An Introduction to the
Industrial Automation
And Robotics

An Introduction to the
Industrial Automation And Robotics

Contributors :
Daniel Pizarro,
Manuel Mazo, *et al.*

AURIS REFERENCE LTD.
London, UK

An Introduction to the Industrial Automation and Robotics
Contributors : Daniel Pizarro *and* Manuel Mazo, *et al.*

Auris Reference Ltd., UK

www.aurisreference.com

United Kingdom

Copyright 2016
Printed in 2017 for Sale in the Indian Subcontinent

Notice

An Introduction to the Industrial Automation and Robotics

ISBN: 978-1-78154-500-3

British Library Cataloguing in Publication Data
A CIP record for this book is available from the British Library

Exclusively distributed by CBS Publishers & Distributors Pvt. Ltd.

Sales & Distribution Rights only for India, Pakistan, Bangladesh, Sri Lanka, Nepal and Bhutan.This book is not to be sold outside these territories.

PREFACE

The purpose of this book is to present an introduction to the multidisciplinary field of automation and robotics for industrial applications. The book initially covers the important concepts of hydraulics and pneumatics and how they are used for automation in an industrial setting. It then moves to a discussion of circuits and using them in hydraulic, pneumatic, and fluidic design. The latter part of the book deals with electric and electronic controls in automation and final chapters are devoted to robotics, robotic programming, and applications of robotics in industry. A companion disc is included with applications and videos. It features: begins with introductory concepts on automation, hydraulics, and pneumatics; covers sensors, PLC's, microprocessors, transfer devices and feeders, robotic sensors, robotic grippers, and robot programming; and, companion DVD has applications and videos from industry.

This page left intentionally blank.

CONTENTS

This page left intentionally blank.

List of Contributors

Daniel Pizarro

Department of Electronics, University of Alcala, NII km 33,600, Alcala de Henares, Spain; E-Mails: mazo@depeca.uah.es (M.M.); santiso@depeca.uah.es (E.S.); marta@depeca.uah.es (M.M.); david.jimenez@depeca.uah.es (D.J.); cobreces@depeca.uah.es (S.C.); losada@depeca.uah.es (C.L.)

Manuel Mazo

Department of Electronics, University of Alcala, NII km 33,600, Alcala de Henares, Spain; E-Mails: mazo@depeca.uah.es (M.M.); santiso@depeca.uah.es (E.S.); marta@depeca.uah.es (M.M.); david.jimenez@depeca.uah.es (D.J.); cobreces@depeca.uah.es (S.C.); losada@depeca.uah.es (C.L.)

Enrique Santiso

Department of Electronics, University of Alcala, NII km 33,600, Alcala de Henares, Spain; E-Mails: mazo@depeca.uah.es (M.M.); santiso@depeca.uah.es (E.S.); marta@depeca.uah.es (M.M.); david.jimenez@depeca.uah.es (D.J.); cobreces@depeca.uah.es (S.C.); losada@depeca.uah.es (C.L.)

Marta Marron

Department of Electronics, University of Alcala, NII km 33,600, Alcala de Henares, Spain; E-Mails: mazo@depeca.uah.es (M.M.); santiso@depeca.uah.es (E.S.); marta@depeca.uah.es (M.M.); david.jimenez@depeca.uah.es (D.J.); cobreces@depeca.uah.es (S.C.); losada@depeca.uah.es (C.L.)

David Jimenez

Department of Electronics, University of Alcala, NII km 33,600, Alcala de Henares, Spain; E-Mails: mazo@depeca.uah.es (M.M.); santiso@depeca.uah.es (E.S.); marta@depeca.uah.es (M.M.); david.jimenez@depeca.uah.es (D.J.); cobreces@depeca.uah.es (S.C.); losada@depeca.uah.es (C.L.)

Santiago Cobreces

Department of Electronics, University of Alcala, NII km 33,600, Alcala de Henares, Spain; E-Mails: mazo@depeca.uah.es (M.M.); santiso@depeca.uah.es (E.S.); marta@depeca.uah.es (M.M.); david.jimenez@depeca.uah.es (D.J.); cobreces@depeca.uah.es (S.C.); losada@depeca.uah.es (C.L.)

Cristina Lo sada

Department of Electronics, University of Alcala, NII km 33,600, Alcala de Henares, Spain; E-Mails: mazo@depeca.uah.es (M.M.); santiso@depeca.uah.es (E.S.); marta@depeca.uah.es (M.M.); david.jimenez@depeca.uah.es (D.J.); cobreces@depeca.uah.es (S.C.); losada@depeca.uah.es (C.L.)

Chapter 1

AUTOMATION

INTRODUCTION

Automation or *automatic control*, is the use of various control systems for operating equipment such as machinery, processes in factories, boilers and heat treating ovens, switching in telephone networks, steering and stabilization of ships, aircraft and other applications with minimal or reduced human intervention. Some processes have been completely automated.

The biggest benefit of automation is that it saves labor, however, it is also used to save energy and materials and to improve quality, accuracy and precision.

The term *automation*, inspired by the earlier word *automatic* (coming from *automaton*), was not widely used before 1947, when General Motors established the automation department. It was during this time that industry was rapidly adopting feedback controllers, which were introduced in the 1930s.

Automation has been achieved by various means including mechanical, hydraulic, pneumatic, electrical, electronic and computers, usually in combination. Complicated systems, such as modern factories, airplanes and ships typically use all these combined techniques.

CONTROL SYSTEM

A **control system** is a device, or set of devices, that manages, commands, directs or regulates the behavior of other device(s) or system(s). Industrial control systems are used in industrial production for controlling an equipment or a machine.

There are two common classes of control systems, open loop control systems and closed loop control systems. In open loop control systems output is generated based on inputs. In closed loop control systems current output is taken into consideration and corrections are made based on feedback. A closed loop system is also called a feedback control system. The human body is a classic example of feedback control system. Fuzzy logic is also used in control systems.

Overview

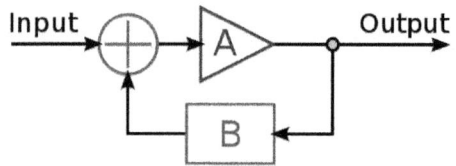

A basic feedback loop

The term "control system" may be applied to the essentially manual controls that allow an operator, for example, to close and open a hydraulic press, perhaps including logic so that it cannot be moved unless safety guards are in place.

An automatic sequential control system may trigger a series of mechanical actuators in the correct sequence to perform a task. For example various electric and pneumatic transducers may fold and glue a cardboard box, fill it with product and then seal it in an automatic packaging machine. Programmable logic controllers are used in many cases such as this, but several alternative technologies exist.

In the case of linear feedback systems, a **control loop**, including sensors, control algorithms and actuators, is arranged in such a fashion as to try to regulate a variable at a setpoint or reference value. An example of this may increase the fuel supply to a furnace when a measured temperature drops. PID controllers are common and effective in cases such as this. Control systems that include some sensing of the results they are trying to achieve are making use of feedback and so can, to some extent, adapt to varying circumstances. Open-loop control systems do not make use of feedback, and run only in pre-arranged ways.

Logic Control

Logic control systems for industrial and commercial machinery were historically implemented at mains voltage using interconnected relays, designed using ladder logic. Today, most such systems are constructed with programmable logic controllers (PLCs) or microcontrollers. The notation of ladder logic is still in use as a programming idiom for PLCs.

Logic controllers may respond to switches, light sensors, pressure switches, *etc.*, and can cause the machinery to start and stop various operations. Logic systems are used to sequence mechanical operations in many applications. PLC software can be written in many different ways – ladder diagrams, SFC – sequential function charts or in language terms known as statement lists.

Examples include elevators, washing machines and other systems with interrelated stop-go operations.

Logic systems are quite easy to design, and can handle very complex operations. Some aspects of logic system design make use of Boolean logic.

On–off Control

A thermostat is a simple negative feedback controller: when the temperature (the "process variable" or PV) goes below a set point (SP), the heater is switched on. Another example could be a pressure switch on an air compressor: when the pressure (PV) drops below the threshold (SP), the pump is powered. Refrigerators and vacuum pumps contain similar mechanisms operating in reverse, but still providing negative feedback to correct errors.

Simple on–off feedback control systems like these are cheap and effective. In some cases, like the simple compressor example, they may represent a good design choice.

In most applications of on–off feedback control, some consideration needs to be given to other costs, such as wear and tear of control valves and perhaps other start-up costs when power is reapplied each time the PV drops. Therefore, practical on–off control systems are designed to include hysteresis: there is a deadband, a region around the setpoint value in which no control action occurs. The width of deadband may be adjustable or programmable.

Linear Control

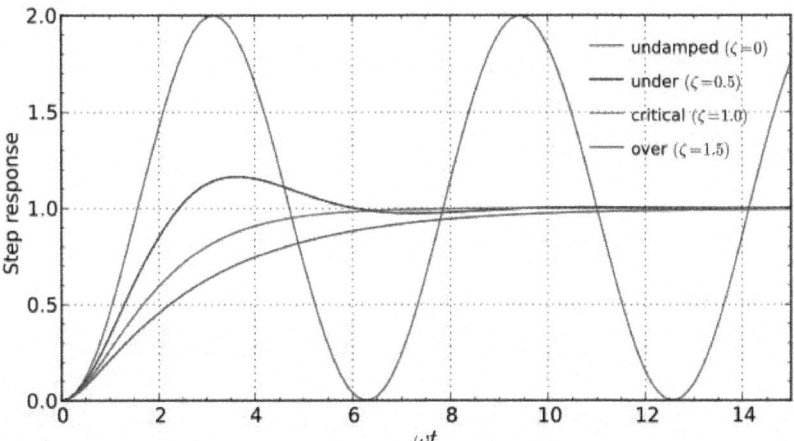

Linear control systems use linear negative feedback to produce a control signal mathematically based on other variables, with a view to maintain the controlled process within an acceptable operating range.

The output from a linear control system into the controlled process may be in the form of a directly variable signal, such as a valve that may be 0 or 100% open or anywhere in between. Sometimes this is not feasible and so, after calculating the current required corrective signal, a linear control system may repeatedly switch an actuator, such as a pump, motor or heater, fully on and then fully off again, regulating the duty cycle using pulse-width modulation.

Proportional Control

When controlling the temperature of an industrial furnace, it is usually better to control the opening of the fuel valve *in proportion to* the current needs of the furnace. This helps avoid thermal shocks and applies heat more effectively.

Proportional negative-feedback systems are based on the difference between the required set point (SP) and process value (PV). This difference is called the *error*. Power is applied in direct proportion to the current measured error, in the correct sense so as to tend to reduce the error (and so avoid positive feedback). The amount of corrective action that is applied for a given error is set by the gain or sensitivity of the control system.

At low gains, only a small corrective action is applied when errors are detected: the system may be safe and stable, but may be sluggish in response to changing conditions; errors will remain uncorrected for relatively long periods of time: it is over-damped. If the proportional gain is increased, such systems become more responsive and errors are dealt with more quickly. There is an optimal value for the gain setting when the overall system is said to be critically damped. Increases in loop gain beyond this point will lead to oscillations in the PV; such a system is under-damped.

In real systems, there are practical limits to the range of the manipulated variable (MV). For example, a heater can be off or fully on, or a valve can be closed or fully open. Adjustments to the gain simultaneously alter the range of error values over which the MV is between these limits. The width of this range, in units of the error variable and therefore of the PV, is called the *proportional band* (PB). While the gain is useful in mathematical treatments, the proportional band is often used in practical situations. They both refer to the same thing, but the PB has an inverse relationship to gain – higher gains result in narrower PBs, and *vice versa*.

Under-damped Furnace Example

In the furnace example, suppose the temperature is increasing towards a set point at which, say, 50% of the available power will be required for steady-state. At low temperatures, 100% of available power is applied. When the PV is within, say 10° of the SP the heat input begins to be reduced by the proportional controller. (Note that this implies a 20° proportional band (PB) from full to no power input, evenly spread around the setpoint value). At the setpoint the controller will be applying 50% power as required, but stray stored heat within the heater sub-system and in the walls of the furnace will keep the measured temperature rising beyond what is required. At 10° above SP, we reach the top of the proportional band (PB) and no power is applied, but the temperature may continue to rise even further before beginning to fall back. Eventually as the PV falls back into the PB, heat is applied again, but now the heater and the furnace walls are too cool and the temperature falls too low before its fall is arrested, so that the oscillations continue.

Over-damped Furnace Example

The temperature oscillations that an under-damped furnace control system produces are unacceptable for many reasons, including the waste of fuel and time (each oscillation cycle may take many minutes), as well as the likelihood of seriously overheating both the furnace and its contents.

Suppose that the gain of the control system is reduced drastically and it is restarted. As the temperature approaches, say 30° below SP (60° proportional band or PB now), the heat input begins to be reduced, the rate of heating of the furnace has time to slow and, as the heat is still further reduced, it eventually is brought up to set point, just as 50% power input is reached and the furnace is operating as required. There was some wasted time while the furnace crept to its final temperature using only 52% then 51% of available power, but at least no harm was done. By carefully increasing the gain this over-damped and sluggish behavior can be improved until the system is critically damped for this SP temperature. Doing this is known as 'tuning' the control system. A well-tuned proportional furnace temperature control system will usually be more effective than on-off control, but will still respond more slowly than the furnace could under skillful manual control.

PID Control

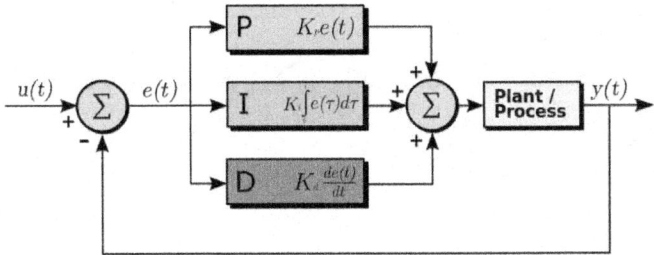

A block diagram of a PID controller

Apart from sluggish performance to avoid oscillations, another problem with proportional-only control is that power application is always in direct proportion to the error. In the example above we assumed that the set temperature could be maintained with 50% power. What happens if the furnace is required in a different application where a higher set temperature will require 80% power to maintain it? If the gain was finally set to a 50° PB, then 80% power will not be applied unless the furnace is 15° below setpoint, so for this other application the operators will have to remember always to set the setpoint temperature 15° higher than actually needed. This 15° figure is not completely constant either: it will depend on the surrounding ambient temperature, as well as other factors that affect heat loss from or absorption within the furnace.

To resolve these two problems, many feedback control schemes include mathematical extensions to improve performance. The most common extensions lead to proportional-integral-derivative control, or PID control.

Derivative Action

The derivative part is concerned with the rate-of-change of the error with time: If the measured variable approaches the setpoint rapidly, then the actuator is backed off early to allow it to coast to the required level; conversely if the measured value begins to move rapidly away from the setpoint, extra effort is applied — in proportion to that rapidity — to try to maintain it.

Derivative action makes a control system behave much more intelligently. On control systems like the tuning of the temperature of a furnace, or perhaps the motion-control of a heavy item like a gun or camera on a moving vehicle, the derivative action of a well-tuned PID controller can allow it to reach and maintain a setpoint better than most skilled human operators could.

If derivative action is over-applied, it can lead to oscillations too. An example would be a PV that increased rapidly towards SP, then halted early and seemed to "shy away" from the setpoint before rising towards it again.

Integral Action

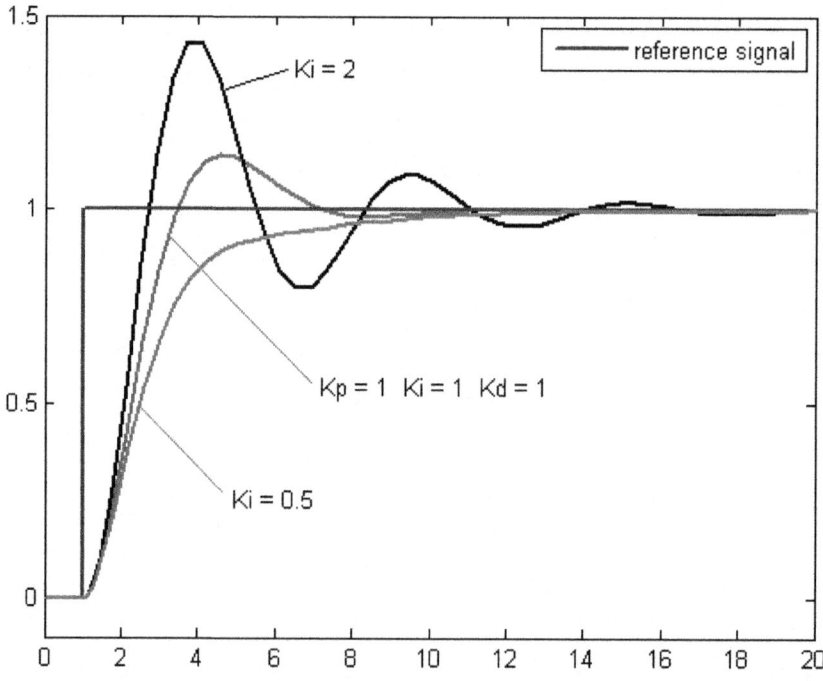

The integral term magnifies the effect of long-term steady-state errors, applying ever-increasing effort until they reduce to zero. In the example of the furnace above working at various temperatures, if the heat being applied does not bring the furnace up to setpoint, for whatever reason, integral action increasingly *moves* the proportional band relative to the setpoint until the PV error is reduced to zero and the setpoint is achieved.

Some controllers include the option to limit the "ramp up % per minute". This option can be very helpful in stabilizing small boilers (3 MBTUH), especially during the summer, during light loads. A utility boiler "unit may be required to change load at a rate of as much as 5% per minute".

Other Techniques

It is possible to filter the PV or error signal. Doing so can reduce the response of the system to undesirable frequencies, to help reduce instability or oscillations. Some feedback systems will oscillate at just one frequency. By filtering out that frequency, more "stiff" feedback can be applied, making the system more responsive without shaking itself apart.

Feedback systems can be combined. In cascade control, one control loop applies control algorithms to a measured variable against a setpoint, but then provides a varying setpoint to another control loop rather than affecting process variables directly. If a system has several different measured variables to be controlled, separate control systems will be present for each of them.

Control engineering in many applications produces control systems that are more complex than PID control. Examples of such fields include fly-by-wire aircraft control systems, chemical plants, and oil refineries. Model predictive control systems are designed using specialized computer-aided-design software and empirical mathematical models of the system to be controlled.

Fuzzy Logic

Fuzzy logic is an attempt to apply the easy design of logic controllers to the control of complex continuously-varying systems. Basically, a measurement in a fuzzy logic system can be partly true, that is if yes is 1 and no is 0, a fuzzy measurement can be between 0 and 1.

The rules of the system are written in natural language and translated into fuzzy logic. For example, the design for a furnace would start with: "If the temperature is too high, reduce the fuel to the furnace. If the temperature is too low, increase the fuel to the furnace."

Measurements from the real world (such as the temperature of a furnace) are converted to values between 0 and 1 by seeing where they fall on a triangle. Usually the tip of the triangle is the maximum possible value which translates to "1."

Fuzzy logic, then, modifies Boolean logic to be arithmetical. Usually the "not" operation is "output = 1 - input," the "and" operation is "output = input.1 multiplied by input.2," and "or" is "output = 1 - ((1 - input.1) multiplied by (1 - input.2))". This reduces to Boolean arithmetic if values are restricted to 0 and 1, instead of allowed to range in the unit interval [0,1].

The last step is to "defuzzify" an output. Basically, the fuzzy calculations make a value between zero and one. That number is used to select a value on a line whose slope and height converts the fuzzy value to a real-world output number. The number then controls real machinery.

If the triangles are defined correctly and rules are right the result can be a good control system.

When a robust fuzzy design is reduced into a single, quick calculation, it begins to resemble a conventional feedback loop solution and it might appear that the fuzzy design was unnecessary. However, the fuzzy logic paradigm may provide scalability for large control systems where conventional methods become unwieldy or costly to derive.

Fuzzy electronics is an electronic technology that uses fuzzy logic instead of the two-value logic more commonly used in digital electronics.

Physical Implementations

Since modern small microprocessors are so cheap, it's very common to implement control systems, including feedback loops, with computers, often in an embedded system. The feedback controls are simulated by having the computer make periodic measurements and then calculate from this stream of measurements.

Computers emulate logic devices by making measurements of switch inputs, calculating a logic function from these measurements and then sending the results out to electronically-controlled switches.

Logic systems and feedback controllers are usually implemented with programmable logic controllers which are devices available from electrical supply houses. They include a little computer and a simplified system for programming. Most often they are programmed with personal computers.

Logic controllers have also been constructed from relays, hydraulic and pneumatic devices, and electronics using both transistors and vacuum tubes (feedback controllers can also be constructed in this manner).

HISTORY

The earliest feedback control mechanism was used to tent the sails of windmills. It was patented by Edmund Lee in 1745.

The centrifugal governor, which dates to the last quarter of the 18th century, was used to adjust the gap between millstones. The centrifugal governor was also used in the automatic flour mill developed by Oliver Evans in 1785, making it the first completely automated industrial process. The governor was adopted by James Watt for use on a steam engine in 1788 after Watt's partner Boulton saw one at a flour mill Boulton & Watt were building.

The governor could not actually hold a set speed; the engine would assume a new constant speed in response to load changes. The governor was able to handle smaller variations such as those caused by fluctuating heat load to the boiler. Also, there was a tendency for oscillation whenever there was a speed change. As a consequence, engines equipped with this governor were not suitable for operations requiring constant speed, such as cotton spinning.

Several improvements to the governor, plus improvements to valve cut-off timing on the steam engine, made the engine suitable for most industrial uses before the end of the 19th century. Advances in the steam engine stayed well ahead of science, both thermodynamics and control theory.

The governor received relatively little scientific attention until James Clerk Maxwell published a paper that established the beginning of a theoretical basis for understanding control theory. Development of the electronic amplifier during the 1920s, which was important for long distance telephony, required a higher signal to noise ratio, which was solved by negative feedback noise cancellation. This and other telephony applications contributed to control theory. Military applications during the Second World War that contributed to and benefited from control theory were fire-control systems and aircraft controls. The so-called classical theoretical treatment of control theory dates to the 1940s and 1950s.

Relay logic was introduced with factory electrification, which underwent rapid adaption from 1900 though the 1920s. Central electric power stations were also undergoing rapid growth and operation of new high pressure boilers, steam turbines and electrical substations created a large demand for instruments and controls.

Central control rooms became common in the 1920s, but as late as the early 1930s, most process control was on-off. Operators typically monitored charts drawn by recorders that plotted data from instruments. To make corrections, operators manually opened or closed valves or turned switches on or off. Control rooms also used color coded lights to send signals to workers in the plant to manually make certain changes.

Controllers, which were able to make calculated changes in response to deviations from a set point rather than on-off control, began being introduced the 1930s. Controllers allowed manufacturing to continue showing productivity gains to offset the declining influence of factory electrification.

In 1959 Texaco's Port Arthur refinery became the first chemical plant to use digital control. Conversion of factories to digital control began to spread rapidly in the 1970s as the price of computer hardware fell.

Significant Applications

The automatic telephone switchboard was introduced in 1892 along with dial telephones. By 1929, 31.9% of the Bell system was automatic. Automatic telephone switching originally used vacuum tube amplifiers and electro-mechanical switches, which consumed a large amount of electricity. Call volume eventually grew so fast that it was feared the telephone system would consume all electricity production, prompting Bell Labs to begin research on the transistor.

The logic performed by telephone switching relays was the inspiration for the digital computer.

The first commercially successful glass bottle blowing machine was an automatic model introduced in 1905. The machine, operated by a two man crew working 12 hour shifts, could produce 17,280 bottles in 24 hours, compared to 2,880 bottles made by a crew of six men and boys working in a shop for a day. The cost of making bottles by machine was 10 to 12 cents per gross compared to $1.80 per gross by the manual glassblowers and helpers.

Sectional electric drives were developed using control theory. Sectional electric drives are used on different sections of a machine where a precise differential must be maintained between the sections. In steel rolling, the metal elongates as it passes through pairs of rollers, which must run at successively faster speeds. In paper making the paper sheet shrinks as it passes around steam heated drying arranged in groups, which must run at successively slower speeds. The first application of a sectional electric drive was on a paper machine in 1919. One of the most important developments in the steel industry during the 20th century was continuous wide strip rolling, developed by Armco in 1928.

Before automation many chemicals were made in batches. In 1930, with the widespread use of instruments and the emerging use of controllers, the founder of Dow Chemical Co. was advocating continuous production.

Self-acting machine tools that displaced hand dexterity so they could be operated by boys and unskilled laborers were developed by James Nasmyth in the 1840s. Machine tools were automated with Numerical control (NC) using punched paper tape in the 1950s. This soon evolved into computerized numerical control (CNC).

Today extensive automation is practiced in practically every type of manufacturing and assembly process. Some of the larger processes include electrical power generation, oil refining, chemicals, steel mills, plastics, cement plants, fertilizer plants, pulp and paper mills, automobile and truck assembly, aircraft production, glass manufacturing, natural gas separation plants, food and beverage processing, canning and bottling and manufacture of various kinds of parts. Robots are especially useful in hazardous applications like automobile spray painting. Robots are also used to assemble electronic circuit boards. Automotive welding is done with robots and automatic welders are used in applications like pipelines.

ADVANTAGES AND DISADVANTAGES

The main advantages of automation are:
- Increased throughput or productivity.
- Improved quality or increased predictability of quality.
- Improved robustness (consistency), of processes or product.
- Increased consistency of output.
- Reduced direct human labor costs and expenses.

The following methods are often employed to improve productivity, quality, or robustness.

- Install automation in operations to reduce cycle time.
- Install automation where a high degree of accuracy is required.
- Replacing human operators in tasks that involve hard physical or monotonous work.
- Replacing humans in tasks done in dangerous environments (*i.e.* fire, space, volcanoes, nuclear facilities, underwater, *etc.*)
- Performing tasks that are beyond human capabilities of size, weight, speed, endurance, *etc.*
- Economic improvement: Automation may improve in economy of enterprises, society or most of humanity. For example, when an enterprise invests in automation, technology recovers its investment; or when a state or country increases its income due to automation like Germany or Japan in the 20th Century.
- Reduces operation time and work handling time significantly.
- Frees up workers to take on other roles.
- Provides higher level jobs in the development, deployment, maintenance and running of the automated processes.

The main disadvantages of automation are:

- Causing unemployment and poverty by replacing human labor.
- Security Threats/Vulnerability: An automated system may have a limited level of intelligence, and is therefore more susceptible to committing errors outside of its immediate scope of knowledge (*e.g.*, it is typically unable to apply the rules of simple logic to general propositions).
- Unpredictable/excessive development costs: The research and development cost of automating a process may exceed the cost saved by the automation itself.
- High initial cost: The automation of a new product or plant typically requires a very large initial investment in comparison with the unit cost of the product, although the cost of automation may be spread among many products and over time.

In manufacturing, the purpose of automation has shifted to issues broader than productivity, cost, and time.

Lights Out Manufacturing

Lights out manufacturing is when a production system is 100% or near to 100% automated (not hiring any workers). In order to eliminate the need for labor costs all together.

Health and Environment

The costs of automation to the environment are different depending on the technology, product or engine automated. There are automated engines that consume more energy resources from the Earth in comparison with previous engines and those that do the opposite too. Hazardous operations, such as oil refining, the manufacturing of industrial chemicals, and all forms of metal working, were always early contenders for automation.

Convertibility and Turnaround Time

Another major shift in automation is the increased demand for flexibility and convertibility in manufacturing processes. Manufacturers are increasingly demanding the ability to easily switch from manufacturing Product A to manufacturing Product B without having to completely rebuild the production lines. Flexibility and distributed processes have led to the introduction of Automated Guided Vehicles with Natural Features Navigation.

Digital electronics helped too. Former analogue-based instrumentation was replaced by digital equivalents which can be more accurate and flexible, and offer greater scope for more sophisticated configuration, parametrization and operation. This was accompanied by the fieldbus revolution which provided a networked (*i.e.* a single cable) means of communicating between control systems and field level instrumentation, eliminating hard-wiring.

Discrete manufacturing plants adopted these technologies fast. The more conservative process industries with their longer plant life cycles have been slower to adopt and analogue-based measurement and control still dominates. The growing use of Industrial Ethernet on the factory floor is pushing these trends still further, enabling manufacturing plants to be integrated more tightly within the enterprise, *via* the internet if necessary. Global competition has also increased demand for Reconfigurable Manufacturing Systems.

Automation Tools

Engineers can now have numerical control over automated devices. The result has been a rapidly expanding range of applications and human activities. Computer-aided technologies (or CAx) now serve the basis for mathematical and organizational tools used to create complex systems. Notable examples of CAx include Computer-aided design (CAD software) and Computer-aided manufacturing (CAM software). The improved design, analysis, and manufacture of products enabled by CAx has been beneficial for industry.

Information technology, together with industrial machinery and processes, can assist in the design, implementation, and monitoring of control systems. One example of an industrial control system is a programmable logic controller (PLC). PLCs are specialized hardened computers which are frequently used to synchronize the flow of inputs from (physical) sensors and events with the flow of outputs to actuators and events.

Human-machine interfaces (HMI) or computer human interfaces (CHI), formerly known as *man-machine interfaces*, are usually employed to communicate with

PLCs and other computers. Service personnel who monitor and control through HMIs can be called by different names. In industrial process and manufacturing environments, they are called operators or something similar. In boiler houses and central utilities departments they are called stationary engineers.

Different types of automation tools exist:

- ANN - Artificial neural network
- DCS - Distributed Control System
- HMI - Human Machine Interface
- SCADA - Supervisory Control and Data Acquisition
- PLC - Programmable Logic Controller
- Instrumentation
- Motion control
- Robotics

Limitations to Automation

- Current technology is unable to automate all the desired tasks.
- Many operations using automation have large amounts of invested capital and produce high volumes of product, making malfunctions extremely costly and potentially hazardous. Therefore, some personnel are needed to insure that the entire system functions properly and that safety and product quality are maintained.
- As a process becomes increasingly automated, there is less and less labor to be saved or quality improvement to be gained. This is an example of both diminishing returns and the logistic function.
- As more and more processes become automated, there are fewer remaining non-automated processes. This is an example of exhaustion of opportunities. New technological paradigms may however set new limits that surpass the previous limits.

Current Limitations

Many roles for humans in industrial processes presently lie beyond the scope of automation. Human-level pattern recognition, language comprehension, and language production ability are well beyond the capabilities of modern mechanical and computer systems. Tasks requiring subjective assessment or synthesis of complex sensory data, such as scents and sounds, as well as high-level tasks such as strategic planning, currently require human expertise. In many cases, the use of humans is more cost-effective than mechanical approaches even where automation of industrial tasks is possible. Overcoming these obstacles is a theorized path to post-scarcity economics.

AUTOMATION *VS.* MECHANIZATION

Mechanization is normally defined as the replacement of a human task with a machine. Automatic transplanters are an example of mechanization. But, true automation encompasses more than mechanization. Automation involves the entire process, including bringing material to and from the mechanized equipment. It normally involves integrating several operations and ensuring that the different pieces of equipment talk to one another to ensure smooth operation. Many times, true automation requires reevaluating and changing current processes rather than simply mechanizing them.

One of the key concepts in automation is differentiating between mechanization and automation. At a recent trade show in Europe, I saw an interesting example of mechanization. On display was a piece of equipment that was trimming boxwood into a ball shape. It accomplished this by mechanical hedge clippers that replicated the motion of a human. While clever, this device is not automation.

Mechanization is normally defined as the replacement of a human task with a machine. Automatic transplanters are an example of mechanization. But, true automation encompasses more than mechanization. Automation involves the entire process, including bringing material to and from the mechanized equipment. It normally involves integrating several operations and ensuring that the different pieces of equipment talk to one another to ensure smooth operation. Many times, true automation requires reevaluating and changing current processes rather than simply mechanizing them.

A Complete Package

An automatic transplanting line can be used as an example of automation. A typical line would consist of automatic destackers for trays and automatic dispensers for pots. These two machines would be connected by means of a conveyor to a flat filling machine that fills the destination trays with soil and levels the soil. The flat filler would be connected by conveyor to an automatic transplanter. A second conveyor could be used to feed source trays into the transplanter. After transplanting, the destination trays would move onto another conveyor, which could feed an automatic tagging machine, and then go through a watering tunnel before being placed onto a final conveyor to be staged for delivery to the greenhouse, a process that can be automated as well. The transplanter itself will normally account for 50-60 percent of the total cost of the line.

Because every piece of equipment in such a line is linked to every other piece not only physically but electronically, the line is truly integrated and will allow for maximum efficiency. Such a line can be run by one or two people, replacing the nine or 10 people required for a manual line. Typically, the cost per flat can be reduced by 50 percent. A future column will deal with cost justifications and payback periods, but obviously, the savings are significant. Because such lines are typically modular, the line can be installed over the course of a few seasons to spread out the cost.

The Next Step

To fully realize the potential savings, simply installing the equipment is not enough. Considerable planning is necessary. Since each change of source or destination trays requires some changeover of the machine, even if only a simple program change, scheduling production to minimize such changeovers is critical. Since humans are less sensitive to changes in products or trays, changing to more rigorous scheduling may require a significant cultural shift in your operation.

Human transplanters are also able to adapt to variable moisture content in plugs. Automatic transplanters, like all automation equipment, require consistency in the product being handled to achieve maximum efficiency. A learning curve will be necessary to determine the proper moisture. Again, this may require a cultural shift in your organization. This change will be rewarded not only by cost savings, but also by a marked improvement in the consistency and quality of your finished product. To some growers, this improvement in quality and predictability may be even more important than the cost savings.

While fewer people will be needed to operate a line, these people will need a higher skill level than current employees. For example, it will be critical to have a maintenance person who understands the equipment and takes ownership of it. Most manufacturers will offer training for these maintenance people. Most manufacturers, of course, also offer service personnel, but downtime is very costly during the peak season. The manufacturer's in-house support people should be able to troubleshoot 75-80 percent of all problems over the telephone when dealing directly with a skilled-grower maintenance person.

But again, this requires a shift in thinking. True automation requires not only mechanization but also integration of multiple operations, rethinking your scheduling and production plan, ensuring consistency in product and evaluating the people skills necessary to achieve your goals. One of the most important considerations before automation is whether or not you want to truly automate your production; you might simply need to automate, or mechanize, certain parts of the process.

RECENT AND EMERGING APPLICATIONS

Automated Retail

Food and Drink

The food retail industry has started to apply automation to the ordering process; McDonald's has introduced touch screen ordering and payment systems in many of its restaurants, reducing the need for as many cashier employees. University of Texas has introduced fully automated cafe retail locations. Some Cafes and restaurants have utilized mobile and tablet "apps" to make the ordering process more efficient by customers ordering and paying on their device. Some restaurants have automated food delivery to customers tables using a Conveyor belt system. The use of robots is sometimes employed to replace waiting staff.

Stores

Many Supermarkets and even smaller stores are rapidly introducing Self checkout systems reducing the need for employing checkout workers.

Online shopping could be considered a form of automated retail as the payment and checkout are through an automated Online transaction processing system. Other forms of automation can also be an integral part of online shopping, for example the deployment of automated warehouse robotics such as that applied by Amazon using Kiva Systems.

Automated Mining

Involves the removal of human labor from the mining process. The mining industry is currently in the transition towards Automation. Currently it can still require a large amount of human capital, particularly in the third world where labor costs are low so there is less incentive for increasing efficiency through automation.

Automated Video Surveillance

The Defense Advanced Research Projects Agency (DARPA) started the research and development of automated visual surveillance and monitoring (VSAM) program, between 1997 and 1999, and airborne video surveillance (AVS) programs, from 1998 to 2002. Currently, there is a major effort underway in the vision community to develop a fully automated tracking surveillance system. Automated video surveillance monitors people and vehicles in real time within a busy environment. Existing automated surveillance systems are based on the environment they are primarily designed to observe, *i.e.*, indoor, outdoor or airborne, the amount of sensors that the automated system can handle and the mobility of sensor, *i.e.*, stationary camera vs. mobile camera. The purpose of a surveillance system is to record properties and trajectories of objects in a given area, generate warnings or notify designated authority in case of occurrence of particular events.

Automated Highway Systems

As demands for safety and mobility have grown and technological possibilities have multiplied, interest in automation has grown. Seeking to accelerate the development and introduction of fully automated vehicles and highways, the United States Congress authorized more than $650 million over six years for intelligent transport systems (ITS) and demonstration projects in the 1991 Intermodal Surface Transportation Efficiency Act (ISTEA). Congress legislated in ISTEA that "the Secretary of Transportation shall develop an automated highway and vehicle prototype from which future fully automated intelligent vehicle-highway systems can be developed. Such development shall include research in human factors to ensure the success of the man-machine relationship. The goal of this program is to have the first fully automated highway roadway or an automated test track in operation by 1997. This system shall accommodate installation of equipment in new and existing motor vehicles." [ISTEA 1991, part B, Section 6054(b)].

Full automation commonly defined as requiring no control or very limited control by the driver; such automation would be accomplished through a combination of sensor, computer, and communications systems in vehicles and along the roadway. Fully automated driving would, in theory, allow closer vehicle spacing and higher speeds, which could enhance traffic capacity in places where additional road building is physically impossible, politically unacceptable, or prohibitively expensive. Automated controls also might enhance road safety by reducing the opportunity for driver error, which causes a large share of motor vehicle crashes. Other potential benefits include improved air quality, increased fuel economy, and spin-off technologies generated during research and development related to automated highway systems.

Automated Waste Management

Automated waste collection trucks prevent the need for as many workers as well as easing the level of Labor required to provide the service.

Home Automation

Home automation (also called domotics) designates an emerging practice of increased automation of household appliances and features in residential dwellings, particularly through electronic means that allow for things impracticable, overly expensive or simply not possible in recent past decades.

Industrial Automation

Industrial automation deals primarily with the automation of manufacturing, quality control and material handling processes. General purpose controllers for industrial processes include Programmable logic controllers and computers. One trend is increased use of Machine vision to provide automatic inspection and robot guidance functions, another is a continuing increase in the use of robots.

Energy efficiency in industrial processes has become a higher priority. Semiconductor companies like Infineon Technologies are offering 8-bit micro-controller applications for example found in motor controls, general purpose pumps, fans, and ebikes to reduce energy consumption and thus increase efficiency.

Agriculture

Now that we're moving towards automated orange-sorting and autonomous tractors, the next step in automated agriculture is robotic strawberry pickers.

Agent-assisted Automation refers to automation used by call center agents to handle customer inquiries. There are two basic types: desktop automation and automated voice solutions. Desktop automation refers to software programming that makes it easier for the call center agent to work across multiple desktop tools. The automation would take the information entered into one tool and populate it across the others so it did not have to be entered more than once, for example. Automated voice solutions allow the agents to remain on the line while disclosures and other important information is provided to customers in the form of pre-recorded audio files. Specialized applications of these automated voice solutions enable the agents to process credit cards without ever seeing or hearing the credit card numbers or CVV codes

The key benefit of agent-assisted automation is compliance and error-proofing. Agents are sometimes not fully trained or they forget or ignore key steps in the process. The use of automation ensures that what is supposed to happen on the call actually does, every time.

RELATIONSHIP TO UNEMPLOYMENT

Based on a formula by Gilles Saint-Paul, an economist at Toulouse 1 University, the demand for unskilled human capital declines at a slower rate than the demand for skilled human capital increases. In the long run and for society as a whole it has led to cheaper products, lower average work hours, and new industries forming (I.e, robotics industries, computer industries, design industries). These new industries provide many high salary skill based jobs to the economy.

LOW COST AUTOMATION – WHAT DOES IT ACTUALLY MEAN?

The notion of low cost automation (LCA) has become an indispensable part of the philosophy of lean production. As the name suggests, **low cost automation** may best be paraphrased as 'simple automation'. But what does this mean in practice?

The source of this kind of automation can be found in the 1970s – at the time of the first Ölund investment crisis. **Low cost automation** was, in its time, the result of the kaizen activities of shop-floor co-workers in Japanese world-class enterprises.

LCA arose partly on account of a product-oriented kind of manufacturing (**lean manufacturing**) in which products were manufactured in the smallest possible number in cells according to the one-piece-pull principle. This was markedly different to the kind of production which, even today, is widespread in the west and marked by a separation of processes according to their functions and the technology involved.

In other words, here in the West, according to the rules of mass production or the principle of volume digression, products are pushed in the greatest possible amounts quite literally through the manufacturing process.

These two different kinds of production, depending respectively on the principles of pull and push, make quite different demands on the flexibility of manufacturing and thus on its degree of automation.

It is interesting to note that recent developments have shown, especially in the West, that the time of quantitative growth is past. Volumes are increasing only through the acquisition of firms or predatory competition. For the sake of stimulating buyers' interest and of serving the increasingly heterogeneous needs of customers, new products are being launched onto the market in continually briefer cycles. This is leading to an increase in the product variants and thus to a rise in the degree of complexity in manufacture.

Where Does the Term Come From?

The source of this kind of automation can be found in the 1970s – at the time of the first Ölund investment crisis. **Low cost automation** was, in its time, the result of the kaizen activities of shop-floor co-workers in Japanese world-class enterprises.

LCA arose partly on account of a product-oriented kind of manufacturing (**lean manufacturing**) in which products were manufactured in the smallest possible number in cells according to the one-piece-pull principle. This was markedly different to the kind of production which, even today, is widespread in the west and marked by a separation of processes according to their functions and the technology involved.

In other words, here in the West, according to the rules of mass production or the principle of volume digression, products are pushed in the greatest possible amounts quite literally through the manufacturing process.

These two different kinds of production, depending respectively on the principles of pull and push, make quite different demands on the flexibility of manufacturing and thus on its degree of automation.

Why LCA?

The highly automated manufacturing equipment of many enterprises often lacks the flexibility to react quickly and inexpensively to new needs of the market, though it may have been introduced for the sake of more flexibility.

As an index for this development we find continually sinking availabilities in firms which have fallen for the automation paradigm. In this respect values below 70%, according to the Japanese method of measurement OEE (overall equipment effectiveness), are no rarity.

Many enterprises react to the cost increase due to this situation by shifting production to lands with low wages, but this mostly turns out to be ineffective. Parts of production which are wasteful or contribute nothing to value are certainly less costly abroad but they are mostly taken along.

A further reason why **LCA** plays a key role in top Japanese enterprises is the effort made to get humans and machines to interact as efficiently as possible. For this sake workers are still given manual tasks. In western enterprises workers stay mostly at only one point or supervise processes, but those in Japanese firms have several functions and look after several processes.

Often they are involved in only simple activities like loading and unloading machines, assembling parts, attaching cables or screwing things in, but these manual activities are precisely what can be lessened by applying simple means and clever ideas. As a rule this is guided and furthered by the team-leader (hancho) with the aim of lessening the number of personnel. In Japanese this is known as shoshinga. The workers who are freed in this way are not sacked but mostly redeployed on other, more important tasks, thanks to which only the best are mostly taken out of the process.

Efforts of this kind lead to many small and simple automations. Mostly this happens within the framework of a wide-reaching campaign within all areas of an enterprise. Hence a few months ago we saw in a Japanese factory run by Denso an **LCA** campaign with the aim of making all manual activities so simple that workers have only to load a utility then the fully processed part is tossed off automatically. Another instance is the gear-maker Jatco Fuji, who is presently carrying out a 'one motion campaign'. The campaign's aim is to optimize all movements to such an extent that they call for only a touch and take no longer than 0.1 minutes.

A further element to be mentioned in connection with **LCA** is called jidoka or autonomation. This refers to a workshop without workers and to a process supervised by a machine on its own. If there are deviations from the normal production process, owing for instance to defects or a tool's breaking or a lack of supplies, or if faults have to be avoided, the machine stops of its own accord and beckons a worker to intervene.

The underlying philosophy of production calls for lean or simple resources, and since these can seldom be supplied from outside, they are often made by the enterprise itself. Since this calls for planning engineers with notable experience in production within the enterprise, no one without this experience is ever called on by Japanese enterprises to take part in planning. Moreover planning and production are mostly brought to together in terms of space and organization.

In contrast to this, planning in the West is mostly based on specialist knowledge from technical colleges, and planning and production are spatially and functionally separate. This often leads to redundancy and to plants' being oversized.

The introduction of **LCA** thus presupposes global thinking or a change of traditional notions of manufacturing.

How Can We Establish LCA?

For **LCA** there are no fixed rules, since in Japan the method is not described as such in concrete terms and its scope is generally unknown. The tendency to simplify processes and utilities is taken for granted, since to the Japanese a belief in the basic simplicity of things is common sense. Nonetheless, on visiting Japan many times, I have found a certain standardization within enterprises. However different the enterprises or branches may be, certain priorities are shared by all factories:

Simple

- Avoidance of complicated motion sequences, functions and structures
- Rotational and linear motions, which are carried out by simple mechanical elements
- Use of basic physical laws (gravitation, force-parallelograms and so on)

Economic

- Use of simple and commercially available materials and components
- Frequent reuse of dismantled utilities

Easily Put Together and Taken to Pieces

- Use of model kit systems able to make many different products

Modular Structure

- Low degree of complexity
- Not too many processing steps in a module
- Can be swiftly changed: modules are set up anew according to the product *i.e.* cells are continually changed. Hence the modules are mostly equipped with wheels and a flexible maintenance structure.

Internal Development and Production

- Compact and lean

Since appliances and resources are often integrated in cells in which workers see to various processes and have to move from module to module, attention is paid to width for the sake of shortening the distances. Efforts are also made to match the size of utilities to the width of work-pieces.

TYPES OF AUTOMATION

General Purpose

Home Automation

Home automation is the residential extension of building automation. It is automation of the home, housework or household activity. Home automation may include centralized control of lighting, HVAC (heating, ventilation and air conditioning), appliances, security locks of gates and doors and other systems, to provide improved convenience, comfort, energy efficiency and security. Home automation for the elderly and disabled can provide increased quality of life for persons who might otherwise require caregivers or institutional care.

The popularity of home automation has been increasing greatly in recent years due to much higher affordability and simplicity through smartphone and tablet connectivity. The concept of the "Internet of Things" has tied in closely with the popularization of home automation.

A home automation system integrates electrical devices in a house with each other. The techniques employed in home automation include those in building automation as well as the control of domestic activities, such as home entertainment systems, houseplant and yard watering, pet feeding, changing the ambiance "scenes" for different events, and the use of domestic robots. Devices may be connected through a home network to allow control by a personal computer, and may allow remote access from the internet. Through the integration of information technologies with the home environment, systems and appliances are able to *communicate* in an integrated manner which results in convenience, energy efficiency, and safety benefits.

Automated *homes of the future* have been staple exhibits for World's Fairs and popular backgrounds in science fiction. However, problems with complexity, competition between vendors, multiple incompatible standards, and the resulting expense have limited the penetration of home automation to homes of the wealthy, or ambitious hobbyists. Possibly the first "home computer" was an experimental home automation system in 1966.

Overview and Benefits

Home automation refers to the use of computer and information technology to control home appliances and features (such as windows or lighting). Systems can range from simple remote control of lighting through to complex computer/microcontroller based networks with varying degrees of intelligence and automation. Home automation is adopted for reasons of ease, security and energy efficiency.

In modern construction in industrialized nations, most homes have been wired for electrical power, telephones, TV outlets, and a doorbell. Many household tasks were automated by the development of specialized automated appliances. For instance, automatic washing machines were developed to reduce the manual labor of cleaning clothes, and water heaters reduced the labor necessary for bathing.

The use of gaseous or liquid fuels, and later the use of electricity enabled increased automation in heating, reducing the labor necessary to manually refuel heaters and stoves. Development of thermostats allowed more automated control of heating, and later cooling.

As the number of controllable devices in the home rises, interconnection and communication becomes a useful and desirable feature. For example, a furnace can send an alert message when it needs cleaning, or a refrigerator when it needs service. If no one is supposed to be home and the alarm system is set, the home automation system could call the owner, or the neighbors, or an emergency number if an intruder is detected.

In simple installations, automation may be as straightforward as turning on the lights when a person enters the room. In advanced installations, rooms can sense not only the presence of a person inside but know who that person is and perhaps set appropriate lighting, temperature, music levels or television channels, taking into account the day of the week, the time of day, and other factors.

Other automated tasks may include reduced setting of the heating or air conditioning when the house is unoccupied, and restoring the normal setting when an occupant is about to return. More sophisticated systems can maintain an inventory of products, recording their usage through bar codes, or an RFID tag, and prepare a shopping list or even automatically order replacements.

Home automation can also provide a remote interface to home appliances or the automation system itself, to provide control and monitoring on a smartphone or web browser.

An example of remote monitoring in home automation could be triggered when a smoke detector detects a fire or smoke condition, causing all lights in the house to blink to alert any occupants of the house to the possible emergency. If the house is equipped with a home theater, a home automation system can shut down all audio and video components to avoid distractions, or make an audible announcement. The system could also call the home owner on their mobile phone to alert them, or call the fire department or alarm monitoring company.

History

Home automation has been a feature of science fiction writing for many years, but has only become practical since the early 20th Century following the widespread introduction of electricity into the home, and the rapid advancement of information technology. Early remote control devices began to emerge in the late 1800s. For example, Nikola Tesla patented an idea for the remote control of vessels and vehicles in 1898.

The emergence of electrical home appliances began between 1915 and 1920; the decline in domestic servants meant that households needed cheap, mechanical replacements. Domestic electricity supply, however, was still in its infancy — meaning this luxury was afforded only the more affluent households.

Ideas similar to modern home automation systems originated during the World's Fairs of the 1930s. Fairs in Chicago, New York and (1964–65), depicted electrified and automated homes. In 1966 Jim Sutherland, an engineer working for Westinghouse Electric, developed a home automation system called "ECHO IV"; this was a private project and never commercialized. The first "wired homes" were built by American hobbyists during the 1960s, but were limited by the technology of the times. The term "smart house" was first coined by the American Association of Housebuilders in 1984.

With the invention of the microcontroller, the cost of electronic control fell rapidly. Remote and intelligent control technologies were adopted by the building services industry and appliance manufacturers.

By the end of the 1990s, "domotics" was commonly used to describe any system in which informatics and telematics were combined to support activities in the home. The phrase is a portmanteau word formed from *domus* (Latin, meaning house) and *informatics*, and refers to the application of computer and robot technologies to domestic appliances.

Despite interest in home automation, by the end of the 1990s there was not a widespread uptake, with such systems still considered the domain of hobbyists or the rich. The lack of a single, simplified, protocol and high cost of entry has put off consumers.

While there is still much room for growth, according to ABI Research, 1.5 million home automation systems were installed in the US in 2012, and a sharp uptake could see shipments topping over 8 million in 2017

System Elements

Elements of a home automation system include sensors (such as temperature, daylight, or motion detection), controllers (such as a general-purpose personal computer or a dedicated automation controller) and actuators, such as motorized valves, light switches, motors, and others.

One or more human-machine interface devices are required, so that the residents of the home can interact with the system for monitoring and control; this may be a specialized terminal or, increasingly, may be an application running on a smart phone or tablet computer. Devices may communicate over dedicated wiring, or over a wired network, or wirelessly using one or more protocols. Building automation networks developed for institutional or commercial buildings may be adapted to control in individual residences. A centralized controller can be used, or multiple intelligent devices can be distributed around the home.

Networks

There have been many attempts to standardize the forms of hardware, electronic and communication interfaces needed to construct a home automation system. Some standards use additional communication and control wiring, some embed signals in the existing power circuit of the house, some use radio frequency (RF) signals, and some use a combination of several methods. Control wiring is hardest to retrofit into an existing house. Some appliances include a USB port that is used for control and connection to a domotics network. Protocol bridges translate information from one standard to another, for example, from X10 to European Installation Bus (EIB now KNX).

Tasks

HVAC

Heating, ventilation and air conditioning (HVAC) systems can include temperature and humidity control, including fresh air heating and natural cooling.

An Internet-controlled thermostat allows the homeowner to control the building's heating and air conditioning systems remotely. The system may automatically open and close windows to cool the house.

Lighting

Lighting control systems can be used to control household electric lights. Lights can be controlled on a time cycle, or arranged to automatically go out when a room is unoccupied. Electronically controlled lamps can be controlled for brightness or color to provide different light levels for different tasks. Lighting can be controlled remotely by a wireless control or over the Internet. Natural lighting (daylighting) can be used to automatically control window shades and draperies to make best use of natural light.

Audio-visual

This category includes audio and video switching and distribution. Multiple audio or video sources can be selected and distributed to one or more rooms and can be linked with lighting and blinds to provide mood settings.

Shading

Automatic control of blinds and curtains can be used for:

- Presence simulation
- Privacy
- Temperature control
- Brightness control
- Glare control
- Security (in case of shutters)

Security

A household security systems integrated with a home automation system can provide additional services such as remote surveillance of security cameras over the Internet, or central locking of all perimeter doors and windows.

With home automation, the user can select and watch cameras live from an Internet source to their home or business. Security systems can include motion sensors that will detect any kind of unauthorized movement and notify the user through the security system or *via* cell phone.

The automation system can simulate the appearance of an occupied home by automatically adjusting lighting or window coverings. Detection systems such as fire alarm, gas leak, carbon monoxide, or water leaks can be integrated. Personal medical alarm systems allow an injured home occupant to summon help.

Intercoms

An intercom system allows communication *via* a microphone and loud speaker between multiple rooms. Integration of the intercom to the telephone, or of the video door entry system to the television set, allowing the residents to view the door camera automatically.

Domestic Robotics (Domotics)

Journalist Bruno de Latour coined the term **domotic** in 1984. **Domotic** has been recently introduced in vocabulary as a composite word of Latin word *domus* and *informatics*, or a contraction of *domestic robotics*, and it refers to *intelligent houses* meaning the use of the automation technologies and computer science applied to the home.

The term covers a range of applications of information technology to the problems of home automation.

Domotics is the study of the realization of an intelligent home environment.

Digital Home includes home automation, multimedia, telecommunications, e-commerce, *etc.* through home networks Domotics and home automation means that systems *talk to* each other for improved convenience, efficiency and safety.

Other Systems

Using special hardware, almost any household appliance can be monitored and controlled automatically or remotely, including cooking appliances, swimming pool systems, and others.

Well and Booster Pump Automation.

Costs

Costs mainly include equipment, components, furniture, and custom installation.

Ongoing costs include electricity to run the control systems, maintenance costs for the control and networking systems, including troubleshooting, and eventual cost of upgrading as standards change. Increased complexity may also increase maintenance costs for networked devices. Cloud-based services supporting an installation may also entail fees for setup, usage, or both.

Learning to use a complex system effectively may take significant time and training.

Control system security may be difficult and costly to maintain, especially if the control system extends beyond the home, for instance by wireless or by connection to the internet or other networks.

Smart Grid

Home automation technologies are viewed as integral additions to the smart grid. Communication between a home automation system and the grid would allow applications like load shedding during system peaks, or would allow the homeowner to automatically defer energy use to periods of low grid cost. Green automation or "demand response" are termns that refer to energy management strategies in home automation when data from smart grids is combined with home automation systems to use resources at either their lowest prices or highest availability, taking advantage, for instance, of high solar panel output in the middle of the day to automatically run washing machines.

Laboratory Automation

Laboratory automation is a multi-disciplinary strategy to research, develop, optimize and capitalize on technologies in the laboratory that enable new and improved processes. Laboratory automation professionals are academic, commercial and government researchers, scientists and engineers who conduct research and develop new technologies to increase productivity, elevate experimental data quality, reduce lab process cycle times, or enable experimentation that otherwise would be impossible.

The most widely known application of laboratory automation technology is laboratory robotics. More generally, the field of laboratory automation comprises many different automated laboratory instruments, devices, software algorithms, and methodologies used to enable, expedite and increase the efficiency and effectiveness of scientific research in laboratories.

The application of technology in today's laboratories is required to achieve timely progress and remain competitive. Laboratories devoted to activities such as high-throughput screening, combinatorial chemistry, automated clinical and analytical testing, diagnostics, large scale biorepositories, and many others, would not exist without advancements in laboratory automation.

Some universities offer entire programs that focus on lab technologies. For example, Indiana University-Purdue University at Indianapolis offers a graduate program devoted to Laboratory Informatics. Also, the Keck Graduate Institute in California offers a graduate degree with an emphasis on development of assays, instrumentation and data analysis tools required for clinical diagnostics, high-throughput screening, genotyping, microarray technologies, proteomics, imaging and other applications.

History

In 1993, Dr. Rod Markin at the University of Nebraska Medical Center created one of the world's first clinical automated laboratory management systems. In the mid-1990s, he chaired a standards group called the Clinical Testing Automation Standards Steering Committee (CTASSC) of the American Association for Clinical Chemistry, which later evolved into an area committee of the Clinical and Laboratory Standards Institute.

In 2004, the National Institutes of Health (NIH) and more than 300 nationally recognized leaders in academia, industry, government, and the public completed the NIH Roadmap to accelerate medical discovery to improve health.

Integrated Library System

An **integrated library system (ILS)**, also known as a **library management system (LMS)**, is an enterprise resource planning system for a library, used to track items owned, orders made, bills paid, and patrons who have borrowed.

An ILS usually comprises a relational database, software to interact with that database, and two graphical user interfaces. Most ILSes separate software functions into discrete programs called modules, each of them integrated with a unified interface. Examples of modules might include:

- acquisitions (ordering, receiving, and invoicing materials)
- cataloging (classifying and indexing materials)
- circulation (lending materials to patrons and receiving them back)
- serials (tracking magazine and newspaper holdings)
- the OPAC (public interface for users)

Each patron and item has a unique ID in the database that allows the ILS to track its activity.

Larger libraries use an ILS to order and acquire, receive and invoice, catalog, circulate, track and shelve materials. Smaller libraries, such as those in private homes or non-profit organizations (like churches or synagogues, for instance), often forgo the expense and maintenance required to run an ILS, and instead use a library computer system.

History

Pre-computerization

Prior to computerization, library tasks were performed manually and independently from one another. Selectors ordered materials with ordering slips, cataloguers manually catalogued items and indexed them with the card catalog system (in which all bibliographic data was kept on a single index card), fines were collected by local bailiffs, and users signed books out manually, indicating their name on cue cards which were then kept at the circulation desk. Early mechanization came in 1936, when the University of Texas began using a punch card system to manage library circulation. While the punch card system allowed for more efficient tracking of loans, library services were far from being integrated, and no other library task was affected by this change.

1960s: The Influence of Computer Technologies

Following this, the next big innovation came with the advent of MARC standards in the 1960s which coincided with the growth of computer technologies – *library automation* was born. From this point onwards, libraries began experimenting with computers, and, starting in the late 1960s and continuing into the 1970s, bibliographic services utilizing new online technology and the shared MARC vocabulary entered the market; these included OCLC, Research Libraries Group (which has since merged with OCLC), and Washington Library Network (which became Western Library Network and is also now part of OCLC).

1970s–1980s: The Early Integrated Library System

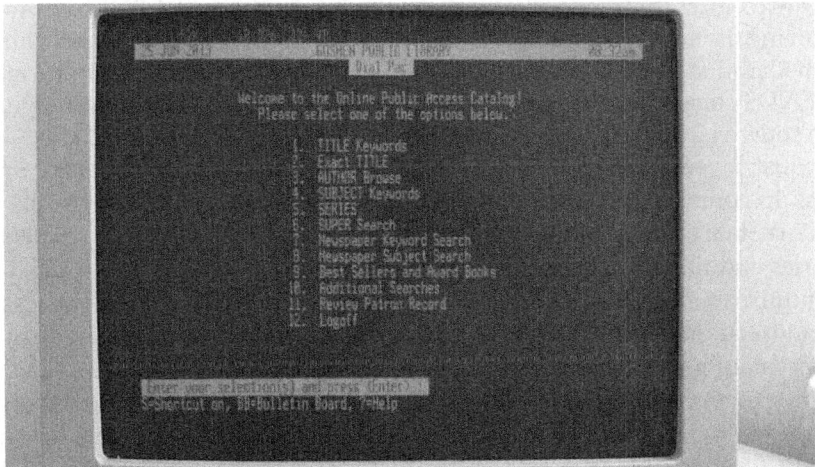

Screenshot of a Dynix menu

The 1970s can be characterized by improvements in computer storage as well as in telecommunications. As a result of these advances, 'turnkey systems on microcomputers,' known more commonly as *integrated library systems* (ILS) finally appeared. These systems included necessary hardware and software which allowed the connection of major circulation tasks, including circulation control and overdue notices. As the technology developed, other library tasks could be accomplished through ILS as well, including acquisition, cataloguing, reservation of titles, and monitoring of serials.

1990s–2000s: The Growth of the Internet

With the evolution of the Internet throughout the 1990s and into the 2000s, ILSs began allowing users to more actively engage with their libraries through OPACs and online web-based portals. Users could log into their library accounts to reserve or renew books, as well as authenticate themselves for access to library-subscribed online databases. Inevitably, during this time, the ILS market grew exponentially. By 2002, the ILS industry averaged sales of approximately US$500 million annually, compared to just US$50 million in 1982.

Mid 2000s–Present: Increasing Costs and Customer Dissatisfaction

By the mid to late 2000s, ILS vendors had increased not only the number of services offered but also their prices, leading to some dissatisfaction among many smaller libraries. At the same time, open source ILS was in its early stages of testing. Some libraries began turning to such open source ILSs as Koha and Evergreen. Common reasons noted were to avoid vendor lock in, avoid license fees, and participate in software development. Freedom from vendors also allowed libraries to prioritize needs according to urgency, as opposed to what their vendor can offer. Libraries which have moved to open source ILS have found that vendors are now more likely to provide quality service in order to continue a partnership since they no longer have the power of owning the ILS software and tying down libraries to strict contracts. This has been the case with the SCLENDS consortium. Following the success of Evergreen for the Georgia PINES library consortium, the South Carolina State Library along with some local public libraries formed the SCLENDS consortium in order to share resources and to take advantage of the open source nature of the Evergreen ILS to meet their specific needs. By October 2011, just 2 years after SCLENDS began operations, 13 public library systems across 15 counties had already joined the consortium, in addition to the South Carolina State Library. Librarytechnology.org does an annual survey of over 2,400 libraries and noted in 2008 2% of those surveyed used open source ILS, in 2009 the number increased to 8%, in 2010 12%, and in 2011 11% of the libraries polled had adopted open source ILSs. The following year's survey reported an increase to 14%, stating that "open source ILS products, including Evergreen and Koha, continue to represent a significant portion of industry activity. Of the 794 contracts reported in the public and academic arena, 113, or 14 percent, were for support services for these open source systems."

2010s–Present: The Rise of Cloud Based Solutions

The use of cloud based library management systems has increased drastically since the rise of "cloud" technology started. Some common management systems include Libramatic, Aura Software and Librarika. Many modern cloud based solutions allow automated cataloging by scanning a book's ISBN. This technology was pioneered by Libramatic, although it is currently in use by systems such as LibraryWorld. Librarika has a method called "Smart Add" that lets Librarians to add book automatically by just inputing the ISBNs. In addition, Librarika offers free library creation with built-in OPAC for up to 10,000 book titles with the ability to get more free titles for non-profit and charities, which serves well for many school, college and other small libraries.

Database Administration and Automation

Database administration is the function of managing and maintaining database management systems (DBMS) software. Mainstream DBMS software such as Oracle, IBM DB2 and Microsoft SQL Server need ongoing management. As such, corporations that use DBMS software often hire specialized IT (Information Technology) personnel called Database Administrators or DBAs.

DBA Responsibilities

- Installation, configuration and upgrading of Database server software and related products.
- Evaluate Database features and Database related products.
- Establish and maintain sound backup and recovery policies and procedures.
- Take care of the Database design and implementation.
- Implement and maintain database security (create and maintain users and roles, assign privileges).
- Database tuning and performance monitoring.
- Application tuning and performance monitoring.
- Setup and maintain documentation and standards.
- Plan growth and changes (capacity planning).
- Work as part of a team and provide 24x7 support when required.
- Do general technical troubleshooting and give cons.
- Database recovery.

Types of Database Administration

There are three types of DBAs:

1. Systems DBAs (also referred to as Physical DBAs, Operations DBAs or Production Support DBAs): focus on the physical aspects of database ad-

ministration such as DBMS installation, configuration, patching, upgrades, backups, restores, refreshes, performance optimization, maintenance and disaster recovery.

2. Development DBAs: focus on the logical and development aspects of database administration such as data model design and maintenance, DDL (data definition language) generation, SQL writing and tuning, coding stored procedures, collaborating with developers to help choose the most appropriate DBMS feature/functionality and other pre-production activities.

3. Application DBAs: usually found in organizations that have purchased 3rd party application software such as ERP (enterprise resource planning) and CRM (customer relationship management) systems. Examples of such application software includes Oracle Applications, Siebel and PeopleSoft and SAP. Application DBAs straddle the fence between the DBMS and the application software and are responsible for ensuring that the application is fully optimized for the database and vice versa. They usually manage all the application components that interact with the database and carry out activities such as application installation and patching, application upgrades, database cloning, building and running data cleanup routines, data load process management, *etc.*

While individuals usually specialize in one type of database administration, in smaller organizations, it is not uncommon to find a single individual or group performing more than one type of database administration.

Nature of Database Administration

The degree to which the administration of a database is automated dictates the skills and personnel required to manage databases. On one end of the spectrum, a system with minimal automation will require significant experienced resources to manage; perhaps 5-10 databases per DBA. Alternatively an organization might choose to automate a significant amount of the work that could be done manually therefore reducing the skills required to perform tasks. As automation increases, the personnel needs of the organization splits into highly skilled workers to create and manage the automation and a group of lower skilled "line" DBAs who simply execute the automation.

Database administration work is complex, repetitive, time-consuming and requires significant training. Since databases hold valuable and mission-critical data, companies usually look for candidates with multiple years of experience. Database administration often requires DBAs to put in work during off-hours (for example, for planned after hours downtime, in the event of a database-related outage or if performance has been severely degraded). DBAs are commonly well compensated for the long hours

One key skill required and often overlooked when selecting a DBA is database recovery. It is not a case of "if" but a case of "when" a database suffers a failure, ranging from a simple failure to a full catastrophic failure. The failure may be data corruption, media failure, or user induced errors. In either situation the DBA must have the skills to recover the database to a given point in time to prevent a loss of data. A highly skilled DBA can spend a few minutes or exceedingly long hours to get the database back to the operational point.

Database Administration Tools

Often, the DBMS software comes with certain tools to help DBAs manage the DBMS. Such tools are called native tools. For example, Microsoft SQL Server comes with SQL Server Enterprise Manager and Oracle has tools such as SQL*Plus and Oracle Enterprise Manager/Grid Control. In addition, 3rd parties such as BMC, Quest Software, Embarcadero Technologies, EMS Database Management Solutions and SQL Maestro Group offer GUI tools to monitor the DBMS and help DBAs carry out certain functions inside the database more easily.

Another kind of database software exists to manage the provisioning of new databases and the management of existing databases and their related resources. The process of creating a new database can consist of hundreds or thousands of unique steps from satisfying prerequisites to configuring backups where each step must be successful before the next can start. A human cannot be expected to complete this procedure in the same exact way time after time - exactly the goal when multiple databases exist. As the number of DBAs grows, without automation the number of unique configurations frequently grows to be costly/difficult to support. All of these complicated procedures can be modeled by the best DBAs into database automation software and executed by the standard DBAs. Software has been created specifically to improve the reliability and repeatability of these procedures such as Stratavia's Data Palette and GridApp Systems Clarity.

The Impact of IT Automation on Database Administration

Recently, automation has begun to impact this area significantly. Newer technologies such as Stratavia's Data Palette suite and GridApp Systems Clarity have begun to increase the automation of databases causing the reduction of database related tasks. However at best this only reduces the amount of mundane, repetitive activities and does not eliminate the need for DBAs. The intention of DBA automation is to enable DBAs to focus on more proactive activities around database architecture, deployment, performance and service level management.

Every database requires a database owner account that can perform all schema management operations. This account is specific to the database and cannot log in to Data Director. You can add database owner accounts after database creation. Data Director users must log in with their database-specific credentials to view the database, its entities, and its data or to perform database management tasks. Database administrators and application developers can manage databases only if they have appropriate permissions and roles granted to them by the organization administrator. The permissions and roles must be granted on the database group or on the database, and they only apply within the organization in which they are granted.

Learning Database Administration

There are several education institutes that offer professional courses, including late-night programs, to allow candidates to learn database administration. Also, DBMS vendors such as Oracle, Microsoft and IBM offer certification programs to help companies to hire qualified DBA practitioners. College degree in Computer Science or related field is helpful but not necessarily a prerequisite.

Broadcast Automation

Broadcast automation incorporates the use of broadcast programming technology to automate broadcasting operations. Used either at a broadcast network, radio station or a television station, it can run a facility in the absence of a human operator. They can also run in a "live assist" mode when there are on-air personnel present at the master control, television studio or control room.

The radio transmitter end of the airchain is handled by a separate automatic transmission system (ATS).

History

Originally, in the USA, many (if not most) broadcast licensing authorities required a licensed board operator to run every station at all times, meaning that every DJ had to pass an exam to obtain a license to be on-air, if their duties also required them to ensure proper operation of the transmitter. This was often the case on overnight and weekend shifts when there was no broadcast engineer present, and all of the time for small stations with only a contract engineer on call.

In the U.S., it was also necessary to have an operator on duty at all times in case the Emergency Broadcast System (EBS) was used, as this had to be triggered manually. While there has not been a requirement to relay any other warnings, any mandatory messages from the U.S. president would have had to first be authenticated with a code word sealed in a pink envelope sent annually to stations by the Federal Communications Commission (FCC).

Gradually, the quality and reliability of electronic equipment improved, regulations were relaxed, and no operator had to be present while a station was operating. In the U.S., this came about when the EAS replaced the EBS, starting the movement toward automation to assist, and sometimes take the place of, the live disc jockeys (DJs) and radio personalities.

Early Analog Systems

Early automation systems were electromechanical systems which used relays. Later systems were "computerized" only to the point of maintaining a schedule, and were limited to radio rather than TV. Music would be stored on reel-to-reel audio tape. Subaudible tones on the tape marked the end of each song. The computer would simply rotate among the tape players until the computer's internal clock matched that of a scheduled event. When a scheduled event would be encountered, the computer would finish the currently-playing song and then execute the scheduled block of events. These events were usually advertisements, but could also include the station's top-of-hour station identification, news, or a bumper promoting the station or its other shows. At the end of the block, the rotation among tapes resumed.

Harris automation system used at the former WWJQ (now WPNW) in 1993.

Advertisements, jingles, and the top-of-hour station identification required by law were often on "carts". Short for cartridges, these were endless tapes similar to 8-track tapes, and looked nearly identical as well. A primary difference between carts and 8 track deals with the pinch roller and capstan. The roller was self-contained in an 8-track; carts had a slot for a pinch roller on a spindle which was activated by solenoid upon pressing the start button on the cart machine. This allowed for nearly instantaneous playback start without artifacts. Mechanical carousels would rotate the carts in and out of multiple tape players as dictated by the computer. Time announcements were provided by a pair of dedicated cart players, with the even minutes stored on one and the odd minutes on the other. This meant an announcement would always be ready to play, even if the minute was changing when the announcement was triggered. The system did require attention throughout the day to change reels as they ran out and reload carts. It became obsolete when a method was developed to automatically rewind and re-cue the reel tapes when they ran out, extending 'walk-away' time indefinitely.

Radio station WIRX may have been one of the world's first completely automated radio stations, built and designed by Brian Jeffrey Brown in 1963 when Brown was only 10 years old. The station broadcast in a classical format, called "More Good Music (MGM)" and featured five minute bottom of the hour news feeds from the Mutual Broadcasting System. The heart of the automation was an 8 x 24 telephone stepping relay which controlled two reel-to-reel tape decks, one twelve inch Ampex machine providing the main program audio and a second RCA seven inch machine providing "fill" music. The tapes played by these machines were originally produced in the MWF's Madison, Wisconsin production facility by WSJM Chief Engineer Richard E. McLemore (and later in-house at WSJM) with sub-audible tones used to signal the end of a song. The stepping relay was programmed by slide switches in the front of the two relay racks which housed the equipment. The news feeds were triggered by a microswitch which was attached to a Western Union clock and tripped by the minute hand of the clock. and then reset the stepping relay. Originally, 30-minute station identification was accomplished by a simulcast switch in the control booth for sister station WSJM (AM), whereupon the disc jockey in the booth would announce "This is WSJM-AM and... (then pressing the momentary contact button)...WSJM-FM, St. Joseph, Michigan." This only lasted about six months, however, and a standard tape cartridge player was wired in to announce the station identification and triggered by the Western Union clock.

A different technology appeared in 1980 with the analog recorders made by Solidyne, which used a computer-controlled tape positioning system. Four GMS 204 units were controlled from a 6809 microprocessor, with the program stored in a solid-state plug-in memory module. This system has a limited programming time of about eight hours.

Satellite programming often used audible dual-tone multi-frequency (DTMF) signals to trigger events at affiliate stations. This allowed the automatic local insertion of ads and station IDs. Because there are 12 (or 16) tone pairs, and typically four tones were sent in rapid succession (less than one second), more events could be triggered than by sub-audible tones (usually 25 Hz and 35 Hz).

Modern Digital Systems

Modern systems run on hard disk, where all of the music, jingles, advertisements, voice tracks, and other announcements are stored. These audio files may be either compressed or uncompressed, or often with only minimal compression as a compromise between file size and quality. For radio software, these disks are usually in computers, sometimes running their own custom operating systems, but more often running as an application on a stable OS like GNU/Linux, Windows NT or others.

Scheduling was an important advance of these systems, allowing for exact timing. Some systems use GPS satellite receivers to obtain exact atomic time, for perfect synchronization with satellite-delivered programming. Reasonablyaccurate timekeeping can also be obtained with the use of Internet protocols (IP) like Network Time Protocol (NTP).

Automation systems are also more interactive than ever before with digital audio workstation (DAW) with console automation and can even record from a telephone hybrid to play back an edited conversation with a telephone caller. This is part of a system's live-assist mode.

The use of automation software and voice tracks to replace live DJs is a current trend in radio broadcasting, done by many Internet radio and adult hits stations. Stations can even be voice-tracked from another city far away, now often delivering sound files over the Internet. In the U.S., this is a common practice under controversy for making radio more generic and artificial. Having local content is also touted as a way for traditional stations to compete with satellite radio, where there may be no radio personality on the air at all.

A commercially-available, for-sale product named Audicom was introduced by Oscar Bonello in 1989. It is based on psychoacoustic lossy compression, the same principle being used in most modern lossy audio encoders such as MP3 and Advanced Audio Coding (AAC), and it allowed both broadcast automation and recording to hard drives.

Television

In television, playout automation is also becoming more practical as the storage space of hard drives increases. Television shows and television commercials, as well as digital on-screen graphics, can all be stored on video servers remotely controlled by computers utilizing the 9-Pin Protocol and the Video Disk Control Protocol (VDCP). These systems can be very extensive, tied-in with parts that allow the "ingest" of video from satellite networks and electronic news gathering (ENG) operations and management of the video library, including archival of footage for later use. In ATSC, Programming Metadata Communication Protocol (PMCP) is then used to pass information about the video through the airchain to Program and System Information Protocol (PSIP), which transmits the current electronic program guide (EPG) information over digital television to the viewer.

Mix Automation

Modern digital audio consoles or mixers use automation. Automation allows the console to remember the audio engineer's adjustment of faders during the post-production editing process. A timecode is necessary for synchronization of automation.

Types of Automation

- Voltage Controlled Automation: fader levels are regulated by voltage-controlled amplifiers (VCA). VCAs control the audio level and not the actual fader.

- Moving Fader Automation: a motor is attached to the fader, which then can be controlled by the console, digital audio workstation (DAW), or user.
- Software Controlled Automation: the software can be internal to the console, or external as part of a DAW. The virtual fader can be adjusted in the software by the user.
- MIDI Automation: the communications protocol MIDI can be used to send messages to the console to control automation.

Modes of Automation

- Auto Write: used the first time automation is created or when writing over existing automation
- Auto Touch: writes automation data only while a fader is touched/faders return to any previously automated position after release
- Auto Latch: starts writing automation data when a fader is touched/stays in position after release
- Auto Read: digital Audio Workstation performs the written automation
- Auto Off: automation is temporarily disabled

All of these include the mute button. If mute is pressed during writing of automation, the audio track will be muted during playback of that automation. Depending on software, other parameters such as panning, sends, and plug-in controls can be automated as well. In some cases, automation can be written using a digital potentiometer (d-pot) instead of a fader.

Building Automation

Building automation is the goal that a Building Management System or a Building Automation System (BAS) attempts to achieve. Both are examples of a distributed control system - the computer networking of electronic devices designed to monitor and control the mechanical, security, fire and flood safety, lighting (especially emergency lighting), HVAC and humidity control and ventilation systems in a building.

BAS core functionality keeps building climate within a specified range, lights rooms based on an occupancy schedule (in the absence of overt switches to the contrary), monitors performance and device failures in all systems, provides malfunction alarms (*via* typically email and/or text notifications) to building engineering/maintenance staff and contractors. BAS reduce building energy and maintenance costs compared to a non-controlled building. Typically they are financed through energy and insurance savings, and other savings associated with pre-emptive maintenance and quick detection of issues.

A building controlled by a BAS is often referred to as an intelligent building, "smart building", or (if a residence) a "smart home". Commercial and industrial buildings have historically relied on robust proven protocols (like BACnet) while proprietary and poorly integrated purpose-specific protocols (like X-10 or those from Honeywell, Siemens or other major manufacturers of smart thermostats, *etc.*)) were used in homes. Recent IEEE standards (notably IEEE 802.15.4, IEEE 1901 and IEEE 1905.1, IEEE 802.21, IEEE 802.11ac, IEEE 802.3at) and consortia efforts like nVoy or QIVICON have provided a standards-based foundation for heterogeneous networking of many devices on many physical networks for diverse purposes, and quality of service and failover guarantees appropriate to support human health and safety. Accordingly commercial, industrial, military and other institutional users now use systems that differ from home systems mostly in scale.

Almost all multi-story green buildings are designed to accommodate a BAS for the energy, air and water conservation characteristics. Electrical device demand response is a typical function of a BAS, as is the more sophisticated ventilation and humidity monitoring required of "tight" insulated buildings. Most green buildings also use as many low-power DC devices as possible, typically integrated with power over Ethernet wiring, so by definition always accessible to a BAS through the Ethernet connectivity. Even a passivhaus design intended to consume no net energy whatsoever will typically require a BAS to manage heat capture, shading and venting, and scheduling device use.

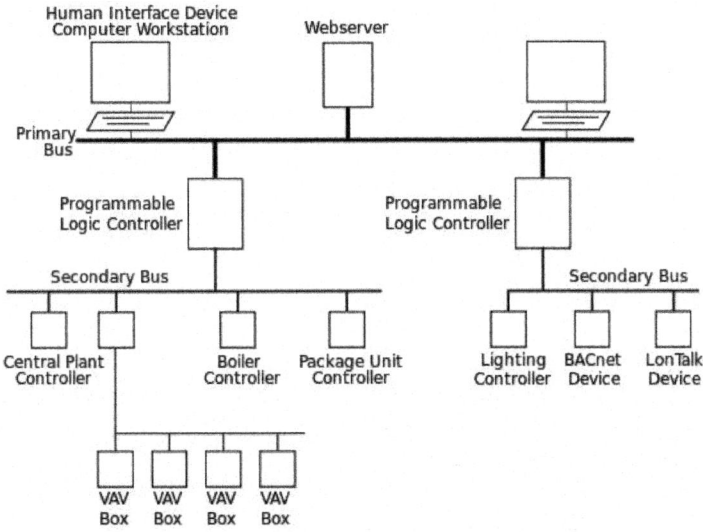

History

The term "Building Automation System", loosely used, refers to any electrical control system that is used to controls a buildings heating, cooling, and ventilation system (HVAC). Modern BAS can also control indoor and outdoor lighting as well as security, fire alarms, and basically everything else that is electrical in the building, on either AC or DC wiring. Old HVAC control systems, such as 24VDC wired thermostats, technically qualify as a building automation system but most such buildings could benefit greatly from the new control systems that are available. There is little resemblance between the IPv6-based technology used in the 2010s and the single purpose systems installed in the 2000s.

Buses and Protocols

Most building automation networks consist of a *primary* and *secondary* bus which connect high-level controllers (generally specialized for building auto-mation, but may be generic programmable logic controllers) with lower-level controllers, input/output devices and a user interface (also known as a human interface device). ASHRAE's open protocol BACnet or the open protocol Lon-Talk specify how most such devices interoperate. Modern systems use SNMP to track events, building on decades of history with SNMP-based protocols in the computer networking world.

Physical connectivity between devices was historically provided by dedicated optical fiber, ethernet, ARCNET, RS-232, RS-485 or a low-bandwidth special pur-pose wireless network. Modern systems rely on standards-based multi-protocol heterogeneous networking such as that specified in the IEEE 1905.1 standard and verified by the nVoy auditing mark. These accommodate typically only IP-based networking but can make use of any existing wiring, and also integrate powerline networking over AC circuits, power over Ethernet low power DC circuits, high-bandwidth wireless networks such as LTE and IEEE 802.11n and IEEE 802.11ac and often integrate these using the building-specific wireless mesh open standard ZigBee).

Proprietary hardware dominates the controller market. Each company has controllers for specific applications. Some are designed with limited controls and no interoperability, such as simple packaged roof top units for HVAC. Software will typically not integrate well with packages from other vendors. Cooperation is at the Zigbee/BACnet/LonTalk level only.

Current systems provide interoperability at the application level, allowing users to mix-and-match devices from different manufacturers, and to provide in-tegration with other compatible building control systems. These typically rely on SNMP, long used for this same purpose to integrate diverse computer networking devices into one coherent network.

Types of Inputs and Outputs

Analog inputs are used to read a variable measurement. Examples are tem-perature, humidity and pressure sensors which could be thermistor, 4-20 mA, 0-10 volt or platinum resistance thermometer (resistance temperature detector), or wireless sensors.

A digital input indicates if a device is turned on or not. Some examples of an inherently digital input would be a 24VDC/AC signal, current switch, an air flow switch, or a volta-free relay contact. Digital inputs could also be pulse type inputs counting the frequency of pulses over a given period of time. An example is a turbine flow meter transmitting rotation data as a frequency of pulses to an input.

Analog outputs control the speed or position of a device, such as a variable frequency drive, an I-P (current to pneumatics) transducer, or a valve or damper actuator. An example is a hot water valve opening up 25% to maintain a setpoint. Another example is a variable frequency drive ramping up a motor slowly to avoid a hard start.

Digital outputs are used to open and close relays and switches as well as drive a load upon command. An example would be to turn on the parking lot lights when a photocell indicates it is dark outside. Another example would be to open a valve by allowing 24VDC/AC to pass through the output powering the valve. Digital outputs could also be pulse type outputs emitting a frequency of pulses over a given period of time. An example is an energy meter calculating kWh and emitting a frequency of pulses accordingly.

Infrastructure

Controller

Controllers are essentially small, purpose-built computers with input and output capabilities. These controllers come in a range of sizes and capabilities to control devices commonly found in buildings, and to control sub-networks of controllers.

Inputs allow a controller to read temperatures, humidity, pressure, current flow, air flow, and other essential factors. The outputs allow the controller to send command and control signals to slave devices, and to other parts of the system. Inputs and outputs can be either digital or analog. Digital outputs are also sometimes called discrete depending on manufacturer.

Controllers used for building automation can be grouped in 3 categories. Programmable Logic Controllers (PLCs), System/Network controllers, and Terminal Unit controllers. However an additional device can also exist in order to integrate 3rd party systems (*i.e.* a stand-alone AC system) into a central Building automation system).

PLC's provide the most responsiveness and processing power, but at a unit cost typically 2 to 3 times that of a System/Network controller intended for BAS applications. Terminal Unit controllers are usually the least expensive and least powerful.

PLC's may be used to automate high-end applications such as clean rooms or hospitals where the cost of the controllers is less of a concern.

In office buildings, supermarkets, malls, and other common automated buildings the systems will use System/Network controllers rather than PLC's. Most System controllers provide general purpose feedback loops, as well as digital circuits, but lack the millisecond response time that PLC's provide.

System/Network controllers may be applied to control one or more mechanical systems such as an Air Handler Unit (AHU), boiler, chiller, *etc.*, or they may supervise a sub-network of controllers. In the diagram above, System/Network controllers are often used in place of PLCs.

Terminal Unit controllers usually are suited for control of lighting and/or simpler devices such as a package rooftop unit, heat pump, VAV box, or fan coil, *etc.* The installer typically selects 1 of the available pre-programmed personalities best suited to the device to be controlled, and does not have to create new control logic.

Occupancy

Occupancy is one of two or more operating modes for a building automation system. Unoccupied, Morning Warmup, and Night-time Setback are other common modes.

Occupancy is usually based on time of day schedules. In Occupancy mode, the BAS aims to provides a comfortable climate and adequate lighting, often with zone-based control so that users on one side of a building have a different thermostat (or a different system, or sub system) than users on the opposite side.

A temperature sensor in the zone provides feedback to the controller, so it can deliver heating or cooling as needed.

If enabled, Morning Warmup (MWU) mode occurs prior to Occupancy. During Morning Warmup the BAS tries to bring the building to setpoint just in time for Occupancy. The BAS often factors in outdoor conditions and historical experience to optimize MWU. This is also referred to as Optimised Start.

An override is a manually initiated command to the BAS. For example, many wall-mounted temperature sensors will have a push-button that forces the system into Occupancy mode for a set number of minutes. Where present, web interfaces allow users to remotely initiate an override on the BAS.

Some buildings rely on occupancy sensors to activate lighting and/or climate conditioning. Given the potential for long lead times before a space becomes sufficiently cool or warm, climate conditioning is not often initiated directly by an occupancy sensor.

Lighting

Lighting can be turned on, off, or dimmed with a building automation or lighting control system based on time of day, or on occupancy sensor, photosensors and timers. One typical example is to turn the lights in a space on for a half hour since the last motion was sensed. A photocell placed outside a building can sense darkness, and the time of day, and modulate lights in outer offices and the parking lot.

Lighting is also a good candidate for Demand response, with many control systems providing the ability to dim (or turn off) lights to take advantage of DR incentives and savings.

In newer buildings, the lighting control is based on the field bus DALI. Lamps with DALI ballasts are fully dimmable. DALI can also detect lamp and ballast failures on DALI luminaires and signals failures.

New components such as WoodTouch by Tecnalia have provided touch interfaces over wooden surfaces for lighting control.

Air Handlers

Most air handlers mix return and outside air so less temperature/humidity conditioning is needed. This can save money by using less chilled or heated water (not all AHUs use chilled/hot water circuits). Some external air is needed to keep the building's air healthy. To optimize energy efficiency while maintaining healthy indoor air quality (IAQ), demand control (or controlled) ventilation (DCV) adjusts the amount of outside air based on measured levels of occupancy.

Analog or digital temperature sensors may be placed in the space or room, the return and supply air ducts, and sometimes the external air. Actuators are placed on the hot and chilled water valves, the outside air and return air dampers. The supply fan (and return if applicable) is started and stopped based on either time of day, temperatures, building pressures or a combination.

Constant Volume Air-handling Units

The less efficient type of air-handler is a "constant volume air handling unit," or CAV. The fans in CAVs do not have variable-speed controls. Instead, CAVs open and close dampers and water-supply valves to maintain temperatures in the building's spaces. They heat or cool the spaces by opening or closing chilled or hot water valves that feed their internal heat exchangers. Generally one CAV serves several spaces

Variable Volume Air-handling Units

A more efficient unit is a "variable air volume (VAV) air-handling unit," or VAV. VAVs supply pressurized air to VAV boxes, usually one box per room or area. A VAV air handler can change the pressure to the VAV boxes by changing the speed of a fan or blower with a variable frequency drive or by moving inlet guide vanes to a fixed-speed fan. The amount of air is determined by the needs of the spaces served by the VAV boxes.

Each VAV box supply air to a small space, like an office. Each box has a damper that is opened or closed based on how much heating or cooling is required in its space. The more boxes are open, the more air is required, and a greater amount of air is supplied by the VAV air-handling unit.

Some VAV boxes also have hot water valves and an internal heat exchanger. The valves for hot and cold water are opened or closed based on the heat demand for the spaces it is supplying. These heated VAV boxes are sometimes used on the perimeter only and the interior zones are cooling only.

A minimum and maximum CFM must be set on VAV boxes to assure adequate ventilation and proper air balance.

VAV Hybrid Systems

Another variation is a hybrid between VAV and CAV systems. In this system, the interior zones operate as in a VAV system. The outer zones differ in that the heating is supplied by a heating fan in a central location usually with a heating coil fed by the building boiler. The heated air is ducted to the exterior dual duct mixing boxes and dampers controlled by the zone thermostat calling for either cooled or heated air as needed.

Central Plant

A central plant is needed to supply the air-handling units with water. It may supply a chilled water system, hot water system and a condenser water system, as well as transformers and auxiliary power unit for emergency power. If well managed, these can often help each other. For example, some plants generate electric power at periods with peak demand, using a gas turbine, and then use the turbine's hot exhaust to heat water or power an absorptive chiller.

Chilled Water System

Chilled water is often used to cool a building's air and equipment. The chilled water system will have chiller(s) and pumps. Analog temperature sensors measure the chilled water supply and return lines. The chiller(s) are sequenced on and off to chill the chilled water supply.

A chiller is a refrigeration unit designed to produce cool (chilled) water for space cooling purposes. The chilled water is then circulated to one or more cooling coils located in air handling units, fan-coils, or induction units. Chilled water distribution is not constrained by the 100 foot separation limit that applies to DX systems, thus chilled water-based cooling systems are typically used in larger buildings. Capacity control in a chilled water system is usually achieved through modulation of water flow through the coils; thus, multiple coils may be served from a single chiller without compromising control of any individual unit. Chillers may operate on either the vapor compression principle or the absorption principle. Vapor compression chillers may utilize reciprocating, centrifugal, screw, or rotary compressor configurations. Reciprocating chillers are commonly used for capacities below 200 tons; centrifugal chillers are normally used to provide higher capacities; rotary and screw chillers are less commonly used, but are not rare. Heat rejection from a chiller may be by way of an air-cooled condenser or a cooling tower. Vapor compression chillers may be bundled with an air-cooled condenser to provide a packaged chiller, which would be installed outside of the building envelope. Vapor compression chillers may also be designed to be installed separate from the condensing unit; normally such a chiller would be installed in an enclosed central plant space. Absorption chillers are designed to be installed separate from the condensing unit.

Condenser Water System

Cooling tower(s) and pumps are used to supply cool condenser water to the chillers. Because the condenser water supply to the chillers has to be constant, variable speed drives are commonly used on the cooling tower fans to control temperature. Proper cooling tower temperature assures the proper refrigerant head pressure in the chiller. The cooling tower set point used depends upon the refrigerant being used. Analog temperature sensors measure the condenser water supply and return lines.

Hot Water System

The hot water system supplies heat to the building's air-handling unit or VAV box heating coils, along with the domestic hot water heating coils (Calorifier). The hot water system will have a boiler(s) and pumps. Analog temperature sensors are placed in the hot water supply and return lines. Some type of mixing valve is usually used to control the heating water loop temperature. The boiler(s) and pumps are sequenced on and off to maintain supply.

The installation and integration of variable frequency drives can lower the energy consumption of the building's circulation pumps to about 15% of what they had been using before. If that sounds hard to believe, I'll explain, and we can do the math. A variable frequency drive functions by modulating the frequency of the electricity provided to the motor that it powers. In the USA, the electrical grid uses a frequency of 60 Hertz or 60 cycles per second. Variable frequency drives are able to decrease the output and energy consumption of motors by lowering the frequency of the electricity provided to the motor, however the relationship between motor output and energy consumption is not a linear one. If the variable frequency drive provides electricity to the motor at 30 Hertz, the output of the motor will be 50% because 30 Hertz divided by 60 Hertz is 0.5 or 50%. The energy consumption of a motor running at 50% or 30 Hertz will not be 50%, but will instead be something like 18% because the relationship between motor output and energy consumption are not linear. The exact ratios of motor output or Hertz provided to the motor (which are effectively the same thing), and the actual energy consumption of the variable frequency drive / motor combination depend on the efficiency of the variable frequency drive. For example, because the variable frequency drive needs power itself to communicate with the building automation system, run its cooling fan, *etc.*, if the motor always ran at 100% with the variable frequency drive installed the cost of operation or electricity consumption would actually go up with the new variable frequency drive installed. The amount of energy that variable frequency drives consume is nominal and is hardly worth consideration when calculating savings, however it did need to be noted that VFD's do consume energy themselves. Due to the fact that the variable frequency drives rarely ever run at 100% and spend most of their time in the 40% output range, and the fact that now the pumps completely shut down when not needed, the variable frequency drives have reduced the energy consumption of the pumps to around 15% of what they had been using before.

Alarms and Security

All modern building automation systems have alarm capabilities. It does little good to detect a potentially hazardous or costly situation if no one who can solve the problem is notified. Notification can be through a computer, pager, cellular phone voice call, audible alarm, or all of these. For insurance and liability purposes all systems keep logs of who was notified, when and how.

Alarms may immediately notify someone or only notify when alarms build to some threshold of seriousness or urgency. At sites with several buildings, momentary power failures can cause hundreds or thousands of alarms from equipment that has shut down — these should be suppressed and recognized as symptoms of a larger failure. Some sites are programmed so that critical alarms are automatically re-sent at varying intervals. For example, a repeating critical alarm (of an [uninterruptible power supply] in 'by pass') might resound at 10 minutes, 30 minutes, and every 2 to 4 hours thereafter until the alarms are resolved.

- Common temperature alarms are: space, supply air, chilled water supply, hot water supply.
- Pressure, humidity, biological and chemical sensors can determine if ventilation systems have failed mechanically or become infected with contaminants that affect human health.
- Differential pressure switches can be placed on a filter to determine if it is dirty or otherwise not performing.
- Status alarms are common. If a mechanical device like a pump is requested to start, and the status input indicates it is off, this can indicate a mechanical failure. Or, worse, an electrical fault that could represent a fire or shock hazard.
- Some valve actuators have end switches to indicate if the valve has opened or not.
- Carbon monoxide and carbon dioxide sensors can tell if concentration of these in the air are too high, either due to fire or ventilation problems in garages or near roads.
- Refrigerant sensors can be used to indicate a possible refrigerant leak.
- Current sensors can be used to detect low current conditions caused by slipping fan belts, clogging strainers at pumps, or other problems.

Security systems can be interlocked to a building automation system. If occupancy sensors are present, they can also be used as burglar alarms. Because security systems are often deliberately sabotaged, at least some detectors or cameras should have battery backup and wireless connectivity and the ability to trigger alarms when disconnected. Modern systems typically use power-over-Ethernet (which can operate a pan-tilt-zoom camera and other devices up to 30-90 watts) which is capable of charging such batteries and keeps wireless networks free for genuinely wireless applications, such as backup communication in outage.

Fire alarm panels and their related smoke alarm systems are usually hard-wired to override building automation. For example: if the smoke alarm is activated, all the outside air dampers close to prevent air coming into the building, and an exhaust system can isolate the blaze. Similarly, electrical fault detection systems can turn entire circuits off, regardless of the number of alarms this triggers or persons this distresses. Fossil fuel combustion devices also tend to have their own over-rides, such as natural gas feed lines that turn off when slow pressure drops are detected (indicating a leak), or when excess methane is detected in the building's air supply.

Good BAS are aware of these overrides and recognize complex failure conditions. They do not send excessive alerts, nor do they waste precious backup power on trying to turn back on devices that these safety over-rides have turned off. A poor BAS, almost by definition, sends out one alarm for every alert, and does not recognize any manual, fire or electric or fuel safety override. Accordingly good BAS are often built on safety and fire systems.

Room Automation

Room automation is a subset of building automation and with a similar purpose, it is the consolidation of one or more systems under centralized control, though in this case in one room.

The most common example of *room automation* is corporate boardroom, presentation suites, and lecture halls, where the operation of the large number of devices that define the room function (such as videoconferencing equipment, video projectors, lighting control systems, public address systems *etc.*) would make manual operation of the room very complex. It is common for room automation systems to employ a touchscreen as the primary way of controlling each operation.

Domotronics

Domotronics Dom-o-tronics "Domus" (Latin for house) -o- "tronics" (Electronics) deals with the interdisciplinary interaction and intelligent networking of building, energy and communications technology in modern especially larger buildings with complex requirements. The aim of the Domotronics is to create buildings, which guarantee optimal comfort, provide a healthy environment, integrate numerous services such as security, communication, HVAC and assistance systems and respond to the occupants and environment through specific sensors for light and heat *etc.* and automated controls. Especially suited to large buildings where passive controls may be inadequate or impractical they have also become popular in modern "luxury" houses.

Specific Purpose

Automated Attendant

In telephony, an **automated attendant** (also **auto attendant, auto-attendant, autoattendant** or **AA**, or **virtual receptionist**) allows callers to be automatically transferred to an extension without the intervention of an operator/receptionist). Many AAs will also offer a simple menu system. An auto attendant may also allow a caller to reach a live operator by dialing a number, usually "0". Typically the auto attendant is included in a business's phone system such as a PBX, but some services allow businesses to use an AA without such a system. Modern AA services (which now overlap with more complicated interactive voice response or IVR systems) can route calls to mobile phones, VoIP virtual phones, other AAs/IVRs, or other locations using traditional land-line phones.

Feature Description

Telephone callers will recognize an automated attendant system as one that greets calls incoming to an organization with a recorded greeting of the form, "Thank you for calling.... If you know your party's extension, you may dial it any time during this message." Callers who have a touch tone (DTMF) phone can dial an extension number or, in most cases, wait for operator ("attendant") assistance. Since the telephone network does not transmit the DC signals from rotary dial telephones, callers who have rotary dial phones have to wait for assistance.

On a purely technical level it could be argued that an automated attendant is a very simple kind of IVR, however in the telecom industry the terms **IVR** and **Auto Attendant** are generally considered distinct. An Automated Attendant serves a very specific purpose (replace live operator and route calls), whereas an IVR can perform all sorts of functions (telephone banking, account inquiries, *etc.*).

An AA will often include a directory which will allow a caller to dial by name in order to find user on a system. There is no standard format to these directories, and they can use combinations of first name, last name, or both.

The following lists common routing steps that are components of an automated attendant (any other routing steps would probably be more suitable to an IVR):

- Transfer to Extension
- Transfer to Voicemail
- Play Message (*i.e.*, "our address is...")
- Go To a Sub Menu
- Repeat Choices

In addition, an Automated Attendant would be expected to have values for the following

- '0' - where to go when the caller dials '0'
- Timeout - what to do if the caller does nothing (usually go to the same place as '0')
- Default mailbox - where to send calls if '0' is not answered (or is not pointing to a live person)

Background

PBXs (Private Branch Exchanges) or PABXs (Private Automatic Branch Exchanges) are telephone systems that serve an organization that has many telephone extensions but fewer telephone lines (sometimes called "trunks") that connect that organization to the rest of the global telecommunications network.

While persons within an enterprise served by a PBX can call each other by dialing their extension numbers, incoming calls, *i.e.*, calls originating from a telephone not served by the PBX but intended for a party served by the PBX, required assistance from a switchboard operator or a telephone service called DID ("Direct Inward Dialing"). Direct Inward Dialing has advantages such as rapid connection to the destination party and disadvantages including cost, lack of identification of the called organization and use of ten-digit telephone numbers.

Automated attendants provide, among many other things, a way for an external caller to be directed to an extension or department served by a PBX system without using Direct Inward Dialing or without switchboard attendant assistance.

History

While many users may think an automated attendant is part of voice mail, voice messaging technology was around since the late 1970s but in the early 1980s companies provided voice prompting systems that allowed callers to reach the intended party, not necessarily to leave a message for the intended party.

Time-based Routing

Many auto attendants will have options to allow for time of day routing, as well as weekend and holiday routing. The specifics of these features will depend entirely on the particular automated attendant, but typically there would be a normal greeting and routing steps that would take place during normal business hours, and a different greeting and routing for non-business hours.

Automatic Painting (Robotic)

Automatic painting is also used to describe painting using a machine or robot.

Industrial robots have been used for decades in automotive applications, including painting, from the first hydraulic versions, which are still in use today but cannot match the quality or safety of the electric robots, to the latest electric offerings from the robot Original Equipment Manufacturers. The newest robots are more accurate and deliver better results with uniform film builds and precise thicknesses.

Originally, industrial paint robots were big and expensive, but today the price of the robots, new and used, have come down to the point that general industry can now afford to have the same level of automation that only the big automotive manufacturers could only once afford.

The selection of today's paint robots is much greater; they vary in size and payload to allow many configurations for painting big items like Boeing 747s and small items like door handles. The prices vary as well, as the new robot market becomes more competitive and the used robot market continues to expand. It is possible to purchase a good used paint robot for as little as $25K.

Painting robots are generally equipped with five to six degrees of freedom, three for the base motions and up to three for applicator orientation. These robots can be used in any explosion hazard Class 1 Division 1 environment.

Robotic Lawn Mower

A **robotic lawn mower** is an autonomous robot used to cut lawn grass. A typical robotic lawn mower requires the user to set up a border wire around the lawn that defines the area to be mowed. The robot uses this wire to locate the boundary of the area to be trimmed and in some cases to locate a recharging dock. Robotic mowers are capable of maintaining up to 20,000 m² (220,000 sq ft) of grass.

Robotic lawn mowers are increasingly sophisticated, are self-docking and some contain rain sensors if necessary, nearly eliminating human interaction. Robotic lawn mowers represented the second largest category of domestic robots used by the end of 2005.

Possibly the first commercial robotic lawn mower was the MowBot, introduced and patented in 1969 and already showing many features of today's most popular products.

In 2012, the growth of robotic lawn mower sales was 15 times that of the traditional styles. With the emergence of smart phones some robotic mowers have integrated features within custom apps to adjust settings or scheduled mowing times and frequency, as well as manually control the mower with a digital joystick.

Technology

In 1995, the first fully solar powered robotic mower became available.

The mower can find its charging station *via* radio frequency emissions, by following a boundary wire, or by following an optional guide wire. This can eliminate wear patterns in the lawn caused by the mower only being able to follow one wire back to the station.

To get to remote and areas only accessible through narrow passages the mower can follow a guide wire or a boundary wire out of the station.

Batteries used varies from NiMH, Li-ion and lead-acid.

Telephone Switchboard

A **telephone switchboard** is a telecommunications system used in the public switched telephone network or in enterprises to interconnect circuits of telephones to establish telephone calls between the subscribers or users, or between other exchanges. The switchboard was an essential component of a manual telephone exchange, and was operated by one or more persons, called operators who either used electrical cords or switches to establish the connections.

PBX switchboard, 1975

The electromechanical automatic telephone exchange, invented by Almon Strowger in 1888, gradually replaced manual switchboards in central telephone exchanges starting in 1919 when the Bell System adopted automatic switching, but many manual branch exchanges remained operational during the last half of the 20th century in offices, hotels, or other enterprises. Later electronic devices and computer technology gave the operator access to an abundance of features. In modern businesses, a private branch exchange (PBX) often has an attendant console for the operator, or an auto-attendant, which bypasses the operator entirely.

Operation

The switchboard is usually designed to accommodate the operator, who sits facing it. It has a high back panel, which consists of rows of female jacks, each jack designated and wired as a local extension of the switchboard (which serves an individual subscriber) or as an incoming or outgoing trunk line. The jack is also associated with a lamp.

On the table or desk area in front of the operator are columns of keys, lamps and cords. Each column consists of a front key and a rear key, a front lamp and a rear lamp, followed by a front cord and a rear cord, making up together a cord circuit. The front key is the "talk" key allowing the operator to speak with that particular cord pair. The rear key on older "manual" boards and PBXs is used to physically ring a telephone. On newer boards, the back key is used to collect (retrieve) money from coin telephones. Each of the keys has three positions: back, normal and forward. When a key is in the normal position an electrical talk path connects the front and rear cords. A key in the forward position connects the operator to the cord pair, and a key in the back position sends a ring signal out on the cord (on older manual exchanges). Each cord has a three-wire TRS phone connector: tip and ring for testing, ringing and voice; and a sleeve wire for busy signals.

When a call is received, a jack lamp lights on the back panel and the operator responds by placing the rear cord into the corresponding jack and throwing the front key forward. The operator then converses with the caller, who informs the operator to whom he or she would like to speak. If it is another extension, the operator places the front cord in the associated jack and pulls the front key backwards to ring the called party. After connecting, the operator leaves both cords "up" with the keys in the normal position so the parties can converse. The supervision lamps light to alert the operator when the parties finish their conversation and go on-hook. Either party could "flash" the operator's supervision lamps by depressing their switch hook for a second and releasing it, in case they needed assistance with a problem. When the operator pulls down a cord, a pulley weight behind the switchboard pulls it down to prevent it from tangling.

On a trunk, on-hook and off-hook signals must pass in both directions. In a one-way trunk, the originating or A board sends a short for off-hook, and an open for on-hook, while the terminating or B board sends normal polarity or reverse polarity. This "reverse battery" signaling was carried over to later automatic exchanges.

History

The first telephones in the 1870s were rented in pairs which were limited to conversation between those two instruments. The use of a central exchange was soon found to be even more advantageous than in telegraphy. In January 1878 the Boston Telephone Dispatch company had started hiring boys as telephone operators. Boys had been very successful as telegraphy operators, but their attitude and behaviour was unacceptable for live phone contact, so the company began hiring women operators instead. Thus, on September 1, 1878, Boston Telephone Dispatch hired Emma Nutt as the first woman operator. Small towns typically had the switchboard installed in the operator's home so that he or she could answer calls on a 24 hour basis. In 1894, New England Telephone and Telegraph Company installed the first battery-operated switchboard on January 9 in Lexington, Massachusetts.

Early switchboards in large cities usually were mounted floor to ceiling in order to allow the operators to reach all the lines in the exchange. The operators were boys who would use a ladder to connect to the higher jacks. Late in the 1890s this measure failed to keep up with the increasing number of lines, and Milo G. Kellogg devised the **Divided Multiple Switchboard** for operators to work together, with a team on the "A board" and another on the "B." These operators were almost always women until the early 1970s, when men were once again hired. Cord switchboards were often referred to as "cordboards" by telephone company personnel. Conversion to Panel switch and other automated switching systems first eliminated the "B" operator and then, usually years later, the "A". Rural and suburban switchboards for the most part remained small and simple. In many cases, customers came to know their operator by name.

As telephone exchanges converted to automatic (dial) service, switchboards continued to serve specialized purposes. Before the advent of direct-dialed long distance calls, a subscriber would need to contact the long-distance operator in order to place a toll call. In large cities, there was often a special number, such as

112, which would ring the long-distance operator directly. Elsewhere, the subscriber would ask the local operator to ring the long-distance operator.

The long distance operator would record the name and city of the person to be called, and the operator would advise the calling party to hang up and wait for the call to be completed. Each toll center had only a limited number of trunks to distant cities, and if those circuits were busy, the operator would try alternate routings through intermediate cities. The operator would plug into a trunk for the destination city, and the inward operator would answer. The inward operator would obtain the number from the local information operator, and ring the call. Once the called party answered, the originating operator would advise him or her to stand by for the calling party, whom she'd then ring back, and record the starting time, once the conversation began.

In the 1940s, with the advent of dial pulse and multi-frequency operator dialing, the operator would plug into a tandem trunk and dial the NPA and operator code for the information operator in the distant city. For instance, the New York City information operator was 212-131. If the customer knew the number, and the point was direct-dialable, the operator would dial the call. If the distant city did not have dialable numbers, the operator would dial the code for the inward operator serving the called party, and ask her to ring the number.

In the 1960s, once most phone subscribers had direct long-distance dialing, a single type of operator began to serve both the local and long distance functions. A customer might call to request a collect call, a call billed to a third number, or a person-to-person call. All toll calls from coin phones required operator assistance. The operator was also available to help complete a local or long-distance number which did not complete. For example, if a customer encountered a reorder tone (a fast busy signal), it could indicate "all circuits busy," or a problem in the destination exchange. The operator might be able to use a different routing to complete the call. If the operator could not get through by dialing the number, she could call the inward operator in the destination city, and ask her to try the number, or to test a line to see if it was busy or out of order.

Cord switchboards used for these purposes were replaced in the 1970s and 1980s by TSPS and similar systems, which greatly reduced operator involvement in calls. The customer would, instead of simply dialing "0" for the operator, dial 0+NPA+7digits, after which an operator would answer and provide the desired service (coin collection, obtaining acceptance on a collect call, *etc.*), and then release the call to be automatically handled by the TSPS.

Before the late 1970s and early 1980s, it was common for many smaller cities to have their own operators. An NPA would usually have its largest city as its primary toll center, with smaller toll centers serving the secondary cities scattered throughout the NPA. TSPS allowed telephone companies to close smaller toll centers and consolidate operator services in regional centers which might be hundreds of miles from the subscriber.

Virtual Switchboard

A virtual switchboard is an automated system used to connect an incoming caller with an agent or staff member. The virtual switchboard user normally has the option of controlling how incoming calls are routed *via* a web interface. For example calls could be routed to different destinations according to certain criteria such as the time of day *etc.*

Interactive voice response (IVR) functionality is also a common feature with virtual switchboards. IVR enables incoming callers to a virtual switchboard to hear prerecorded announcements. Popular announcements instruct callers to press a number on their key pad to select which department they want to reach (for example, "press 1 for sales, 2 for accounts, 3 for support, and so on").

Automated Teller Machine

An NCR Personas 75-Series interior, multi-function ATM in the United States

An **automated teller machine** or **automatic teller machine** (**ATM**, American, Australian, Singaporean, Indian, and Hiberno-English), also known as an **automated banking machine** (**ABM**, Canadian English), **cash machine, cashpoint, cashline,** or colloquially **hole in the wall** (Australian, British, South African, and Sri Lankan English), is an electronic telecommunications device that enables the customers of a financial institution to perform financial transactions without the need for a human cashier, clerk or bank teller.

On most modern ATMs, the customer is identified by inserting a plastic ATM card with a magnetic stripe or a plastic smart card with a chip that contains a unique card number and some security information such as an expiration date or CVVC (CVV). Authentication is provided by the customer entering a personal identification number (PIN).

Using an ATM, customers can access their bank deposit or credit accounts in order to make a variety of transactions such as cash withdrawals, check balances, or credit mobile phones. If the currency being withdrawn from the ATM is different from that in which the bank account is denominated the money will be converted at an official exchange rate. Thus, ATMs often provide the best possible exchange rates for foreign travellers, and are widely used for this purpose.History

The idea of self-service in retail banking developed through independent and simultaneous efforts in Japan, Sweden, the United Kingdom and the United States. In the US patent record, Luther George Simjian has been credited with developing a "prior art device". Specifically his 132nd patent was first filed on 30 June 1960. The roll-out of this machine, called Bankograph, was delayed by a couple of years, due in part to Simjian's Reflectone Electronics Inc. being acquired by Universal Match Corporation. An experimental Bankograph was installed in New York City in 1961 by the City Bank of New York, but removed after six months due to the lack of customer acceptance. The Bankograph was an automated envelope deposit machine and did not have cash dispensing features.

In simultaneous and independent efforts, engineers in Japan, Sweden, and Britain developed their own cash machines during the early 1960s. The first of these that was put into use was by Barclays Bank in Enfield Town in North London, United Kingdom, on 27 June 1967. This machine was the first in the world and was used by English comedy actor Reg Varney, at the time so as to ensure maximum publicity for the machines that were to become mainstream in the UK. This instance of the invention is credited to John Shepherd-Barron of printing firm De La Rue, who was awarded an OBE in the 2005 New Year Honours. This design used paper cheques issued by a teller, marked with carbon-14 for machine readability and security, which in a latter model were matched with a personal identification number.

The Barclays-De La Rue machine (called De La Rue Automatic Cash System or DACS) beat the Swedish saving banks' and a company called Metior's machine (a device called Bankomat) by a mere nine days and Westminster Bank's-Smith Industries-Chubb system by a month. The online version of the Swedish machine is listed to been operational on 6 May 1968, while claiming to be the first online cash machine in the world (ahead of a similar claim by IBM and Lloyds Bank in 1971). The collaboration of a small start-up called Speytec and Midland Bank developed a third machine which was marketed after 1969 in Europe and the USA by the Burroughs Corporation. The patent for this device was filed on September

1969 by John David Edwards, Leonard Perkins, John Henry Donald, Peter Lee Chappell, Sean Benjamin Newcombe & Malcom David Roe.

Both the DACS and MD2 accepted only a single-use token or voucher which was retained by the machine while the Speytec worked with a card with a magnetic strip at the back. They used principles including Carbon-14 and low-coercivity magnetism in order to make fraud more difficult. The idea of a PIN stored on the card was developed by a British engineer working on the MD2 named James Goodfellow in 1965 (patent GB1197183 filed on 2 May 1966 with Anthony Davies). The essence of this system was that it enabled the verification of the customer with the debited account without human intervention. This patent is also the earliest instance of a complete "currency dispenser system" in the patent record. This patent was filed on 5 March 1968 in the USA and granted on 1 December 1970. It had a profound influence on the industry as a whole.

Docutel United States 1969

After looking first hand at the experiences in Europe, in 1968 the networked ATM was pioneered in the US, in Dallas, Texas, by Donald Wetzel, who was a department head at an automated baggage-handling company called Docutel. Recognised by the United States Patent Office for having invented the ATM network are Fred J. Gentile and Jack Wu Chang, under US Patent # 3,833,885. On September 2, 1969, Chemical Bank installed the first ATM in the U.S. at its branch in Rockville Centre, New York. The first ATMs were designed to dispense a fixed amount of cash when a user inserted a specially coded card. A Chemical Bank advertisement boasted "On Sept. 2 our bank will open at 9:00 and never close again." Chemical's ATM, initially known as a Docuteller was designed by Donald Wetzel and his company Docutel. Chemical executives were initially hesitant about the electronic banking transition given the high cost of the early machines. Additionally, executives were concerned that customers would resist having machines handling their money. In 1995, the Smithsonian National Museum of American History recognised Docutel and Wetzel as the inventors of the networked ATM.

The first modern ATM was an IBM 2984 and came into use at Lloyd Bank, Brentwood High Street, Essex, England in December 1972. The IBM 2984 was designed at the request of Lloyds Bank. The 2984 Cash Issuing Terminal was the first true ATM, similar in function to today's machines and named a by Lloyds Bank: Cashpoint; Cashpoint is still a registered trademark of Lloyds TSB in the UK. All were online and issued a variable amount which was immediately deducted from the account. A small number of 2984s were supplied to a US bank. A couple of well known historical models of ATMs include the IBM 3614, IBM 3624 and 473x series, Diebold 10xx and TABS 9000 series, NCR 1780 and earlier NCR 770 series.

The newest ATM at Royal Bank of Scotland allows customers to withdraw cash up to £100 without a card by inputting a six-digit code requested through their smartphones.

Location

ATMs are placed not only near or inside the premises of banks, but also in locations such as shopping centers/malls, airports, grocery stores, petrol/gas stations, restaurants, or anywhere frequented by large numbers of people. There are two types of ATM installations: on- and off-premise. On-premise ATMs are typically more advanced, multi-function machines that complement a bank branch's capabilities, and are thus more expensive. Off-premise machines are deployed by financial institutions and Independent Sales Organisations (ISOs) where there is a simple need for cash, so they are generally cheaper single function devices. In Canada, ATMs (also known there as ABMs) not operated by a financial institution are known as "white-label ABMs".

In the U.S., Canada and some Gulf countries, banks often have drive-thru lanes providing access to ATMs using an automobile.

Many ATMs have a sign above them, indicating the name of the bank or organisation owning the ATM and possibly including the list of ATM networks to which that machine is connected.

ATMs can also be found in train stations and metro stations. In recent times, countries like India and some countries in Africa are installing ATM's in rural areas as well, which are solar powered. These ATM's also do not require air conditioning.

Financial Networks

Most ATMs are connected to interbank networks, enabling people to withdraw and deposit money from machines not belonging to the bank where they have their accounts or in the countries where their accounts are held. Some examples of interbank networks include NYCE, PULSE, PLUS, Cirrus, AFFN, Interac, Interswitch, STAR, LINK, MegaLink and BancNet.

ATMs rely on authorisation of a financial transaction by the card issuer or other authorising institution *via* the communications network. This is often performed through an ISO 8583 messaging system.

Many banks charge ATM usage fees. In some cases, these fees are charged solely to users who are not customers of the bank where the ATM is installed; in other cases, they apply to all users.

In order to allow a more diverse range of devices to attach to their networks, some interbank networks have passed rules expanding the definition of an ATM to be a terminal that either has the vault within its footprint or utilises the vault or cash drawer within the merchant establishment, which allows for the use of a scrip cash dispenser.

ATMs typically connect directly to their host or ATM Controller *via* either ADSL or dial-up modem over a telephone line or directly *via* a leased line. Leased lines are preferable to plain old telephone service (POTS) lines because they require less time to establish a connection. Less-trafficked machines will usually rely on a dial-up modem on a POTS line rather than using a leased line, since a leased line may be comparatively more expensive to operate versus a POTS line. That dilemma may be solved as high-speed Internet VPN connections become more ubiquitous. Common lower-level layer communication protocols used by ATMs to communicate back to the bank include SNA over SDLC, TC500 over Async, X.25, and TCP/IP over Ethernet.

In addition to methods employed for transaction security and secrecy, all communications traffic between the ATM and the Transaction Processor may also be encrypted *via* methods such as SSL.

Global Use

There are no hard international or government-compiled numbers totaling the complete number of ATMs in use worldwide. Estimates developed by ATMIA place the number of ATMs in use currently at over 2.2 million, or approximately 1 ATM per 3000 people in the world.

To simplify the analysis of ATM usage around the world, financial institutions generally divide the world into seven regions, due to the penetration rates, usage statistics, and features deployed. Four regions have high numbers of ATMs per million people. Despite the large number of ATMs, there is additional demand for machines in the Asia/Pacific area as well as in Latin America. ATMs have yet to reach high numbers in the Near East and Africa.

One of the world's most northerly installed ATMs is located at Longyearbyen, Svalbard, Norway.

The world's most southerly installed ATM is located at McMurdo Station, located in New Zealand's Ross Dependency, in Antarctica.

According to international statistics, the highest installed ATM in the world is located at Nathu La Pass, in India, installed by the Indian Axis Bank at 4023 metres (13200 ft). According to the Mainland Chinese media and CPC statistics, the highest installed ATM in the world is located in Nagchu County, Tibet, China, at 4500 metres, allegedly installed by the Agricultural Bank of China.

Israel has the world's lowest installed ATM at Ein Bokek at the Dead Sea, installed independently by a grocery store at 421 metres below sea level.

While ATMs are ubiquitous on modern cruise ships, ATMs can also be found on some US Navy ships.

Hardware

An ATM is typically made up of the following devices:
- CPU (to control the user interface and transaction devices)
- Magnetic or chip card reader (to identify the customer)
- PIN pad EEP4 (similar in layout to a touch tone or calculator keypad), manufactured as part of a secure enclosure
- Secure cryptoprocessor, generally within a secure enclosure

- Display (used by the customer for performing the transaction)
- Function key buttons (usually close to the display) or a touchscreen (used to select the various aspects of the transaction)
- Record printer (to provide the customer with a record of the transaction)
- Vault (to store the parts of the machinery requiring restricted access)
- Housing (for aesthetics and to attach signage to)
- Sensors and indicators

Due to heavier computing demands and the falling price of personal computer-like architectures, ATMs have moved away from custom hardware architectures using microcontrollers or application-specific integrated circuits and have adopted the hardware architecture of a personal computer, such as USB connections for peripherals, Ethernet and IP communications, and use personal computer operating systems.

Business owners often lease ATM terminals from ATM service providers, however based on the economies of scale, the price of equipment has dropped to the point where many business owners are simply paying for ATMs using a credit card.

New ADA voice and text-to-speech guidelines imposed in 2010, but required by March 2012 have forced many ATM owners to either upgrade non-compliant machines or dispose them if they are not up-gradable, and purchase new compliant equipment. This has created an avenue for hackers and thieves to obtain ATM hardware at junkyards from improperly disposed decommissioned ATMs.

The vault of an ATM is within the footprint of the device itself and is where items of value are kept. Scrip cash dispensers do not incorporate a vault.

Mechanisms found inside the vault may include:

- Dispensing mechanism (to provide cash or other items of value)
- Deposit mechanism including a check processing module and bulk note acceptor (to allow the customer to make deposits)
- Security sensors (magnetic, thermal, seismic, gas)
- Locks (to ensure controlled access to the contents of the vault)
- Journaling systems; many are electronic (a sealed flash memory device based on in-house standards) or a solid-state device (an actual printer) which accrues all records of activity including access timestamps, number of notes dispensed, *etc.* This is considered sensitive data and is secured in similar fashion to the cash as it is a similar liability.

ATM vaults are supplied by manufacturers in several grades. Factors influencing vault grade selection include cost, weight, regulatory requirements, ATM type, operator risk avoidance practices and internal volume requirements. Industry standard vault configurations include Underwriters Laboratories UL-291 "Business Hours" and Level 1 Safes, RAL TL-30 derivatives, and CEN EN 1143-1 - CEN III and CEN IV.

ATM manufacturers recommend that an ATM vault be attached to the floor to prevent theft, though there is a record of a theft conducted by tunnelling into an ATM floor.

Software

With the migration to commodity Personal Computer hardware, standard commercial "off-the-shelf" operating systems, and programming environments can be used inside of ATMs. Typical platforms previously used in ATM development include RMX or OS/2.

Today the vast majority of ATMs worldwide use a Microsoft Windows operating system, primarily Windows XP Professional or Windows XP Embedded. A small number of deployments may still be running older versions of Windows OS such as Windows NT, Windows CE, or Windows 2000.

There is a computer industry security view that general public desktop operating systems have greater risks as operating systems for cash dispensing machines than other types of operating systems like (secure) real-time operating systems (RTOS). RISKS Digest has many articles about cash machine operating system vulnerabilities.

Linux is also finding some reception in the ATM marketplace. An example of this is Banrisul, the largest bank in the south of Brazil, which has replaced the MS-DOS operating systems in its ATMs with Linux. Banco do Brasil is also migrating ATMs to Linux. Indian-based Vortex Engineering is Manufacturing ATM's which operates only with Linux. Common application layer transaction protocols, such as Diebold 91x and NCR NDC or NDC+ provide emulation of older generations of hardware on newer platforms with incremental extensions made over time to address new capabilities, although companies like NCR continuously improve these protocols issuing newer versions (*e.g.* NCR's AANDC v3.x.y, where x.y are subversions). Most major ATM manufacturers provide software packages that implement these protocols. Newer protocols such as IFX have yet to find wide acceptance by transaction processors.

With the move to a more standardised software base, financial institutions have been increasingly interested in the ability to pick and choose the application programs that drive their equipment. WOSA/XFS, now known as CEN XFS (or simply XFS), provides a common API for accessing and manipulating the various devices of an ATM. J/XFS is a Java implementation of the CEN XFS API.

While the perceived benefit of XFS is similar to the Java's "Write once, run anywhere" mantra, often different ATM hardware vendors have different interpretations of the XFS standard. The result of these differences in interpretation means that ATM applications typically use a middleware to even out the differences between various platforms.

With the onset of Windows operating systems and XFS on ATM's, the software applications have the ability to become more intelligent. This has created a new breed of ATM applications commonly referred to as programmable applications.

These types of applications allows for an entirely new host of applications in which the ATM terminal can do more than only communicate with the ATM switch. It is now empowered to connected to other content servers and video banking systems.

Notable ATM software that operates on XFS platforms include Triton PRISM, Diebold Agilis EmPower, NCR APTRA Edge, Absolute Systems AbsoluteINTER-ACT, KAL Kalignite Software Platform, Phoenix Interactive VISTAatm, Wincor Nixdorf ProTopas and Euronet EFTS.

With the move of ATMs to industry-standard computing environments, concern has risen about the integrity of the ATM's software stack.

Security

Security, as it relates to ATMs, has several dimensions. ATMs also provide a practical demonstration of a number of security systems and concepts operating together and how various security concerns are dealt with.

Physical

Early ATM security focused on making the ATMs invulnerable to physical attack; they were effectively safes with dispenser mechanisms. A number of attacks on ATMs resulted, with thieves attempting to steal entire ATMs by ram-raiding. Since late 1990s, criminal groups operating in Japan improved ram-raiding by stealing and using a truck loaded with a heavy construction machinery to effectively demolish or uproot an entire ATM and any housing to steal its cash.

Another attack method, *plofkraak*, is to seal all openings of the ATM with silicone and fill the vault with a combustible gas or to place an explosive inside, attached, or near the ATM. This gas or explosive is ignited and the vault is opened or distorted by the force of the resulting explosion and the criminals can break in. This type of theft has occurred in the Netherlands, Belgium, France, Denmark, Germany and Australia. This type of attacks can be prevented by a number of gas explosion prevention devices also known as gas suppression system. These systems use explosive gas detection sensor to detect explosive gas and to neutralise it by releasing a special explosion suppression chemical which changes the composition of the explosive gas and renders it ineffective.

Several attacks in the UK (at least one of which was successful) have emulated the traditional WW2 escape from POW camps by digging a concealed tunnel under the ATM and cutting through the reinforced base to remove the money.

Modern ATM physical security, per other modern money-handling security, concentrates on denying the use of the money inside the machine to a thief, by using different types of Intelligent Banknote Neutralisation Systems.

A common method is to simply rob the staff filling the machine with money. To avoid this, the schedule for filling them is kept secret, varying and random. The money is often kept in cassettes, which will dye the money if incorrectly opened.

Transactional Secrecy and Integrity

The security of ATM transactions relies mostly on the integrity of the secure cryptoprocessor: the ATM often uses general commodity components that sometimes are not considered to be "trusted systems".

Encryption of personal information, required by law in many jurisdictions, is used to prevent fraud. Sensitive data in ATM transactions are usually encrypted with DES, but transaction processors now usually require the use of Triple DES. Remote Key Loading techniques may be used to ensure the secrecy of the initialisation of the encryption keys in the ATM. Message Authentication Code (MAC) or Partial MAC may also be used to ensure messages have not been tampered with while in transit between the ATM and the financial network. In some countries a system has been developed that if the ATM card holder is told to withdraw the cash forcefully by the thief then if he entered his card password starting from the last digit to the first digit then the alarm will sound in the nearest police station.

Customer Identity Integrity

There have also been a number of incidents of fraud by Man-in-the-middle attacks, where criminals have attached fake keypads or card readers to existing machines. These have then been used to record customers' PINs and bank card information in order to gain unauthorised access to their accounts. Various ATM manufacturers have put in place countermeasures to protect the equipment they manufacture from these threats.

Alternative methods to verify cardholder identities have been tested and deployed in some countries, such as finger and palm vein patterns, iris, and facial recognition technologies. Cheaper mass-produced equipment has been developed and is being installed in machines globally that detect the presence of foreign objects on the front of ATMs, current tests have shown 99% detection success for all types of skimming devices.

Device Operation Integrity

Openings on the customer-side of ATMs are often covered by mechanical shutters to prevent tampering with the mechanisms when they are not in use. Alarm sensors are placed inside the ATM and in ATM servicing areas to alert their operators when doors have been opened by unauthorised personnel.

Rules are usually set by the government or ATM operating body that dictate what happens when integrity systems fail. Depending on the jurisdiction, a bank may or may not be liable when an attempt is made to dispense a customer's money from an ATM and the money either gets outside of the ATM's vault, or was exposed in a non-secure fashion, or they are unable to determine the state of the money after a failed transaction. Customers often commented that it is difficult to recover money lost in this way, but this is often complicated by the policies regarding suspicious activities typical of the criminal element.

Customer Security

In some countries, multiple security cameras and security guards are a common feature. In the United States, The New York State Comptroller's Office has advised the New York State Department of Banking to have more thorough safety inspections of ATMs in high crime areas.

Consultants of ATM operators assert that the issue of customer security should have more focus by the banking industry; it has been suggested that efforts are now more concentrated on the preventive measure of deterrent legislation than on the problem of ongoing forced withdrawals.

At least as far back as July 30, 1986, consultants of the industry have advised for the adoption of an emergency PIN system for ATMs, where the user is able to send a silent alarm in response to a threat. Legislative efforts to require an emergency PIN system have appeared in Illinois, Kansas and Georgia, but none have succeeded yet. In January 2009, Senate Bill 1355 was proposed in the Illinois Senate that revisits the issue of the reverse emergency PIN system. The bill is again supported by the police and denied by the banking lobby.

In 1998 three towns outside the Cleveland, Ohio, in response to an ATM crime wave, adopted ATM Consumer Security Legislation requiring that an emergency telephone number switch be installed at all outside ATMs within their jurisdiction. In the wake of an ATM Murder in Sharon Hill, Pennsylvania, The City Council of Sharon Hill passed an ATM Consumer Security Bill as well. As of July 2009, ATM Consumer Security Legislation is currently pending in New York, New Jersey, and Washington D.C.

In China and elsewhere, many efforts to promote security have been made. On-premises ATMs are often located inside the bank's lobby which may be accessible 24 hours a day. These lobbies have extensive security camera coverage, a courtesy telephone for consulting with the bank staff, and a security guard on the premises. Bank lobbies that are not guarded 24 hours a day may also have secure doors that can only be opened from outside by swiping the bank card against a wall-mounted scanner, allowing the bank to identify which card enters the building. Most ATMs will also display on-screen safety warnings and may also be fitted with convex mirrors above the display allowing the user to see what is happening behind them.

As of 2013, the only claim available about the extent of ATM connected homicides is that they range from 500 to 1000 nationwide, covering only cases where the victim had an ATM card and the card was used by the killer after the known time of death.

Uses

Although ATMs were originally developed as just cash dispensers, they have evolved to include many other bank-related functions:

- Paying routine bills, fees, and taxes (utilities, phone bills, social security, legal fees, taxes, *etc.*)

- Printing bank statements
- Updating passbooks
- Cash advances
- Cheque Processing Module
- Paying (in full or partially) the credit balance on a card linked to a specific current account.
- Transferring money between linked accounts (such as transferring between checking and savings accounts)
- Deposit currency recognition, acceptance, and recycling

In some countries, especially those which benefit from a fully integrated cross-bank ATM network (*e.g.*: Multibanco in Portugal), ATMs include many functions which are not directly related to the management of one's own bank account, such as:

- Loading monetary value into stored value cards
- Adding pre-paid cell phone / mobile phone credit.
- Purchasing
  - o Postage stamps.
  - o Lottery tickets
  - o Train tickets
  - o Concert tickets
  - o Movie tickets
  - o Shopping mall gift certificates.
  - o Gold
- Donating to charities

Increasingly banks are seeking to use the ATM as a sales device to deliver pre approved loans and targeted advertising using products such as ITM (the Intelligent Teller Machine) from Aptra Relate from NCR. ATMs can also act as an advertising channel for other companies.*

However several different technologies on ATMs have not yet reached worldwide acceptance, such as:

- Videoconferencing with human tellers, known as video tellers
- Biometrics, where authorisation of transactions is based on the scanning of a customer's fingerprint, iris, face, *etc.*
- Cheque/Cash Acceptance, where the ATM accepts and recognise cheques and/or currency without using envelopes Expected to grow in importance in the US through Check 21 legislation.
- Bar code scanning
- On-demand printing of "items of value" (such as movie tickets, traveler's cheques, *etc.*)

- Dispensing additional media (such as phone cards)
- Co-ordination of ATMs with mobile phones
- Integration with non-banking equipment
- Games and promotional features
- CRM at the ATM

E.G. In Canada, ATMs are called *guichets automatiques* in French and some-times "Bank Machines" in English. The Interac shared cash network does not allow for the selling of goods from ATMs due to specific security requirements for PIN entry when buying goods. CIBC machines in Canada, are able to top-up the minutes on certain pay as you go phones.

Reliability

Before an ATM is placed in a public place, it typically has undergone extensive testing with both test money and the backend computer systems that allow it to perform transactions. Banking customers also have come to expect high reliability in their ATMs, which provides incentives to ATM providers to minimise machine and network failures. Financial consequences of incorrect machine operation also provide high degrees of incentive to minimise malfunctions.

ATMs and the supporting electronic financial networks are generally very reliable, with industry benchmarks typically producing 98.25% customer avail-ability for ATMs and up to 99.999% availability for host systems that manage the networks of ATMs. If ATM networks do go out of service, customers could be left without the ability to make transactions until the beginning of their bank's next time of opening hours.

This said, not all errors are to the detriment of customers; there have been cases of machines giving out money without debiting the account, or giving out higher value notes as a result of incorrect denomination of banknote being loaded in the money cassettes. The result of receiving too much money may be influenced by the card holder agreement in place between the customer and the bank.

Errors that can occur may be mechanical (such as card transport mechanisms; keypads; hard disk failures; envelope deposit mechanisms); software (such as operating system; device driver; application); communications; or purely down to operator error.

To aid in reliability, some ATMs print each transaction to a roll paper journal that is stored inside the ATM, which allows both the users of the ATMs and the related financial institutions to settle things based on the records in the journal in case there is a dispute. In some cases, transactions are posted to an electronic journal to remove the cost of supplying journal paper to the ATM and for more convenient searching of data.

Improper money checking can cause the possibility of a customer receiving counterfeit banknotes from an ATM. While bank personnel are generally trained better at spotting and removing counterfeit cash, the resulting ATM money supplies used by banks provide no guarantee for proper banknotes, as the Federal Criminal Police Office of Germany has confirmed that there are regularly incidents of false banknotes having been dispensed through bank ATMs. Some ATMs may be stocked and wholly owned by outside companies, which can further complicate this problem. Bill validation technology can be used by ATM providers to help ensure the authenticity of the cash before it is stocked in an ATM; ATMs that have cash recycling capabilities include this capability.

Fraud

As with any device containing objects of value, ATMs and the systems they depend on to function are the targets of fraud. Fraud against ATMs and people's attempts to use them takes several forms.

The first known instance of a fake ATM was installed at a shopping mall in Manchester, Connecticut in 1993. By modifying the inner workings of a Fujitsu model 7020 ATM, a criminal gang known as The Bucklands Boys were able to steal information from cards inserted into the machine by customers.

WAVY-TV reported an incident in Virginia Beach in September 2006 where a hacker who had probably obtained a factory-default administrator password for a gas station's white label ATM caused the unit to assume it was loaded with US$5 bills instead of $20s, enabling himself—and many subsequent customers—to walk away with four times the money they wanted to withdraw. This type of scam was featured on the TV series *The Real Hustle*.

ATM behavior can change during what is called "stand-in" time, where the bank's cash dispensing network is unable to access databases that contain account information. In order to give customers access to cash, customers may be allowed to withdraw cash up to a certain amount that may be less than their usual daily withdrawal limit, but may still exceed the amount of available money in their accounts, which could result in fraud if the customers intentionally withdraw more money than what they had in their accounts.

Card Fraud

In an attempt to prevent criminals from shoulder surfing the customer's personal identification number (PIN), some banks draw privacy areas on the floor.

For a low-tech form of fraud, the easiest is to simply steal a customer's card along with its PIN. A later variant of this approach is to trap the card inside of the ATM's card reader with a device often referred to as a Lebanese loop. When the customer gets frustrated by not getting the card back and walks away from the machine, the criminal is able to remove the card and withdraw cash from the customer's account, using the card and its PIN.

This type of ATM fraud has spread globally. Although somewhat replaced in terms of volume by ATM skimming incidents, a re-emergence of card trapping has been noticed in regions such as Europe, where EMV chip and PIN cards have increased in circulation.

Another simple form of fraud involves attempting to get the customer's bank to issue a new card and its PIN and stealing them from their mail.

By contrast, a newer high-tech method of operating, sometimes called **card skimming** or **card cloning**, involves the installation of a magnetic card reader over the real ATM's card slot and the use of a wireless surveillance camera or a modified digital camera or a false PIN keypad to observe the user's PIN. Card data is then cloned into a duplicate card and the criminal attempts a standard cash withdrawal. The availability of low-cost commodity wireless cameras, keypads, card readers, and card writers has made it a relatively simple form of fraud, with comparatively low risk to the fraudsters.

In an attempt to stop these practices, countermeasures against card cloning have been developed by the banking industry, in particular by the use of smart cards which cannot easily be copied or spoofed by unauthenticated devices, and by attempting to make the outside of their ATMs tamper evident. Older chip-card security systems include the French Carte Bleue, Visa Cash, Mondex, Blue from American Express and EMV '96 or EMV 3.11. The most actively developed form of smart card security in the industry today is known as EMV 2000 or EMV 4.x.

EMV is widely used in the UK (Chip and PIN) and other parts of Europe, but when it is not available in a specific area, ATMs must fall back to using the easy–to–copy magnetic strip to perform transactions. This fallback behaviour can be exploited. However the fall-back option has been removed on the ATMs of some UK banks, meaning if the chip is not read, the transaction will be declined.

Card cloning and skimming can be detected by the implementation of magnetic card reader heads and firmware that can read a signature embedded in all magnetic strips during the card production process. This signature, known as a "MagnePrint" or "BluPrint", can be used in conjunction with common two-factor authentication schemes used in ATM, debit/retail point-of-sale and prepaid card applications.

The concept and various methods of copying the contents of an ATM card's magnetic strip onto a duplicate card to access other people's financial information was well known in the hacking communities by late 1990.

In 1996, Andrew Stone, a computer security consultant from Hampshire in the UK, was convicted of stealing more than £1 million by pointing high-definition video cameras at ATMs from a considerable distance, and by recording the card numbers, expiry dates, *etc.* from the embossed detail on the ATM cards along with video footage of the PINs being entered. After getting all the information from the videotapes, he was able to produce clone cards which not only allowed him to withdraw the full daily limit for each account, but also allowed him to sidestep withdrawal limits by using multiple copied cards. In court, it was shown that he could withdraw as much as £10,000 per hour by using this method. Stone was sentenced to five years and six months in prison.

In February 2009, a group of criminals used counterfeit ATM cards to steal $9 million from 130 ATMs in 49 cities around the world, all within a period of 30 minutes.

Related Devices

A talking ATM is a type of ATM that provides audible instructions so that people who cannot read an ATM screen can independently use the machine, therefore effectively eliminating the need for assistance from an external, potentially malevolent source. All audible information is delivered privately through a standard headphone jack on the face of the machine. Alternatively, some banks such as the Nordea and Swedbank use a built-in external speaker which may be invoked by pressing the talk button on the keypad. Information is delivered to the customer either through pre-recorded sound files or *via* text-to-speech speech synthesis.

A postal interactive kiosk may also share many of the same components as an ATM, but only dispenses items related to postage.

A scrip cash dispenser may share many of the same components as an ATM, but lacks the ability to dispense physical cash and consequently requires no vault. Instead, the customer requests a withdrawal transaction from the machine, which prints a receipt. The customer then takes this receipt to a nearby sales clerk, who then exchanges it for cash from the till.

A teller assist unit (TAU) may also share many of the same components as an ATM, but they are distinct in that they are designed to be operated solely by trained personnel and not by the general public, they do not integrate directly into interbank networks, and are usually controlled by a computer that is not directly integrated into the overall construction of the unit.

Automated Pool Cleaner

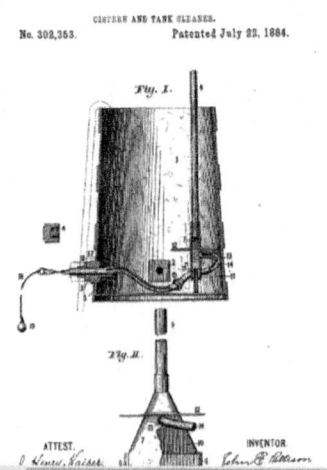

The first patented cistern cleaner, the forerunner of the swimming pool cleaner

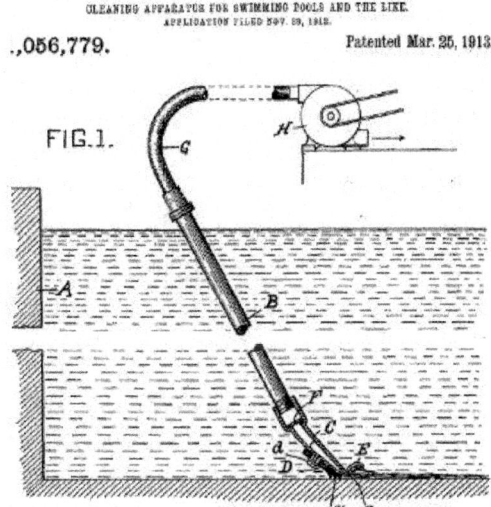

2012 was the Centennial anniversary of the first swimming pool cleaner

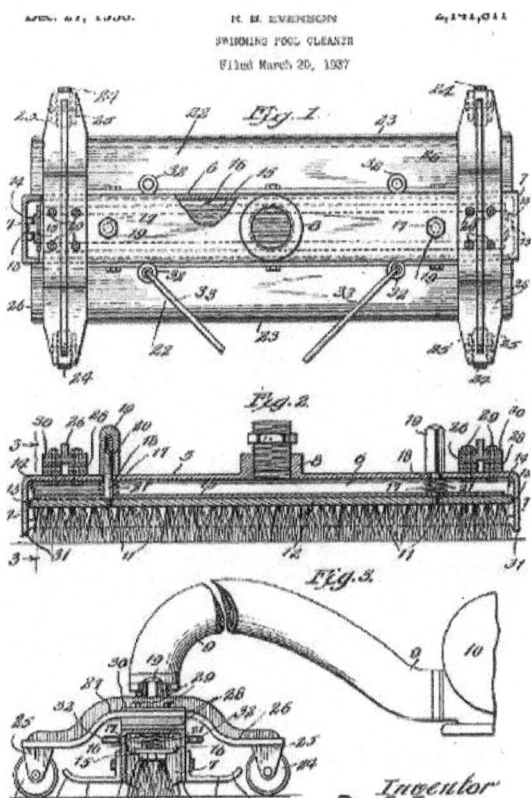

R.B. Everson invented the first suction-side pool vacuum cleaner

An **automated pool cleaner** is a vacuum cleaner intended to collect debris and sediment from swimming pools with minimal human intervention. It is one of several types of swimming pool vacuum cleaners. Other major types are battery-powered or manually powered wands effective only for very small pools, kiddie or wading pools and small spas and hot tubs, and battery-powered, handheld/ extended reach pool and spa vacuums. The latter are powered by rechargeable batteries and can be hand held attached to a telescopic pole used for extended reach. These are used for small to medium-sized pools, larger spas, and to spot clean larger pools.

History

Evolution

Swimming pool cleaners evolved from two areas of science: development of the water filter and early cistern cleaners. The forerunner of today's pool cleaners were cistern cleaners. A cistern (Middle English cisterne, from the Latin cisterna, from cista, box, from Greek kistê, basket) is a waterproof receptacle for holding liquids, usually water. Often cisterns were and still are built to catch and store rainwater. The great palaces of antiquity had both lavish pools and cisterns. They were prevalent in early America as well. United States Patent and Trademark Office makes reference to a cistern cleaner patent filed (though never issued) as early as 1798. Before swimming pools were affordable and fashionable, many swam in their larger cisterns.

In 1883 John E. Pattison of New Orleans filed an application for a "Cistern and Tank Cleaner " and the first discovered patent was issued the following year. It swept and scraped the bottom of a cistern or tank, and through a combination of suction and manipulation of the water pressure was able to separate and remove sediment without removing the water. Over the next 20 years his invention was improved on numerous occasions. Many pool cleaner patents issued in the modern era refer to some of the cistern cleaners as predecessors of their invention.

Early Models

The first swimming pool cleaner was invented in 1912 by Pittsburgh, Pennsylvania local citizen John M. Davison. On November 26, 1912, he submitted a patent application to the United States Patent and Trademark Office entitled "Cleaning Apparatus For Swimming Pools And The Like," patent number 1,056,779 that was issued on March 25, 1913.

The first suction-side pool cleaner was invented by Roy B. Everson of Chicago in 1937, which he entitled "Swimming Pool Cleaner".

Nineteen years later, the first suction-side pool cleaner was the work of Joseph Eistrup of San Mateo, California, who called his invention simply "Pool Cleaner".

Two years later, the first truly automatic, aptly named "Automatic Swimming Pool Cleaner" was created by Andrew L. Pansini of Greenbrae, California, founder of the industry icon Jandy Corporation. Patent Number 3,032,044 was touted by Pansini as "an automatic swimming pool cleaner", which is effective to remove the scum, dirt and other accumulations from both the bottom and side walls of a pool to disperse foreign matter in the, water for removal therefrom by normal pump-filter system of the pool.

The first robotic pool cleaner that used electricity was the work of Robert B. Myers of Boca Raton, Florida in 1967, who filed a patent.

The third and last of the generally accepted pool cleaner technologies, the pressure-side cleaner, was invented by Melvyn L. Henkin of Tarzana, California in 1972. His "Automatic Swimming Pool Cleaner, United States Patent Number 3,822,754 utilized three wheels to allow the machine "to travel underwater along a random path on the pool vessel surface for dislodging debris therefrom". The design is probably familiar to pool owners as the Polaris Pool Cleaner.

Types

There are three main types of automated or automatic swimming pool cleaners, classified by the drive mechanism and source of power used:

Suction-side

In this type, water pumped out of the pool *via* its skimmer or drains is used for locomotion and debris suction and returned after being filtered *via* pool return or outlet valves. This is the least expensive and most popular type. It traces a random course. This type of cleaner is usually attached *via* a 1.5 inch hose to a vacuum plate in the skimmer, or to a dedicated extraction or "vac" line on the side of the pool. The suction action of the pool's pump provides motive force to the machine to randomly traverse the floor and walls of the pool, extracting dirt and debris in its path. The first automatic pool cleaner was a suction cleaner.

These are the least expensive and most widely used pool cleaners with purchase costs ranging in the $100–$300 price range. They are powered solely by the main pump of the pool and utilize the pool's filter system to remove dirt and debris from the water. These machines effectively diminish the suction of the main pump - using them will increase the electricity costs and require the main pump and filter system to be serviced more frequently.

Pressure-side

In this design, pool water inflow is further pressurized using a secondary "booster" pump on most but not all models. This high-pressure water is used for locomotion and debris suction, employing the venturi effect. It traces a random course. The requirement of a booster pump makes this type the highest in electricity use of the three types of pool cleaners.

The pressure causes turbulence in the water, distributing some debris on the floor and walls of the pool, some of which is re-floated to the pool surface and then sucked into the main filter through the skimmer inlets. A portion of the dirt and debris is caught in an attached filter bag. The purchase cost of this type of cleaner range from a minimum of $200 to about $700 plus the costs of the booster pump, usually over $200. Some more sophisticated models can cost over $1,000.

Both suction-side and pressure-side cleaners are dependent on the pool's main pump and filter system to remove contaminants from the pool water cannot remove particles smaller than the pore size of the pool's existing filter element. Such elements can be made of sand, diatomaceous earth, zeolite or other natural or synthetic materials. That particle size ranges from under 5 μm for diatomaceous filters to well in excess of 50 μm for sand filters. Disadvantages of these types of pool cleaners are the additional electricity use, and filtration limitation by the pore size of the main filter element, as well as the time and effort needed to attach the device to the ports that connect to the main pump and filter, and the increased burden of maintenance time and expense on the pool's mechanics.

Electric Robotic

These cleaners are independent from the pool's main filter and pump system and are powered by a separate electricity source, usually in the form of a set-down transformer that is kept at least 10 feet (3.0 m) from the water in the pool, often on the pool deck. They have two internal motors: one to suck in water through a self-contained filter bag and then return the filtered water at a high rate of speed back into the pool water. The second is a drive motor that is connected to tractor-like rubber or synthetic tracks and "brushes" connected by rubber or plastic bands *via* a metal shaft. The brushes, resembling paint rollers, are located on the front and back of the machine and help remove contaminating particles from the pool's floor, walls (and in some designs even the pool steps) depending on size and configuration. They also direct the particles into the internal filter bag.

An internal microchip is pre-programmed to turn on and off and reverse the direction of the drive motors. The chip will cause the machine to change direction when it reaches a wall or the water surface after climbing the pool walls.

These machines may also be directed by sensors located in the bump bars which, on contact with objects such as a wall, cause a reverse in direction, with a small offset allowing it to move one machine's width over on each crossing of the pool. The delay timer is an important feature for many pools, as many switch off a number of circulation pumps during the night to allow suspended particles to settle on the bottom of the pool; after a couple of hours the pool cleaner begins its cleaning cycle. This cleaning cycle is set up to complete before the pumps are turned back on. though not necessary for adequate pool cleaning, this feature saves energy and improves cleaning efficiency.

In order to move forward and backward and negotiate walls and steps electric robotic cleaners rely on three natural principals, traction and movement caused by the drive motor and tracks, buoyancy created by the large areas inside the machine that fills with air, and the force resulting from the high pressure of water being emitted from the top of the machine that pushes it against the floor and walls. Some electric robotic machines use brushes made out of polyvinyl alcohol (PVA) Polyvinyl alcohol that has an adherence quality that allows the unit to almost cling to the walls, steps and floors. They also are resistant to dirt and oil improving lifespan over rubber or other synthetic materials.

The combination of these three natural principles along with an internal mercury switch that tells the microchip that the unit has gone from a horizontal to vertical position as it climbs a wall allows it to change direction from ascending to descending the wall at pre-programmed intervals based on the average height of a pool walls. Some machines have delayed timers that cause the robot to remain at the water line, where more dirt accumulates, for momentarily resulting in a scrubbing action, much like the wheels of a powerful automobile spinning or peeling out.

The major benefits of these machines are efficiency in time, energy, and cleaning ability as well as low maintenance requirements and costs. The major disadvantage is purchase cost which can range from $500 for floor-cleaning-only machines to over $2,000 for the most sophisticated residential units.

Commercial Versions

According to P.K. Data of Duluth, a Georgia a consulting and market research firm that has been retained for many years by the pool and spa industry's internal trade organization, The Association of Pool & Spa Professionals (APSP) there are approximately 14, 000,000 residential pools and spas in the United States and over 400,000 commercial or public pools. As a result this has created a market for larger, more powerful commercial pool cleaners. All commercial pool cleaners are electric robotic and can range in price from a little over $1,000 to upwards of $15,000 or more. They closely resemble residential models but in addition to their addition size they are made with heavy duty components and often more sophisticated computer guidance and on and off systems.

Controlling Legislation

There have been attempts for nearly 100 years to mandate the use of pool cleaners, primarily addressed to public pools. Currently the Center for Disease Control and Prevention located in the Greater Atlanta, Georgia metropolitan area on a grant provided by the National Swimming Pool Foundation (NPSP) is about to publish the first uniform Model Aquatic Health Code (MAHC).. Included is a section on pool filtration proposed regulations directed to the nation's 3200+ state and local agencies that enforce laws and ordinances relating to the operation of swimming pools and spas.

Historical Perspective

The proposed MAHC is not the first attempt to propose a uniform aquatic health code. The credit for that goes to the American Public Health

Association(APHA) which 100 years ago recognized the dangers of improperly maintained aquatic facilities and formed a committee in 1918 to that for the next 66 years issued eleven so-called "Swimming Pools and Other Public Bathing Places Standards For Design, Construction, Equipment And Operation" recommended ordinances and regulations. But for a variety of reasons none of these recommendations were adopted, at least not formally or completely adopted.

Uniform Aquatic Health Code

The APHA has tried to develop a uniform aquatic health code, or what it referred to for years as referenced above, "and published short reports annually from 1920 through 1925 that it simply referred to as "Report Of The Committee On Bathing Places". and finally in 1926 published in its journal its first comprehensive report it called "Standards For Design, Construction, Equipment And Operation" for "Swimming Pools And Other Public Bathing Places". Twelve others were published through 1981, however its lack of authority to enforce them is implied by the changing description of what was limited to their recommendations or suggestions and the expressed purposes in issuing them.

Reports

- In 1957, it referred to its report as "Recommended Practice for Design, Equipment and Operation of Swimming Pools and Other Public Bathing Places".

- In the most comprehensive one since 1926 and until it stopped issuing them in 1981 it referred in 1964 to its report as "Suggested Ordinance and Regulations Covering Public Swimming Pools" and in 1970 one for "Private Swimming Pools". Its last report in 1981 was called "Public Swimming Pools: Recommended Regulations for Design and Construction, Operation and Maintenance". In 1926 appears the committee believed its standards would be adopted by the empowered jurisdictions, although who they were never mentioned in the report.

- In 1964 the boldest move of all took place as it presented its recommendations in the form of proposed ordinances and regulations and in the Forward concluded: "State and local governments who desire to enact this Suggested Ordinance and Regulations Covering Public Swimming Pools a useful resource".

New York City Meeting, 1912

In 1912, coincidentally the same year when the United States Patent and Trademark Office issued the first patent for a swimming pool cleaner, the Sanitary Engineering Section of the American Public Health Association (APHA) convened in New York City to lay the groundwork for the first recommended pool and spa regulations. As reported in the American Journal of Public Health in April 1912 a meeting was held in Havana the previous December and at the New York meeting among the subjects that the committee was to be studying was "Hygiene of swimming pools".

Six years later a committee on swimming pools was appointed at the APHA's annual meeting in Chicago and in 1920 a similar committee was appointed at the meeting in Washington, D.C. In 1921 and periodically over the next seven decades until the work of the APHA on this subject matter went through a series of divisions and consolidations, diverted elsewhere its committees and joint committees with other health-orientated public and quasi-public organizations issued proposed ordinances and regulations in the form of unenforceable recommendations. Despite their intended and published goals, none became law, uniform, much less national.

None of the proposed Standards included more than a passing reference of the need to properly clean a pool. A few, but curiously not all of these recommended ordinances and regulations, related to the use of a vacuum, although the first that included any specificity in 1923 at least required a certain level of clarity. The 1921 report, barely a few pages in length, made this reference to the need to clean the pool.

Pool cleaning is done by completely emptying the pool an average of twice weekly and scrubbing with stiff brushes and soap. Hose flushing follows the scrubbing. After the flushing outlet is opened, the well turned on and clean water allowed to water over the floor of the drains, *etc...*

The 1923 report of the American Journal of Public Health, Sanitary Engineering Section American Public Health Association read before the Sanitary Engineering Section of the American Public Health Association at the Fifty-second Annual Meeting at Boston, Massachusetts, October 8, 1923. slightly longer, but still very brief stated:

Section 3. Clearness: At all times when the pool is in use the water shall be sufficiently clear to permit a black disk six inches in diameter on a white filed, placed on the bottom of the pool at the deepest point, to be clearly visible from both sides of the pool when the water is quiet.

It further stated:

No swimming pool shall be opened to the use of bathers on any day until all visible dirt on the bottom of the pool and any visible scum or floating matter on the surface has been removed. Scum and floating matters may be infectious material and should always be removed as soon as possible after they are observed.

Therefore, in 1921 it was recognized that infectious material, namely pathogens collect in the pool and should be removed.

It was not until 1926 twelve years after the organization recognized the need to address swimming pool "hygiene" and eight years after the committee was organized that the first true report was issued and later published in the Journal of the American Public Health Association. Of all of its reports from 1920 through

1981 the first major report by the APHA the 1926 one, written in narrative form as were the succeeding nine ones though 1957 the committee included the detailed provisions relating to pool cleaning, vacuuming and vacuums:

E. Suction Cleaner: In the opinion of the committee the only satisfactory method of removing the dirt, hair, *etc.*, settling on the bottom of a pool is by means of a suction cleaner. As such cleaners are commonly operated by the circulation pumps; they may be classed as an adjunct to the recirculation system. When a suction cleaner is to be operated by the recirculation pump, a gate with graduated stem or other registering device should be provided for throttling the flow from the pool outlet to permit the pump to operate at maximum efficiency when the suction cleaner is in use. Fixed pipe connections for attachment of suction cleaner to pump suction should be of ample size to reduce friction to a minimum and the cleaner and all removable connections should be designed to provide a maximum velocity at the suction nozzle.

XXVI Cleaning Pool

A. Visible dirt on the bottom of a swimming pool shall not be permitted to remain more than 24 hours. B. Any visible scum or floating material on the surface of a pool shat be removed within 24 hours by flushing or other effective means

The 1964 report included the following language:

A vacuum-cleaning system shall be provided. When an integral part of the recirculation system, sufficient connections shall be located in the walls of the swimming pool, at least eight inches below the water line and "Visible dirt on the bottom of the swimming pool shall be removed every 24 hours or more frequently as required. Visible scum or floating matter on the swimming pool surface shall be removed within 24 hours by flushing or other effective means.

The CDC was founded, followed by the Cabinet-level Department of Health, Education and Welfare, now the Department of Health, and Human Services and its 11 operating divisions, the National Health Service Corps and along the way a variety of private and non-profit aquatic organizations such as the National Spa and Pool Association, now the Association of Pool and Spa Professionals the National Swimming Pool Foundation.

Presently a variety, but not by a long shot the majority, of states and jurisdictions that have codified the requirement of inclusion of an independent vacuum cleaner including the two states with the highest number and concentration of both residential and public pools:

California: 2010 Title 24, Part 2, Vol. 2 California Building Code. Section 3140B, Cleaning Systems:

A vacuum cleaning system shall be available which is capable of removing sediment from all parts of the pool floor. A cleaning system using potable water shall be provided with an approved backflow protection device as required by the California Department of Public Health under Sections 7601 to 7605.

Florida:

Florida Department of Health section 64E-9.007 Recirculation and Treatment System Requirements

(12) Cleaning system – A portable or plumbed in vacuum cleaning system shall be provided. All vacuum pumps shall be equipped with hair and lint strainers. Recirculation or separate vacuum pumps shall not be used for vacuuming purposes when in excess of 3 horsepower. When the system is plumbed in, the vacuum fittings shall be located to allow cleaning the pool with a 50 foot maximum length of hose. Vacuum fittings shall be mounted no more than 15 inches below the water level, flush with the pool walls, and shall be provided with a spring loaded safety cover which shall be in place at all times. Bag type cleaners which operate as ejectors on potable water supply pressure must be protected by a vacuum breaker. Cleaning devices shall not be used while the pool is open to bathers.

Call to Action

In 2005 the CDC in response to a growing concern and feared epidemic with the Pathogen Crytosporidium the Center for Disease Control Center for Disease Control, much like the American Public Health Association did in 1912 convened many of the country's foremost medical and other scientific experts to study the concern for aquatic health. As a result in 2007 they began their quest, again much like the APHC for a uniform aquatic health code.

Each health and safety segment has been assigned to a committee to study it and draft a proposed module open for public comment before being adopted and then recommended to the nation's 3200+ state and local health agencies that enact ordinances and regulations for swimming pools and spa and other aquatic facilities, inspect and monitor them and then enforce the regulations. Since the focus of the MAHC was to respond to the threat of Crytosporidium the Technical Committee of Recirculation Systems and Filtration is a major focus. The University of North Carolina Charlotte Associate Professor James Amburgey is the Chairperson of the Center For Disease Control, Model Aquatic Health Code Technical Committee on Recirculation Systems and Filtration.

Amburgey has conducted many tests to evaluation existing swimming pool filters and his conclusions have been they are extremely ineffective in most cases to help remove Crytosporidium. He is reported to be working with several manufacturers of swimming pool and spa vacuum cleaners to develop a filter bag that will result in exponential advancements in the current filter bags, cleaners and vacuums on the market.

Automated Guided Vehicle

An **automated guided vehicle** or **automatic guided vehicle** (AGV) is a mobile robot that follows markers or wires in the floor, or uses vision, magnets, or lasers for navigation. They are most often used in industrial applications to move materials around a manufacturing facility or a warehouse. Application of the automatic guided vehicle has broadened during the late 20th century.

Forklift AGV picking load from rack

VNA AGV can travel through aisle with very little side clearance safely.

Introduction

Automated guided vehicles (AGVs) increase efficiency and reduce costs by helping to automate a manufacturing facility or warehouse. The first AGV was invented by Barrett Electronics in 1953. The AGV *can tow objects* behind them in trailers to which they can autonomously attach. The trailers can be used to move raw materials or finished product. The AGV can also store objects on a bed. The objects can be placed on a set of motorized rollers (conveyor) and then pushed off by reversing them. AGVs are employed in nearly every industry, including, pulp, paper, metals, newspaper, and general manufacturing. Transporting materials such as food, linen or medicine in hospitals is also done.

An AGV can also be called a laser guided vehicle (LGV). In Germany the technology is also called *Fahrerlose Transportsysteme* (FTS) and in Sweden *förarlösa truckar*. Lower cost versions of AGVs are often called Automated Guided Carts (AGCs) and are usually guided by magnetic tape. AGCs are available in a variety of models and can be used to move products on an assembly line, transport goods throughout a plant or warehouse, and deliver loads.

The first AGV was brought to market in the 1950s, by Barrett Electronics of Northbrook, Illinois, and at the time it was simply a tow truck that followed a wire in the floor instead of a rail. In 1976, Egemin Automation (Holland, MI) started working on the development of an automatic driverless control system for use in several industrial and commercial applications. Out of this technology came a new type of AGV, which follows invisible UV markers on the floor instead of being towed by a chain. The first such system was deployed at the Willis Tower in Chicago, Illinois to deliver mail throughout its offices.

Over the years the technology has become more sophisticated and today automated vehicles are mainly Laser navigated *e.g.* LGV. In an automated process, LGVs are programmed to communicate with other robots to ensure product is moved smoothly through the warehouse, whether it is being stored for future use or sent directly to shipping areas. Today, the AGV plays an important role in the design of new factories and warehouses, safely moving goods to their rightful destination.

Navigation

Wired

A slot is cut in to the floor and a wire is placed approximately 1 inch below the surface. This slot is cut along the path the AGV is to follow. This wire is used to transmit a radio signal. A sensor is installed on the bottom of the AGV close to the ground. The sensor detects the relative position of the radio signal being transmitted from the wire. This information is used to regulate the steering circuit, making the AGV follow the wire.

Guide Tape

AGVs (some known as automated guided carts or AGCs) use tape for the guide path. The tapes can be one of two styles: magnetic or colored. The AGC is fitted with the appropriate guide sensor to follow the path of the tape. One major advantage of tape over wired guidance is that it can be easily removed and relocated if the course needs to change. Colored tape is initially less expensive, but lacks the advantage of being embedded in high traffic areas where the tape may become damaged or dirty. A flexible magnetic bar can also be embedded in the floor like wire but works under the same provision as magnetic tape and so remains unpowered or passive. Another advantage of magnetic guide tape is the dual polarity. small pieces of magnetic tape may be placed to change states of the AGC based on polarity and sequence of the tags.

Laser Target Navigation

The navigation is done by mounting reflective tape on walls, poles or fixed machines. The AGV carries a laser transmitter and receiver on a rotating turret. The laser is transmitted and received by the same sensor. The angle and (sometimes) distance to any reflectors that in line of sight and in range are automatically calculated. This information is compared to the map of the reflector layout stored in the AGV's memory. This allows the navigation system to triangulate the current position of the AGV. The current position is compared to the path programmed in to the reflector layout map. The steering is adjusted accordingly to keep the AGV on track. It can then navigate to a desired target using the constantly updating position.

- Modulated Lasers The use of modulated laser light gives greater range and accuracy over pulsed laser systems. By emitting a continuous fan of modulated laser light a system can obtain an uninterrupted reflection as soon as the scanner achieves line of sight with a reflector. The reflection ceases at the trailing edge of the reflector which ensures an accurate and consistent measurement from every reflector on every scan. By using a modulated laser a system can achieve an angular resolution of ~ 0.1 mrad (0.006°) at 8 scanner revolutions per second.

- Pulsed Lasers A typical pulsed laser scanner emits pulsed laser light at a rate of 14,400 Hz which gives a maximum possible resolution of ~ 3.5 mrad (0.2°) at 8 scanner revolutions per second. To achieve a workable navigation, the readings must be interpolated based on the intensity of the reflected laser light, to identify the centre of the reflector.

Inertial (Gyroscopic) Navigation

Another form of an AGV guidance is inertial navigation. With inertial guidance, a computer control system directs and assigns tasks to the vehicles. Transponders are embedded in the floor of the work place. The AGV uses these transponders to verify that the vehicle is on course. A gyroscope is able to detect the slightest change in the direction of the vehicle and corrects it in order to keep the AGV on its path. The margin of error for the inertial method is ±1 inch.

Inertial can operate in nearly any environment including tight aisles or extreme temperatures. Inertial navigation can include use of magnets embedded in the floor of the facility that the vehicle can read and follow.

Natural Features (Natural Targeting) Navigation

Navigation without retrofitting of the workspace is called Natural Features or Natural Targeting Navigation. One method uses one or more range-finding sensors, such as a laser range-finder, as well as gyroscopes or inertial measurement units with Monte-Carlo/Markov localization techniques to understand where it is as it dynamically plans the shortest permitted path to its goal. The advantage of such systems is that they are highly flexible for on-demand delivery to any location. They can handle failure without bringing down the entire manufacturing operation, since AGVs can plan paths around the failed device. They also are quick to install, with less down-time for the factory.

Steering Control

To help an AGV navigate it can use three different steer control systems. The differential speed control is the most common. In this method there are two independent drive wheels. Each drive is driven at different speeds in order to turn or the same speed to allow the AGV to go forwards or backwards. The AGV turns in a similar fashion to a tank. This method of steering is the simplest as it does not require additional steering motors and mechanism. More often than not, this is seen on an AGV that is used to transport and turn in tight spaces or when the AGV is working near machines. This setup for the wheels is not used in towing applications because the AGV would cause the trailer to jackknife when it turned.

The second type of steering used is steered wheel control AGV. This type of steering can be similar to a cars steering. But this is not very manoeuvrable. It is more common to use a three wheeled vehicle similar to a conventional three wheeled forklift. The drive wheel is the turning wheel. It is more precise in following the programmed path than the differential speed controlled method. This type of AGV has smoother turning. Steered wheel control AGV can be used in all applications; unlike the differential controlled. Steered wheel control is used for towing and can also at times have an operator control it.

The third type is a combination of differential and steered. Two independent steer/drive motors are placed on diagonal corners of the AGV and swivelling castors are placed on the other corners. It can turn like a car (rotating in an arc) in any direction. It can crab in any direction and it can drive in differential mode in any direction.

Vision Guidance

Vision-Guided AGVs can be installed with no modifications to the environment or infrastructure. They operate by using cameras to record features along the route, allowing the AGV to replay the route by using the recorded features to navigate. Vision-Guided AGVs use Evidence Grid technology, an application of probabilistic volumetric sensing, and was invented and initially developed by Dr. Moravec at Carnegie Mellon University. The Evidence Grid technology uses probabilities of occupancy for each point in space to compensate for the uncertainty in the performance of sensors and in the environment. The primary navigation sensors are specially designed stereo cameras. The vision-guided AGV uses 360-degree images and build a 3D map, which allows the vision-guided AGVs to follow a trained route without human assistance or the addition of special features, landmarks or positioning systems.

Geoguidance

A geoguided AGV recognizes its environment to establish its location. Without any infrastructure, the forklift equipped with geoguidance technology detects and identifies columns, racks and walls within the warehouse. Using these fixed references, it can position itself, in real time and determine its route. There are no limitations on distances to cover or number of pick-up or drop-off locations. Routes are infinitely modifiable.

Path Decision

AGVs have to make decisions on path selection. This is done through different methods: frequency select mode, and path select mode (wireless navigation only) or *via* a magnetic tape on the floor not only to guide the AGV but also to issue steering commands and speed commands.

Frequency Select Mode

Frequency select mode bases its decision on the frequencies being emitted from the floor. When an AGV approaches a point on the wire which splits the AGV detects the two frequencies and through a table stored in its memory decides on the best path. The different frequencies are required only at the decision point for the AGV. The frequencies can change back to one set signal after this point. This method is not easily expandable and requires extra cutting meaning more money.

Path Select Mode

An AGV using the path select mode chooses a path based on preprogrammed paths. It uses the measurements taken from the sensors and compares them to values given to them by programmers. When an AGV approaches a decision point it only has to decide whether to follow path 1, 2, 3, *etc.* This decision is rather simple since it already knows its path from its programming. This method can increase the cost of an AGV because it is required to have a team of programmers to program the AGV with the correct paths and change the paths when necessary. This method is easy to change and set up.

Magnetic Tape Mode

The magnetic tape is laid on the surface of the floor or buried in a 10mm channel; not only does it provide the path for the AGV to follow but also strips of the tape in different combos of polarity, sequence, and distance laid alongside the track tell the AGV to change lane, speed up, slow down, and stop.

Traffic Control

Flexible manufacturing systems containing more than one AGV may require it to have traffic control so the AGV's will not run into one another. Traffic control can be carried out locally or by software running on a fixed computer elsewhere in the facility. Local methods include zone control, forward sensing control, and combination control. Each method has its advantages and disadvantages.

Zone Control

Zone control is the favorite of most environments because it is simple to install and easy to expand. Zone control uses a wireless transmitter to transmit a signal in a fixed area. Each AGV contains a sensing device to receive this signal and transmit back to the transmitter. If the area is clear the signal is set at "clear" allowing any AGV to enter and pass through the area. When an AGV is in the area the "stop" signal is sent and all AGV attempting to enter the area stop and wait for their turn. Once the AGV in the zone has moved out beyond the zone the "clear" signal is sent to one of the waiting AGVs. Another way to set up zone control traffic management is to equip each individual robot with its own small transmitter/receiver. The individual AGV then sends its own "do not enter" message to all the AGVs getting to close to its zone in the area. A problem with this method is if one zone goes down all the AGV's are at risk to collide with any other AGV. Zone control is a cost efficient way to control the AGV in an area.

Forward Sensing Control

Forward sensing control uses collision avoidance sensors to avoid collisions with other AGV in the area. These sensors include: sonic, which work like radar; optical, which uses an infrared sensor; and bumper, physical contact sensor. Most AGVs are equipped with a bumper sensor of some sort as a fail safe. Sonic sensors send a "chirp" or high frequency signal out and then wait for a reply from the outline of the reply the AGV can determine if an object is ahead of it and take the necessary actions to avoid collision. The optical uses an infrared transmitter/receiver and sends an infrared signal which then gets reflected back; working on a similar concept as the sonic sensor. The problems with these are they can only protect the AGV from so many sides. They are relatively hard to install and work with as well.

Combination Control

Combination control sensing is using collision avoidance sensors as well as the zone control sensors. The combination of the two helps to prevent collisions in any situation. For normal operation the zone control is used with the collision avoidance as a fail safe. For example, if the zone control system is down, the collision avoidance system would prevent the AGV from colliding.

System Management

Industries with AGVs need to have some sort of control over the AGVs. There are three main ways to control the AGV: locator panel, CRT color graphics display, and central logging and report.

A locator panel is a simple panel used to see which area the AGV is in. If the AGV is in one area for too long, it could mean it is stuck or broken down. CRT color graphics display shows real time where each vehicle is. It also gives a status of the AGV, its battery voltage, unique identifier, and can show blocked spots. Central logging used to keep track of the history of all the AGVs in the system. Central logging stores all the data and history from these vehicles which can be printed out for technical support or logged to check for up time.

AGV is a system often used in FMS to keep up, transport, and connect smaller subsystems into one large production unit. AGVs employ a lot of technology to ensure they do not hit one another and make sure they get to their destination. Loading and transportation of materials from one area to another is the main task of the AGV. AGV require a lot of money to get started with, but they do their jobs with high efficiency. In places such as Japan automation has increased and is now considered to be twice as efficient as factories in America. For a huge initial cost the total cost over time decreases

Vehicle Types

- AGVS Towing Vehicles were the first type introduced and are still a very popular type today. Towing vehicles can pull a multitude of trailer types and have capacities ranging from 8,000 pounds to 60,000 pounds.
- AGVS Unit Load Vehicles are equipped with decks, which permit unit load transportation and often automatic load transfer. The decks can either be lift and lower type, powered or non-powered roller, chain or belt decks or custom decks with multiple compartments.
- AGVS Pallet Trucks are designed to transport palletized loads to and from floor level; eliminating the need for fixed load stands.
- AGVS Fork Truck has the ability to service loads both at floor level and on stands. In some cases these vehicles can also stack loads in rack.
- AGVS Hybrid Vehicles are adapted from a standard CAT-style man-aboard truck so that they can run fully automated or be driven by a fork truck driver. These can be used for trailer loading as well as moving materials around warehouses. Most often, they are equipped with forks, but can be customized to accommodate most load types.

- Light Load AGVS are vehicles which have capacities in the neighborhood of 500 pounds or less and are used to transport small parts, baskets, or other light loads though a light manufacturing environment. They are designed to operate in areas with limited space.
- AGVS Assembly Line Vehicles are an adaptation of the light load AGVS for applications involving serial assembly processes.

Common AGV Applications

Automated Guided Vehicles can be used in a wide variety of applications to transport many different types of material including pallets, rolls, racks, carts, and containers. AGVs excel in applications with the following characteristics:

- Repetitive movement of materials over a distance
- Regular delivery of stable loads
- Medium throughput/volume
- When on-time delivery is critical and late deliveries are causing inefficiency
- Operations with at least two shifts
- Processes where tracking material is important

Raw Materials Handling

AGVs are commonly used to transport raw materials such as paper, steel, rubber, metal, and plastic. This includes transporting materials from receiving to the warehouse, and delivering materials directly to production lines.

Work-in-process Movement

Work-in-Process movement is one of the first applications where automated guided vehicles were used, and includes the repetitive movement of materials throughout the manufacturing process. AGVs can be used to move material from the warehouse to production/processing lines or from one process to another.

Pallet Handling

Pallet handling is an extremely popular application for AGVs as repetitive movement of pallets is very common in manufacturing and distribution facilities. AGVs can move pallets from the palletizer to stretch wrapping to the warehouse/ storage or to the outbound shipping docks.

Finished Product Handling

Moving finished goods from manufacturing to storage or shipping is the final movement of materials before they are delivered to customers. These movements often require the gentlest material handling because the products are complete and subject to damage from rough handling. Because AGVs operate with precisely controlled navigation and acceleration and deceleration this minimizes the potential for damage making them an excellent choice for this type of application

Trailer Loading

Automatic loading of trailers is a relatively new application for automated guided vehicles and becoming increasingly popular. AGVs are used to transport and load pallets of finished goods directly into standard, over-the-road trailers without any special dock equipment. AGVs can pick up pallets from conveyors, racking, or staging lanes and deliver them into the trailer in the specified loading pattern. Some Automatic Trailer Loading AGVs utilize Natural Targeting to view the walls of the trailer for navigation. These types of ATL AGVs can be either completely driverless or hybrid CAT-based vehicles.

Roll Handling

AGVs are used to transport rolls in many types of plants including paper mills, converters, printers, newspapers, steel producers, and plastics manufacturers. AGVs can store and stack rolls on the floor, in racking, and can even automatically load printing presses with rolls of paper.

Container Handling

AGVs are used to move sea containers in some maritime container terminals. The main benefits are reduced labour costs and a more reliable (less variable) performance. This use of AGVs was pioneered by ECT in the Netherlands at the Delta terminal in the Port of Rotterdam.

Primary Application Industries

Efficient, cost effective movement of materials is an important, and common element in improving operations in many manufacturing plants and warehouses. Because automatic guided vehicles (AGVs) can delivery efficient, cost effective movement of materials, AGVs can be applied to various industries in standard or customized designs to best suit an industry's requirements. Industry's currently utilizing AGVs include (but are not limited to):

Pharmaceutical

AGVs are a preferred method of moving materials in the pharmaceutical industry. Because an AGV system tracks all movement provided by the AGVs, it supports process validation and cGMP (current Good Manufacturing Practice).

Chemical

AGVs deliver raw materials, move materials to curing storage warehouses, and provide transportation to other processing cells and stations. Common industries include rubber, plastics, and specialty chemicals.

Manufacturing

AGVs are often used in general manufacturing of products. AGVs can typically be found delivering raw materials, transporting work-in process, moving finished goods, removing scrap materials, and supplying packaging materials.

Automotive

AGV installations are found in Stamping Plants, Power Train Plants, and Assembly Plants delivering raw materials, transporting work-in process, and moving finished goods. AGVs are also used to supply specialized tooling which must be changed.

Paper and Print

AGVs can move paper rolls, pallets, and waste bins to provide all routine material movement in the production and warehousing (storage/retrieval) of paper, newspaper, printing, corrugating, converting, and plastic film.

Food and Beverage

AGVs can be applied to move materials in food processing (such as the loading of food or trays into sterilizers) and at the "end of line," linking the palletizer, stretch wrapper, and the warehouse. AGVs can load standard, over-the-road trailers with finished goods, and unload trailers to supply raw materials or packaging materials to the plant. AGVs can also store and retrieve pallets in the warehouse.

Hospital

AGVs are becoming increasingly popular in the healthcare industry for efficient transport, and are programmed to be fully integrated to automatically operate doors, elevators/lifts, cart washers, trash dumpers, *etc.* AGVs typically move linens, trash, regulated medical waste, patient meals, soiled food trays, and surgical case carts.

Warehousing

AGVs used in Warehouses and Distribution Centers logically move loads around the warehouses and prepare them for shipping/loading or receiving or move them from an induction conveyor to logical storage locations within the warehouse. Often, this type of use is accompanied by customized warehouse management software.

Battery Charging

AGVs utilize a number of battery charging options. Each option is dependent on the users preference. The most commonly used battery charging technologies are *Battery Swap, Automatic/Opportunity Charging,* and *Automatic Battery Swap.*

Battery Swap

"Battery swap technology" requires an operator to manually remove the discharged battery from the AGV and place a fully charged battery in its place after approximately 8 – 12 hours (about one shift) of AGVs operation. 5 – 10 minutes is required to perform this with each AGV in the fleet.

Automatic and Opportunity Charging

"Automatic and opportunity battery charging" allows for continuous operation. On average an AGV charges for 12 minutes every hour for automatic charging and no manual intervention is required. If opportunity is being utilized the AGV will receive a charge whenever the opportunity arises. When a battery pack gets to a predetermined level the AGV will finish the current job that it has been assigned before it goes to the charging station.

Automatic Battery Swap

"Automatic battery swap" is an alternative to manual battery swap. It requires an additional piece of automation machinery, an automatic battery changer, to the overall AGV system. AGVs will pull up to the battery swap station and have their batteries automatically replaced with fully charged batteries. The automatic

battery changer then places the removed batteries into a charging slot for automatic recharging. The automatic battery changer keeps track of the batteries in the system and pulls them only when they are fully charged.

While a battery swap system reduces the manpower required to swap batteries, recent developments in battery charging technology allow batteries to be charged more quickly and efficiently potentially eliminating the need to swap batteries.

REASONS TO AUTOMATION

Automating business processes is often referred to as BPA, namely Business Process Automation which is a way of taking different applications and integrating them, using software applications and restructuring to make business processes happen in a manner that is seamless and effortless. In fact there are very compelling reasons why business processes should be automated; here are the top 10.

1. Technology and the automation of processes can increase productivity by reducing the time taken to perform repetitive tasks. This is indeed a compelling reason for automation of processes: since increased productivity with less labour costs means increased profit.

2. Automated processes will also reduce defects, which in turn saves money and enhances the production process.

3. An automated process will also ensure that each process will run effortlessly and that it will run the same way every time it is run; there will be no room for human errors and as everyone knows, 'To err is human'. So there will be fewer mistakes made and mistakes may not simply be defects, but can be about making mistakes that can impact on the production process as a whole.

4. Automated systems do not get bored. Anyone who has to undertake a repetitive task over and over again will get bored. This will slow down their performance (and increase the risks of defects) however, an automated process will not become bored, no matter how often it is run, so performance is not slowed down; it is run at the same speed every time.

5. The automation of processes also means that they will always ensure compliance with internal or external requirements, such as statutory requirements.

6. Cost savings are also made by the reduction of training that staff will require. Historically it could have taken a long time to train staff to do even the most repetitive of tasks, especially if the task required a high degree of accuracy. However, if they are using an automated process, it takes far less training, so overheads are reduced significantly.

7. Automated systems also make auditing an easier process in itself, so there is less time taken to analyse and audit processes and the production process as a whole.

8. Automated systems are also incredibly flexible, so it is easy to make changes, or at least far easier than it would be to change processes that involve only

humans. This flexibility can be important if changes are required at short notice.

9. The fact that defects are reduced means that customer satisfaction is increased, often quite significantly, so customers will be kept happy whilst profits are increased, which is good business sense and one of the most compelling arguments for automating any business process.

10. A workforce that is not bored by repeatedly undertaking very repetitive tasks or who are not faced with reams of data that they need to analyse, or who don't have to worry about achieving high degrees of accuracy each time, because the technology solves all these problems, will actually be a happier workforce. In a sense this brings about two benefits:

    *   A happier workforce tends to be a more productive workforce because they are not stressed by their work and they have a more relaxed and happier working environment, which helps them to devote their attention to the tasks in hand.

    *   A happier workforce is also a more stable workforce, because they are less likely to leave. This in turn further reduces staff training costs, because staff are not being recruited, trained only for them to leave. Instead they remain in post and so do not require further training.Thus the automation of processes is not a radical concept, it is simply about making sure that businesses can survive within very tough times!

INDUSTRIAL AUTOMATION - WHEN AND WHAT TO AUTOMATE

Over the past decade, *Lean Manufacturing* has been a major topic regarding the manufacturing strategies of progressive industrial companies. According to this approach, plants will initially move from traditional batch manufacturing to *Lean Manufacturing, which is* also known as *Just In Time Manufacturing (JIT)* or *Continuous Flow Manufacturing (CFM)*. Many companies are taking the next step in a plan to move to an automated and integrated manufacturing facility. With this step, the plants invest capital in order to add automated processes and *Computer Integrated Manufacturing (CIM)* to their strategy. Within this strategy, the automation of material conversion, assembly operations, and quality control processes can achieve significant benefits.

When regarding industrial automation, the journey from *Lean Manufacturing* to *Integrated Manufacturing* is sound since this planned approach will eliminate the "waste" processes which otherwise would have been included in the automation. In addition, it is easier and better to automate a flow line of similar part families and similar processes than to attempt to mechanize islands of operations and hope there is some overall benefit. Does this mean that all automation projects should be put on hold until *Lean Manufacturing* is fully implemented? The answer is a definitive No! Although flow lines are a natural candidate for automation, there are many other places where industrial automation would yield meaningful benefits and it would not make sense to delay the projects. This article will deal with the questions of when and what to automate.

When to Automate

The first thing that should be done when considering automation is to review the tactical issues. That is, what are you actually trying to accomplish by automating? Although special circumstances may offer a number of reasons to automate, the five primary reasons that dominate corporate tactical thinking are: reduce cost, improve quality, reduce inventory, improve response time, and improve ergonomics.

Each of these is summarized below:

Reduce Cost

When considering industrial automation as a cost reduction, management traditionally focuses on direct labor. Reducing direct labor is the most obvious method of reducing costs; however, we must be careful when analyzing this scenario. The important thing to remember is that eliminating direct labor results in a cost reduction only when the actual headcount is lower than what would be required to produce the same volume manually. When industrial automation does create an actual headcount reduction, that labor savings is usually specified as the entire cost reduction and other cost benefits are often overlooked.

One frequently overlooked cost saving is the reduction of indirect labor since automation will generally reduce the amount of material handling and orientation time required with manual operations. Reduction of in-process inventory is another significant cost savings that occurs when automation replaces batch operations. Finished goods inventory can often be reduced as well. Another cost benefit of automation is scrap reduction since an automated line will inspect for parts present and will often integrate functional testing. In addition, defective piece parts are rejected by automated stations. Finally, when there is an automated line instead of individual workstations, the cost for training can be effectively reduced when it is necessary to add personnel.

Improve Quality

There are a number of ways that automation will improve the quality of your products. Two methods that were previously mentioned are the inspection of parts present and the use of in-process testing. Any proper industrial automation will verify the presence and position of a part after it has been placed into the assembly. Often the equipment will incorporate in-process testing to verify that the product is correct before sending it to the next operation. Another quality improvement is the elimination of piece part rejects. A defective or out-of-tolerance part will usually not pass through the tooling required to automatically feed and load it. This sorting of rejects will improve the quality of the finished goods, but it will also put pressure on the upstream operations to control the piece part quality.

Industrial automation is often used to improve quality since machines can accomplish tasks that are difficult to do manually. For example, automation is frequently used to gauge and match components to achieve a more accurate fit than the primary processes will allow. When automation is used to test products, the parameters must be quantified and the products are accepted or rejected according to those parameters. This eliminates the subjective decision-making and potential for operator error, which is present when testing is done manually. In addition, it is relatively simple to maintain and document test results when utilizing automatic testers. The tester database can be utilized for SPC and Pareto Analysis in order to control root causes of failure.

Reduce Inventory

Industrial automation reduces inventory in the exactly the same manner as *Lean Manufacturing*. Since product flows from station to station instead of sitting in bulk at each operation, the work-in-process inventory is dramatically reduced. In addition, the finished goods inventory can be reduced as automation allows you to make products when they are needed or *Just In Time*.

Improve Response Time

Response Time is related to *manufacturing cycle time*, which is the time from receipt of an order to shipment of the product. When manufacturing in batch through a number of manual operations, most of that time is spent with the products sitting in "buckets" waiting for value added work to be performed. Again, in a manner similar to *Lean Manufacturing*, automation will improve the response time as products flow from operation to operation. In addition, sub-assemblies can be integrated into the automation line in order to eliminate individual workstations. With the technology available today, flexible automation can further improve response time by processing small lot sizes or incorporating programmable changeovers.

Another way that industrial automation can improve response time is by handling "surge" orders. When the automation project is developed, it should be planned to handle peak capacity. For example, if the equipment can produce the forecast annual quantity in two shifts, five days a week, a "surge" order can be run on third shift or on the weekend. Without the automation, people would have to be added and trained in order to handle the peak volumes.

Improve Ergonomics

The impact of ergonomics is continually increasing when considering industrial automation. As people become aware of the problems associated with cumulative trauma disorders, vibration syndrome, fatigue, and so forth, they also become aware of the need to mechanize the operations that are causing these problems. Automation projects are often initiated when an operation is observed as "carpal tunnel waiting to happen." In most cases, the productivity gained by eliminating the difficult or repetitive operation will justify the expense of automation.

What to Automate

"When to Automate" outlines the primary benefits of industrial automation, but what analysis should be done to indicate that an automation project makes sense in your plant? First, concentrate on the high-volume products. If you only have a few people involved in your annual production, it probably is not worth the effort to investigate automation. Second, look at products with a long product life cycle. If a product is going to be around for less than five years, and your automation project will take two years to implement, it is probably not worth the effort unless the equipment is reconfigurable. Other issues that indicate a potential for automation are excess material handling, operator dependent quality, and operations that are repetitive or difficult to do manually.

When the potential for industrial automation is first recognized, there is a natural tendency to try to automate everything produced in order to increase volumes. This tendency has been the failure of many potentially sound automation projects. Exactly what products to include in the automation have to be analyzed carefully.

KEYS TO INTEGRATING AUTOMATION, MES, AND BUSINESS SYSTEMS

Delivering an integrated manufacturing execution system (MES) is complex; however, by focusing on the key factors outlined in this article, successful projects have been implemented. Folklore among many automation and IT professionals is that MES projects are complex, expensive, and never complete. This article illustrates how to develop a high-level understanding of the MES system implementation process with concrete examples of how automation and IT delivery teams can function together to ensure success.

MES Functionality and Standards

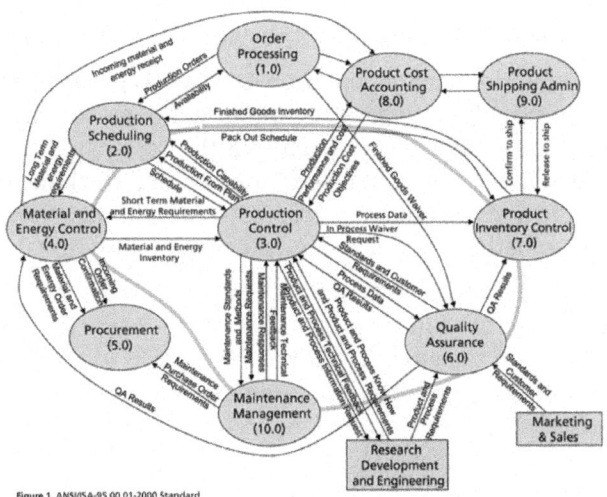

Figure 1. ANSI/ISA-95.00.01-2000 Standard

The ANSI/ISA-95.00.01-2000 standard, shown in Figure, defines ten functions related to the production operations; it also defines the boundaries between enterprise and control (MES/automation systems) domains. One excellent first step to a successful project execution is reviewing the standards documentation as a way to provide a common language related to data structures and interface definitions for discussions between automation and MES execution teams.

For the purposes of this article, we will focus on the following three major MES functions:

- Production operations management
- Quality operations management
- Inventory operations management

Production operations management is, at the core, the sequencing and execution of both operator and automation production steps. The sequencing is based on the product manufacturing requirements (how much is needed) and specifications. Also included is the capture of all product- and production-related data. Benefits include the ability to rapidly modify a "shop floor" production schedule to meet current product demand, the availability of near real-time data to business systems, and a reduction in time for data entry time by interfacing directly with the automation systems.

Quality operations management is the automated monitoring of production and product data to ensure conformance to established specifications. Benefits include rapid identification of either process or quality issues. One major time savings is to be able to "review by exception" the critical parameters necessary to ensure a quality product. With all critical product and process data being tracked in real time against established specifications, the MES is able to present for review only those items outside established limits; leading to a quicker review and approval by quality representatives. Another time-saving aspect is the ability to automate equipment state tracking due to the MES's awareness of what products and batches are associated with each unit of equipment over time.

Inventory operations management is the tracking and movement of material through the production process, including items considered "in-process" or stored as a "warehouse item." Included also is the automated tracking of "weigh and dispense" activities. Benefits include the detailed tracking and reporting of materials produced and consumed in near real-time, allowing for the rapid updating of production schedules to minimize downtime and reduce overall inventory.

Looking to the actual MES components, a key success factor is to design the MES hardware and software architecture to have the same level of availability as the facility automation systems. Avoid the temptation to design a facility based on the belief that the MES system will be unavailable, as this adds unnecessary complexity and cost. Systems designed for high-availability systems are available; simply use them.

No Project is an Island: The Truth of an Mes Implementation and Requirements

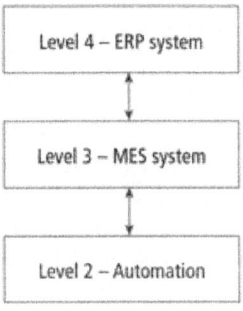

Figure.

When describing the role of an MES, it serves, in part, as the "glue" between the "business" and "automation" systems.

As such, implementations are seldom done in isolation, with critical success factors being the need to understand the overall multi-level requirements definition process and establish effective communications between all organizations. As a start to understand the overall requirements, consider the major functionality of the systems at each level: The enterprise resource planning, or ERP, level systems focus on customer order demand, overall material and inventory planning, and financials. MES level systems focus on multi-production center sequencing conforming to a detailed schedule, material production and consumption, and the capture and analysis of production history. Automation level systems focus on production center equipment operation, control, and reporting. This is a very simplified view of the functionality at each level, but understanding the key items identified is critical for a successful integrated project.

Having now established that there are actually three co-dependent projects occurring during an MES implementation, we will now explore the interaction between each. Let us begin with a simplified model of each project consisting of requirements definition, build and test, and commissioning as shown in Figure. It would be great if projects at the ERP, MES, and automation levels had the same timelines and project implementation process. In looking at the typical project timing for each level, however, the actual integrated project appears more like Figure.

As can be seen, one of the major challenges is that, given the long lead time of certain automation projects, business-related requirements may not be fully understood until the automation team is in the middle of the build-and-test or commissioning phase of their project. So a key success factor is to understand the requirements from each level early enough in the overall project schedule to avoid major changes. Pay close attention to asking questions regarding the long-term information analysis needs of the business, as these usually translate to data capture and interfacing requirements.

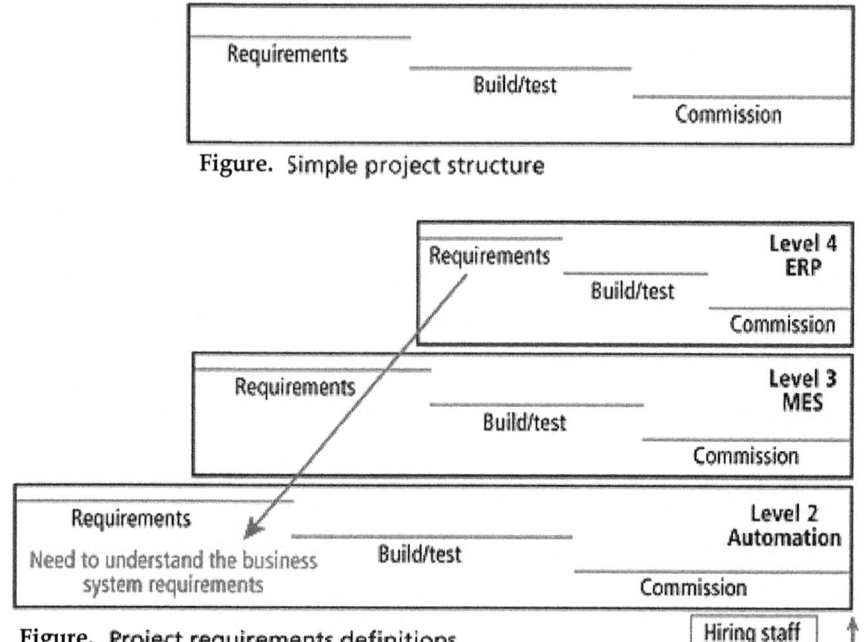

Figure. Simple project structure

Figure. Project requirements definitions

If you are fortunate enough to be implementing at a new facility, typically the operations staff is not completely hired until close to the actual go-live. You then have the fun of seeing experienced individuals review what is being delivered and then ask that it be changed to meet their needs for operational oversight. To minimize the chances of this happening, the key is to find experienced individuals to participate in the definition of the requirements and project delivery. We have been successful in having operations staff from other facilities join our project teams on temporary assignments. If you are implementing at an existing facility, plan on having individuals from operations join the team on a full-time basis. Chosen well, these folks know what is actually happening in the facility for defining the requirements and have the capability to bring the rest of the operations staff along with the implementation. Help these critical individuals in their discussions with coworkers by developing and using a "visual language".

Describing MES and automation actions and operational sequences visually assists in the identification of parallel and co-mingled activities. On one project, we established a "team room" where we filled the walls with large drawings of the visual language and brought all the operations staff in multiple times to review and ensure that we had the process defined correctly prior to committing to coding. One side benefit was that we were able to identify several "unhappy path" scenarios where the production process might have deviated from the intended actions, allowing us to define and code these as alternative paths for the operations staff.

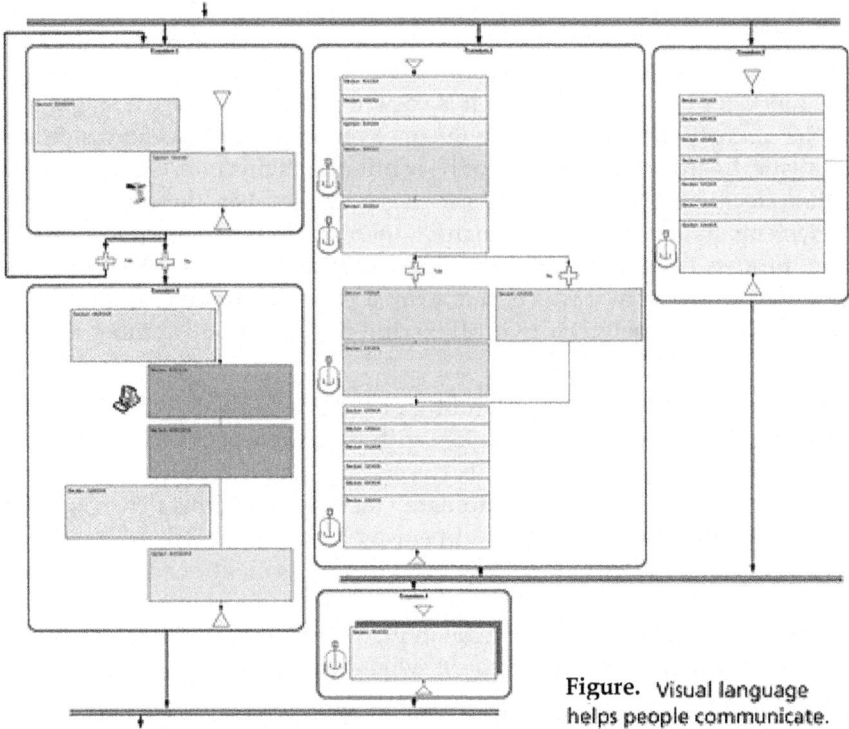

Figure. Visual language helps people communicate.

Additionally, consideration must be given to the organizational change impact of the delivery teams of multiple co-dependent projects. Automation and IT organizations must be closely aligned throughout the requirements definition and project execution processes through the use of robust change control that communicates and considers the impact of a modification at all levels and aided by the use of a common "cross-system" data definitions dictionary. Finally, it is important to understand and accept the implementation philosophy of each organization within the overall project. For example, the system environment and the development processes at the ERP level typically allow for rapid prototyping of several possible solutions prior to establishing the final business process. This leads to a desire for quick responses from the ERP team on questions regarding changes in requirements at the MES and automation levels.

Interfaces: The True Joy of an MES Implementation

Focus on the interfaces – simply stated: what interfaces are required; how is each interface triggered; and what data elements are contained within the interface. Document and reference these in your interactions with other system development teams. Avoid the temptation to try to understand the internal coding and data structures of each system and instead look at the "touch points" between systems. This allows you to build the necessary boundaries for the expectations of each system and avoid "functionality creep," where there is a loss of separation between the ERP, MES, and automation levels. Each level serves a distinct purpose with established business and operations processes; respect these differences by focusing on the interface. In addition, watch the number and level of complexity of the interfaces being established. It is easy to fall into the trap of trying to automate the interactions between systems to cover all possible scenarios, but this is not realistic. Learn to ask: "When (or how often) will this interface be triggered?" and then decide if the interface is necessary. Remember, there is a reason that most systems have a process for manual intervention; be prepared to suggest a manual process to avoid interfacing complexity of seldom-used actions. On one project, we decided that it was best to print a paper document for the movement of certain materials between two MESs due to the limited number of possible times this would occur.

Also seek to gain an understanding of the operational goals for the facility and the degree of operator interaction with the automated processes. Operator interaction can provide a "buffer" between automated actions, and it stands to reason that a greater desire to "automate" facility operations drives greater implementation complexity – usually appearing in an integrated automation/MES project as a need for increased interfacing. Further, as mentioned earlier, you may find yourself in the position of defining "automation" level requirements prior to there being either an MES or ERP team in place. In these cases, it is helpful to look at each system's focus as a predictor of what data interfacing may be required.

As a simple example, since the focus of an ERP system is customer orders, inventory, and financials, plan automation level interfaces to include:

- An interface to the automation system identifying a production job, sequenced to meet customer demand that includes the product to be produced, required raw materials, necessary parameters, and specifications
- An interface from the automation system on the completion of a production job – including start and end times, the amount of material produced and raw materials consumed, any byproducts of the productions process (including scrap and waste), any associated downtime during the production job, and quality indicators

Effective Communications and Accountability do Matter

Consider the potential players in a major MES implementation today. Our current project includes an overall contracting team, seven major equipment/automation suppliers, six teams engaged in the MES development, and two ERP teams. And given the nature of software delivery today, these 16 teams are distributed across the world in seven different time zones. Since co-location is not an option, there is a need for an effective communications and accountability process that allows for rapid issue discovery and resolution. One effective method is to invest in an issues tracking system that is accessible to all individuals engaged in the project. The one we are using is cloud based and allows for the following: Overall statement and detailed description of the issue, the attachment of documents, entry of updates to the issue status with email notification to impacted individuals, and the assignment of an issue to a project milestone and issue owner. Simple in overall functionality, but with quickly customized reporting, we have accountability to an owner and are able to track open items to closure by a required milestone. This allows us to use focused conference calls with document display capability to allow for the rapid design decisions. Finally, we assign an individual to be the "issues moderator" to ensure that items are being addressed.

10 STRATEGIES FOR AUTOMATION AND PROCESS IMPROVEMENT IN CIM?

Introduction

Today manufacturing Industries want to automize their processes to increase their productivity to meet the ever increasing demand in the market. Automation becomes unnecessary after the process has been simplified. When it is found to be feasible solution for improved performance, quality and productivity, it is needed to study about the ten strategies that will give a road map for any manufacturing organization. The 10 points are given below:

- Specialization of operations
- Combined operations
- Simultaneous operations
- Integration of operations
- Increased flexibility
- Improved material handling and storage
- Online inspection
- Process control and optimization
- Plant operations control
- Computer Intergrated Manufacturing

Specialization of Operations

Some special purpose machines or equipments are used to perform particular operation with highest possible efficiency. This is employed to utilize the specialization of labour and improve the labour productivity.

Combined Operations

Production of parts is obtained in a sequence of operations. Several processing steps are required for complex parts. Combined operations eliminate or reduce the processing steps. Just reducing the number of distinct production machines results better. That is, more than one operation are carried out in a given machine so that number of separate machines needed are reduced considerably. By this strategies set up time for each machine can be reduced as much. Apart from this, Ideal time, non-operation time, material handling time and production lead time are also reduced.

Simultaneous Operations

The operations combined in one particular workstations are carried out simultaneously to reduce the total processing time. This will result in greater increase in productivity. This is very common logical strategies in every industries. Assembly operations can also be done using this strategies.

Integration of Operations

Intergration means linking several workstations together into a single integrated mechanism to tranfer parts between them by automatic material handling system. The products are easily scheduled and several parts are processed at a time, thereby resulting in greater increase in overall output of the system.

Increased Flexibility

Flexibility plays important role in a manufacturing system. Use of flexible automation concept is followed here. That when needed, variety of products or parts are processed in a single machine. Here the maximum utilization of equipments is done. The important and primary objective of this strategy is to reduce setup time as much as possible. Less work in process can also be achieved.

Improved Material Handling and Storage

Automated material handling is adopted for reducing non-productive time here. Other benefits included are less work in process and shorter manufacturing lead time.

Online Inspection

After the process is completed, inspection can be performed for measuring quality of work. Any poor quality product is found in this strategy and is processed and again inspected to find improvement on it. Online inspection means that poor quality product is inspected regularly to eliminate the scrap and overall quality can be increased.

Process Control and Optimization

A wide range of control scheme is included here to operate the individual processes and associated equipments. The individual processing time is reduced by this strategy. Product quality can also be improved.

Plant Operations Control

This is applied in control of plant level whereas previous is cocerned with processes only. The functions of this strategy are to manage and coordinate the aggregate oprations in the plant. A high level computer networking may be incorporated to do this task.

Computer Integrated Manufacturing

Engineering design and business functions are integrated in this step. This is higher level activity in the above startegies. The extensive use of computer applications, databases and networking are the parts of this strategy

Conclusion

These 10 strategies consititute a check list on improving the manufacturing system by automation. Processes are improved and optimized to increase productivity. Multiple strategies can be implemented in one project. So we can conclude that these points are very much effective for producing highest level quality product.

Chapter 2

BASIC LAWS AND PRINCIPLE

FLUID POWER

Fluid power is a term describing *hydraulics* and *pneumatics* technologies. Both technologies use a fluid (liquid or gas) to transmit power from one location to another. With hydraulics, the fluid is a liquid (usually oil), whereas pneumatics uses a gas (usually compressed air). Both are forms of *power transmission*, which is the technology of converting power to a more useable form and distributing it to where it is needed. The common methods of power transmission are electrical, mechanical, and fluid power.

Although they sometimes are viewed as competing technologies, no single method of power transmission is the best choice for all applications. In fact, most applications are served by a combination of technologies. However, *fluid power offers important advantages over the other technologies.*

Fluid power systems easily produce linear motion using hydraulic or pneumatic cylinders, whereas electrical and mechanical methods usually must use a mechanical device to convert rotational motion to linear. Fluid power systems generally can transmit equivalent power within a much smaller space than mechanical or electrical drives can, especially when extremely high force or torque is required. Fluid power systems also offer simple and effective *control of direction, speed, force*, and *torque* using simple control valves. Fluid power systems often do not require electrical power, which eliminates the risk of electrical shock, sparks, fire, and explosions.

What is Hydraulics?

To visualize a basic hydraulic system, think of two identical syringes connected together with tubing and filled with water. *Syringe A* represents a pump, and *Syringe B* represents an actuator, in this case a cylinder. Pushing the plunger of *Syringe A* pressurizes the liquid inside. This fluid pressure acts equally in all directions, and causes the water to flow out the bottom, into the tube, and into

Syringe B. If you placed a 5 lb. object on top of the plunger of *Syringe B*, you would need to push on *Syringe A's* plunger with at least 5 lbs. of force to move the weight upward. If the object weighed 10 lbs., you would have to push with at least 10 lbs. of force to move the weight upward.

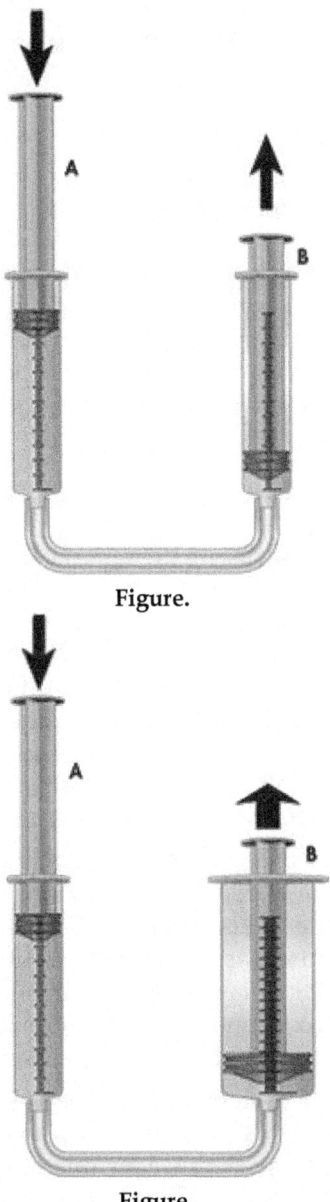

Figure.

Figure.

If the area of the plunger (which is a piston) of *Syringe A* is 1 sq. in., and you push with 5 lbs. of force, the fluid pressure will be 5 lbs./sq. in. (psi). Because fluid pressure acts equally in all directions, if the object on *Syringe B* (which, again has an area of 1 sq. in.) weighs 10 lbs., fluid pressure would have to exceed 10 psi before the object would move upward. If we double the diameter of *Syringe B*, the area of the plunger becomes four times what it was. This means a 10 lb. weight would be supported on 4 sq. in. of fluid. Therefore, fluid pressure would only have to exceed 2.5 psi (10 lbs. ÷ 4 sq. in. = 2.5 psi) to move the 10 lb. object upward. So moving the 10 lb. object would only require 2.5 lbs. of force on the plunger of *Syringe A*, but the plunger on *Syringe B* would only move upward ¼ as far as when both plungers were the same size. This is the essence of fluid power. Varying the sizes of pistons (plungers) and cylinders (syringes) allows multiplying the applied force.

In actual hydraulic systems, pumps contain many pistons or other types of pumping chambers. They are driven by a prime mover (usually an electric motor, diesel engine, or gas engine) that rotates at several hundred revolutions per minute (rpm). Every rotation causes all of the pump's pistons to extend and retract — drawing fluid in and pushing it out to the hydraulic circuit in the process. Hydraulic systems typically operate at fluid pressures of thousands of psi. So a system that can develop 2,000 psi can push with 10,000 lbs. of force from a cylinder about the same size as a can of soda pop.

What is Pneumatics?

The principles of pneumatics are the same as those for hydraulics, but pneumatics transmits power using a gas instead of a liquid. Compressed air is usually used, but nitrogen or other inert gases can be used for special applications. With pneumatics, air is usually pumped into a receiver using a compressor. The receiver holds a large volume of compressed air to be used by the pneumatic system as needed. Atmospheric air contains airborne dirt, water vapor, and other contaminants, so filters and air dryers are often used in pneumatic systems to keep compressed air clean and dry, which improve reliability and service life of the components and system. Pneumatic systems also use a variety of valves for controlling direction, pressure, and speed of actuators.

Most pneumatic systems operate at pressures of about 100 psi or less. Because of the lower pressure, cylinders and other actuators must be sized larger than their hydraulic counterparts to apply an equivalent force. For example, a hydraulic cylinder with a 2 in. diameter piston and fluid pressure of 1,000 psi can push with 3140 lbs. of force. A pneumatic cylinder using 100 psi air would need a bore of almost 6½ in. (33 sq. in.) to develop the same force.

Even though pneumatic systems usually operate at much lower pressure than hydraulic systems do, **pneumatics holds many advantages that make it more suitable for many applications.** Because pneumatic pressures are lower, components can be made of thinner and lighter weight materials, such as aluminum and engineered plastics, whereas hydraulic components are generally made of steel and ductile or cast iron. Hydraulic systems are often considered rigid, whereas pneumatic systems usually offer some cushioning, or "give." Pneumatic systems are generally simpler because air can be exhausted to the atmosphere, whereas hydraulic fluid usually is routed back to a fluid reservoir.

Pneumatics also holds advantages over electromechanical power transmission methods. Electric motors are often limited by heat generation. *Heat generation is usually not a concern with pneumatic motors* because the stream of compressed air running through them carries heat from them. Furthermore, because pneumatic components require no electricity, they don't need the bulky, heavy, and expensive explosion-proof enclosures required by electric motors. In fact, even without special enclosures, electric motors are substantially larger and heavier than pneumatic motors of equivalent power rating. Plus, if overloaded, pneumatic motors will simply stall and not use any power. Electric motors, on the other hand, can overheat and burn out if overloaded. Moreover, torque, force, and speed control with pneumatics often requires simple pressure- or flow-control valves, as opposed to more expensive and complex electrical drive controls. And as with hydraulics, pneumatic actuators can instantly reverse direction, whereas electromechanical components often rotate with high momentum, which can delay changes in direction.

Another advantage of pneumatics is that it allows using vacuum for picking up and moving objects. Vacuum can be thought of as negative pressure — by removing air (evacuating) from the volume between two parts, atmospheric pressure outside the volume pushes the parts together. For example, attempting to pick up a single sheet of paper or a raw egg presents a challenge with conventional grippers. But with a vacuum pneumatic system, evacuating a suction cup in contact with a sheet of paper or eggshell will cause atmospheric pressure to push the paper or egg against the cup, allowing it to be lifted.

FLUID POWER ADVANTAGES

Hydraulic and pneumatic systems share many benefits for the machines in which they are installed. These include:

- high horsepower-to-weight ratio — You could probably hold a 5-hp hydraulic motor in the palm of your hand, but a 5-hp electric motor might weight 40 lb or more.
- safety in hazardous environments because they are inherently spark-free and can tolerate high temperatures.
- force or torque can be held constant — this is unique to fluid power transmission
- high torque at low speed — unlike electric motors, pneumatic and hydraulic motors can produce high torque while operating at low rotational speeds. Some fluid power motors can even maintain torque at zero speed without overheating
- pressurized fluids can be transmitted over long distances and through complex machine configurations with only a small loss in power

- multi-functional control — a single hydraulic pump or air compressor can provide power to many cylinders, motors, or other actuators
- elimination of complicated mechanical trains of gears, chains, belts, cams, and linkages
- motion can be almost instantly reversed

Hydraulic and pneumatic systems are both widely used in stationary (industrial) and off-highway (mobile) equipment. **Hydraulic systems are widely used when heavy force or torque is involved**, such as lifting loads weighing several tons, crushing or pressing strong materials like rock and solid metal, and digging, lifting, and moving large amounts of earth. **And although pneumatics is capable of transmitting high force and torque, it is more widely used for fast-moving, repetitive applications**, such as pick-and-place operations, gripping, and repetitive gripping or stamping. In both cases, electronic controls and sensors have been implemented into fluid power systems for the last few decades. These electronics make hydraulic and pneumatic systems faster, more precise and efficient, more reliable, and allow them to be tied into statistical process control and other factory and mobile equipment control networks.

Fluid Power Components

Fluid power systems consist of multiple components that work together or in sequence to perform some action or work. People well versed in fluid power circuit and system design may purchase individual components and assemble them into a fluid power systems themselves. However, many fluid power systems are designed by distributors, consultants, and other fluid power professionals who may provide the system in whole or in part.

The major components of any fluid power system include:

- a pumping device — a hydraulic pump or air compressor to provide fluid power to the system
- fluid conductors — tubing, hoses, fittings, manifolds and other components that distribute pressurized fluid throughout the system
- valves — devices that control fluid flow, pressure, starting, stopping and direction
- actuators — cylinders, motors, rotary actuators, grippers, vacuum cups and other components that perform the end function of the fluid power system.
- support components — filters, heat exchangers, manifolds, hydraulic reservoirs, pneumatic mufflers, and other components that enable the fluid power system to operate more effectively.

Electronic sensors and switches are also incorporated into many of today's fluid power systems to provide a means for electronic controls to monitor operation of components. Diagnostic instruments are also used for measuring pressure, temperature and flow in assessing the condition of the system and for troubleshooting.

FLUID PROPERTIES

In addition to the properties like mass, velocity, and pressure usually considered in physical problems, the following are the basic properties of a fluid:

Density

The density of a fluid, is generally designated by the Greek symbol ρ(*rho*) is defined as the mass of the fluid over an infinitesimal volume. Density is expressed in the British Gravitational (BG) system as slugs/ft³, and in the SI system kg/m³.

$$\rho = \lim_{\Delta V \to 0} \frac{\Delta m}{\Delta V} = \frac{dm}{dV}$$

If the fluid is assumed to be uniformly dense the formula may be simplified as:

$$\rho = \frac{m}{V}$$

Specific Weight

The specific weight of a fluid is designated by the Greek symbol γ(gamma), and is generally defined as the weight per unit volume. The units for gamma are lb/ft³ and N/m³ in the imperial and SI systems, respectively.

$\gamma = \rho * g$

g = local acceleration of gravity and ρ = density

Note: It is customary to use:

$g = 32.174$ ft/s² = 9.81 m/s²

$\rho = 1000$ kg/m³

Relative Density (Specific Gravity)

The relative density of any fluid is defined as the ratio of the density of that fluid to the density of the standard fluid. For liquids we take water as a standard fluid with density ρ=1000 kg/m³. For gases we take air or O_2 as a standard fluid with density, ρ=1.293 kg/m³.

Viscosity

Viscosity is a material property, unique to fluids, that measures the fluid's resistance to flow. Though a property of the fluid, its effect is understood only when the fluid is in motion. When different elements move with different velocities, each element tries to drag its neighbouring elements along with it. Thus, shear stress occurs between fluid elements of different velocities.

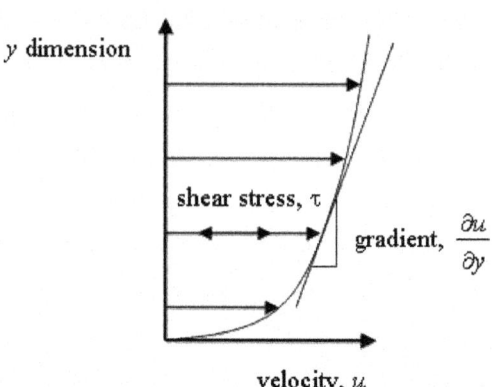

Velocity gradient in laminar shear flow

The relationship between the shear stress and the velocity field was studied by Isaac Newton and he proposed that the shear stresses are directly proportional to the velocity gradient. $\tau = \mu \dfrac{\partial u}{\partial y}$ The constant of proportionality is called the coefficient of dynamic viscosity.

Another coefficient, known as the kinematic viscosity is defined as the ratio of dynamic viscosity and density.

I.e., $\boldsymbol{v = \mu / \rho}$

It is the property of a fluid that quantifies resistance to flow of the fluid.

Dimensionless Parameters

Dimensionless parameters are used to simplify analysis, and describe the physical situation without referring to units. A dimensionless quantity has no physical unit associated with it. They arose from dimensional analysis techniques. These numbers have many applications in fluid mechanics as well as in related subjects like aerodynamics and convective heat transfer.

Reynolds Number

Reynolds number is used in the study of fluid flows. It compares the relative strength of inertial and viscous effects.

The value of the Reynolds number is defined as:

$$\mathrm{Re} = \frac{\rho V L}{\mu} = \frac{V L}{v}$$

where ρ(rho) is the density, μ(mu) is the absolute viscosity, V is the characteristic velocity of the flow, and L is the characteristic length for the flow.

Example 0.1: Reynold's number for flat plate flow

Air at 293K temperature, and 1.225 kg m⁻³ density is flowing past a flat plate at 1 m s⁻¹. What's the Reynold's Number 1 m downstream from the leading the edge of the plate?

Absolute viscosity for air is $1.8 \times 10\text{-}5$ N s m⁻².

$$Re = \frac{\rho V L}{\mu} = \frac{1.225(1)(1)}{1.8\,E10^{-5}} = 68,055$$

Additionally, we define a parameter v(nu) as the *kinematic viscosity.*

Low *Re* indicates creeping flow, medium *Re* is *laminar* flow, and high *Re* indicates *turbulent* flow.

Reynolds number can also be transformed to take account of different flow conditions. For example the Reynolds number for flow within a pipe is expressed as

$$Re = \frac{\rho u d}{\mu}$$

where u is the average fluid velocity within the pipe and d is the inside diameter of the pipe.

Application of dynamic forces (and the Reynolds number) to the real world: sky-diving, where friction forces equal the falling body's weight.

Chapter 3

BASIC PNEUMATIC AND HYDRAULIC SYSTEM

PNEUMATICS

Pneumatics is a section of technology that deals with the study and application of pressurized gas to produce mechanical motion.

Pneumatic systems, that are used extensively in industry, and factories, are commonly plumbed with compressed air or compressed inert gases. This is because a centrally located and electrically powered compressor, that powers cylinders and other pneumatic devices through solenoid valves, can often provide motive power in a cheaper, safer, more flexible, and more reliable way than a large number of electric motors and actuators.

Pneumatics also has applications in dentistry, construction, mining, and other areas.

Examples of Pneumatic Systems and Components

- Air brakes on buses and trucks
- Air brakes on trains
- Air compressors
- Air engines for pneumatically powered vehicles
- Barostat systems used in Neurogastroenterology and for researching electricity
- Cable jetting, a way to install cables in ducts
- Dental drill
- Compressed-air engine and compressed-air vehicles
- Gas-operated reloading

- Holman Projector, a pneumatic anti-aircraft weapon
- HVAC control systems
- Inflatable structures
- Lego pneumatics can be used to build pneumatic models
- Exercise machines

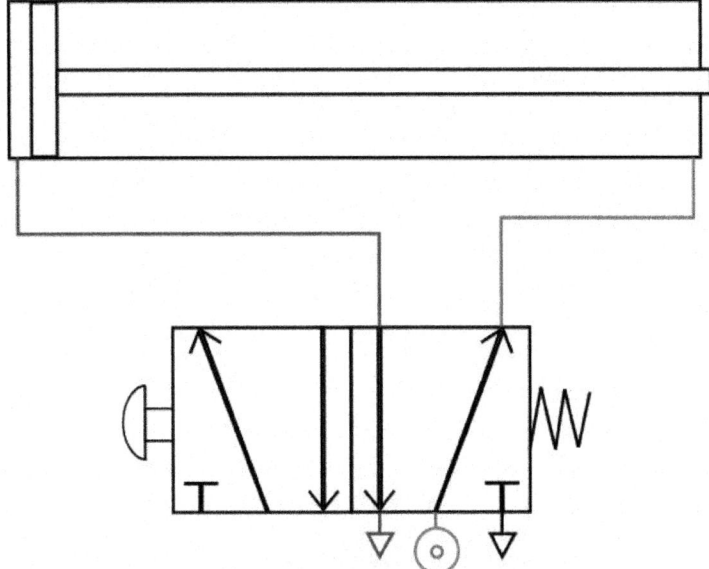

Pneumatic circuitry

- Pipe organs:
  - o Electro-pneumatic action
  - o Tubular-pneumatic action
- Player piano
- Pneumatic actuator
- Pneumatic air guns
- Pneumatic bladder
- Pneumatic cylinder
- Pneumatic Launchers, a type of spud gun
- Pneumatic mail systems
- Pneumatic motor
- Pneumatic tire
- Pneumatic tools:
  - o Jackhammer used by road workers
  - o Pneumatic nailgun

- Pressure regulator
- Pressure sensor
- Pressure switch
- Vacuum pump

Gases Used in Pneumatic Systems

Pneumatic systems in fixed installations, such as factories, use compressed air because a sustainable supply can be made by compressing atmospheric air. The air usually has moisture removed, and a small quantity of oil is added at the compressor to prevent corrosion and lubricate mechanical components.

Factory-plumbed pneumatic-power users need not worry about poisonous leakage, as the gas is usually just air. Smaller or stand-alone systems can use other compressed gases that present an asphyxiation hazard, such as nitrogen – often referred to as OFN (oxygen-free nitrogen) when supplied in cylinders.

Any compressed gas other than air is an asphyxiation hazard – including nitrogen, which makes up 78% of air. Compressed oxygen (approx. 21% of air) would not asphyxiate, but is not used in pneumatically-powered devices because it is a fire hazard, more expensive, and offers no performance advantage over air.

Portable pneumatic tools and small vehicles, such as Robot Wars machines and other hobbyist applications are often powered by compressed carbon dioxide, because containers designed to hold it such as soda stream canisters and fire extinguishers are readily available, and the phase change between liquid and gas makes it possible to obtain a larger volume of compressed gas from a lighter container than compressed air requires. Carbon dioxide is an asphyxiant and can be a freezing hazard if vented improperly.

Comparison to Hydraulics

Both pneumatics and hydraulics are applications of fluid power. Pneumatics uses an easily compressible gas such as air or a suitable pure gas – while hydraulics uses relatively incompressible liquid media such as oil. Most industrial pneumatic applications use pressures of about 80 to 100 pounds per square inch (550 to 690 kPa). Hydraulics applications commonly use from 1,000 to 5,000 psi (6.9 to 34.5 MPa), but specialized applications may exceed 10,000 psi (69 MPa).

Advantages of Pneumatics

- Simplicity of design and control – Machines are easily designed using standard cylinders and other components, and operate *via* simple on-off control.
- Reliability – Pneumatic systems generally have long operating lives and require little maintenance. Because gas is compressible, Equipment is less subject to shock damage. Gas absorbs excessive force, whereas fluid

in hydraulics directly transfers force. Compressed gas can be stored, so machines still run for a while if electrical power is lost.

• Safety — There is a very low chance of fire compared to hydraulic oil. Newer machines are usually overload safe.

Advantages of Hydraulics

• Liquid does not absorb any of the supplied energy.

• Capable of moving much higher loads and providing much higher forces due to the incompressibility.

• The hydraulic working fluid is basically incompressible, leading to a minimum of spring action. When hydraulic fluid flow is stopped, the slightest motion of the load releases the pressure on the load; there is no need to "bleed off" pressurized air to release the pressure on the load.

Pneumatic Logic

Pneumatic logic systems (sometimes called **air logic control**) are often used to control industrial processes, consisting of primary logic units such as:

• And Units

• Or Units

• 'Relay or Booster' Units

• Latching Units

• 'Timer' Units

• Sorteberg relay

• Fluidics amplifiers with no moving parts other than the air itself

Pneumatic logic is a reliable and functional control method for industrial processes. In recent years, these systems have largely been replaced by analog electronic or digital control systems in new installations because of the smaller size, lower cost, greater precision, and more powerful features of digital controls. Many pneumatic devices are still used in processes, however, where the advantages of digital controls are outweighed by such considerations as:

• The cost of upgrading an entire system from pneumatic to digital control is prohibitive

• Safety might be compromised (*e.g.*, potential sparks near explosive gases)

• Compressed air is the most viable energy source available

HYDRAULICS

Hydraulics is a topic in applied science and engineering dealing with the mechanical properties of liquids. At a very basic level hydraulics is the liquid version of pneumatics. Fluid mechanics provides the theoretical foundation for hydraulics, which focuses on the engineering uses of fluid properties. In fluid power, hydraulics is used for the generation, control, and transmission of power by the use of pressurized liquids. Hydraulic topics range through some part of science and most of engineering modules, and cover concepts such as pipe flow, dam design, fluidics and fluid control circuitry, pumps, turbines, hydropower, computational fluid dynamics, flow measurement, river channel behavior and erosion.

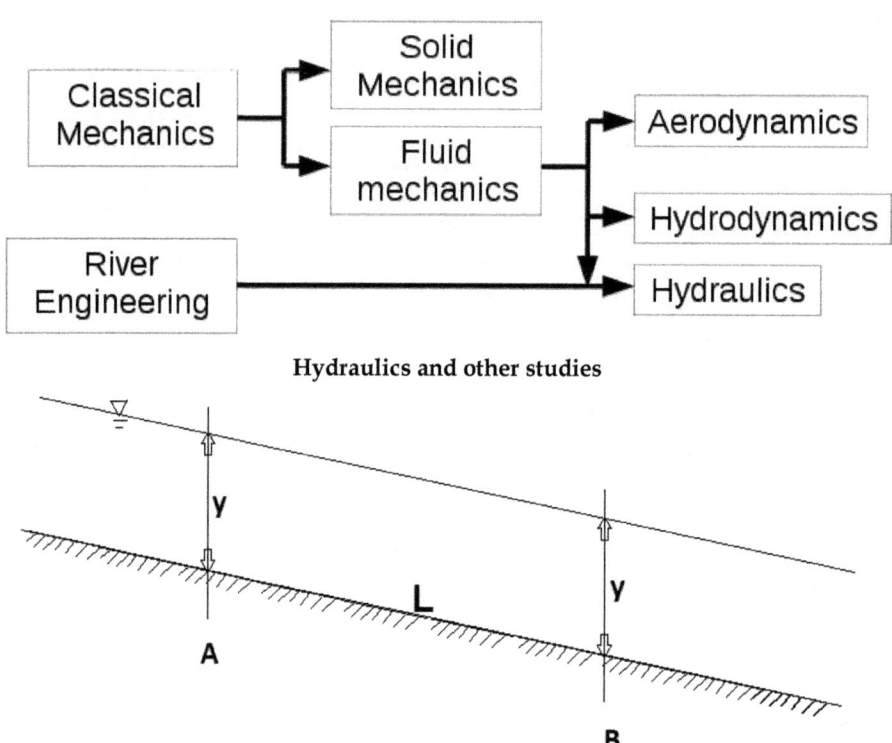

Hydraulics and other studies

An open channel, with a uniform depth, Open Channel Hydraulics deals with uniform and non-uniform streams.

Free surface hydraulics is the branch of hydraulics dealing with free surface flow, such as occurring in rivers, canals, lakes, estuaries and seas. Its sub-field **open channel flow** studies the flow in open channels.

ADVANTAGES OF PNEUMATICS OVER HYDRAULICS

Despite the immense capabilities of hydraulics presented in terms of moving higher loads and in other industrial utilization, pneumatics are still in wide use today. The article discusses some applications and advantages of pneumatics in industry.

Pneumatics is study of mechanical motion caused by pressurized gases and how this motion can be used to perform engineering tasks. Pneumatics is used mainly in mining and general construction works. Pneumatic devices are used frequently in the dentistry industry across the world. On the other hand, hydraulics means use of pressurized fluids to execute a mechanical task. Hydraulics is frequently used in the concepts of turbines, dams, and rivers. Air brakes in buses, air compressors, compressed air engines, jackhammers, and vacuum pumps are some of the most commonly used types of mechanical equipment that are based on pneumatics technology. Commonly seen hydraulics based equipment types are hydraulic presses, hydraulic hoppers, hydraulic cylinders, and hydraulic rams. In the subsequent sections of this article, you will learn how a pneumatic system works, what its best features are, and its major advantages over hydraulic systems.

Operation of a Pneumatic System

In order to affect mechanical motion, pneumatics employs compression of gases, based on the working principles of fluid dynamics in the concept of pressure. Any equipment employing pneumatics uses an interconnecting set of components: a pneumatic circuit consisting of active components such as gas compressor, transition lines, air tanks, hoses, open atmosphere, and passive components. Compressed air is supplied by the compressor and is transmitted through a series of hoses. Air flows are regulated by valves and the pneumatic cylinder transfers the energy provided by the compressed gas to mechanical energy. Aside from compressed air, inert gases are also applied particularly for self-contained systems. Pneumatics is applied in a wide range in industries, even in mining and dentistry. The majority of industries use gas pressures of about 80 to 100 pounds per square inch.

Over pneumatics, hydraulics is capable of moving heavier loads and having greater force, and since its working fluids are incompressible, it has minimum spring actions. But at the same time pneumatics are cleaner; the system uses no return lines and gases are exhausted to the atmosphere. Thus leaks will be of less

concern since the working fluid of pneumatics is air, unlike oil in hydraulics. Its working fluid is also widely available and most factories are pre-plumbed for compressed air distribution, hence pneumatic equipment is easier to set-up. To control the system, only ON and OFF are used and the system consists only of standard cylinders and other components, making it simpler than hydraulics. Pneumatic systems require low maintenance and have long operating lives. Lastly the working fluid of the pneumatic system absorbs excessive force, leading to less frequent damage to equipment. Compressible gases are also easy to store and safer; no fire hazard is presented and machines could be made to be overload safe.

Advantages of Pneumatics over Hydraulics

Like hydraulics, pneumatics is a type of fluid power application where instead of an incompressible liquid, pneumatics employ gas in their system. Hydraulics present certain advantages over pneumatics, but in a given application, pneumatic powered equipment is more suitable, particularly in industries where the factory units are plumbed for compressed air.

The air used in pneumatic devices is dried and free from moisture so that it does not create any problem to the internal parts of the system. Moreover, to avoid corrosive actions, oil or lubricants are added so that friction effects can be reduced. Compressed air is used in most of the machines and in some cases compressed carbon dioxide is used. As most of the pneumatic devices are air based, they have a less complicated design and can be made of inexpensive material. Mass production techniques can be adopted to produce pneumatic systems, which not only save money but save time too.

Other major advantages are listed below.

1. Initial cost is less; hydraulics equipment cost as much as twice the price of pneumatic equipment.

2. A pneumatic water treatment automation system reduces the costs of installation and operation compared with conventional electrical installations. For opening and closing of underwater valves, pneumatic systems work well because they can sustain overload pressure conditions.

3. Pneumatic actuators also have long life and perform well with negligible maintenance requirement throughout their life cycle.

4. Very suitable for power transmission when distance of transmission is more.

The major disadvantage of pneumatic systems is that they cannot be employed for tasks that require working under high pressures. However, modern technology is working on finding better solutions to this address this problem so that heavy engineering tasks can be executed using pneumatic devices. In a nutshell, in order to execute low scale engineering and mechanical tasks, pneumatic devices would be the best suited and a viable alternative over hydraulic systems.

HYDRAULICS *VS* PNEUMATICS

There are almost no significant differences between hydraulics and and for non-engineers but if you examine further, there are lots of technical uniqueness in each system.

By definition alone, hydraulics is very different from pneumatics because it is used in controlling, transmitting and harnessing power using pressured fluids. The latter is dealing more on studying the impact of pressurized gases and how it influences mechanical movement. Hydraulics is frequently used in the concepts of dams, rivers, turbines and even erosion whereas pneumatics is applied in various fields of dentistry, mining and general construction among others.

The material or substance used differs between the two. In hydraulics, the substance used is an incompressible fluid medium wherein the most common example is oil. Pneumatics, on the contrary, makes use of a very compressible gas like air itself or an appropriate pure gas.

Another difference between the two when applied is the strength of the pressures used in their applications. Hydraulic systems use a greater amount of pressure compared to pneumatic applications. In pneumatics, only 80-100 psi (pounds per square inch) of pressure is used for its industrial applications. Hydraulic-based applications frequently use pressures that range from 1,000-5,000 psi. Nevertheless, other more advanced hydraulic systems even use pressures of up to 10,000 psi. Because of this high power demand, hydraulic systems chiefly use bigger components while pneumatic systems use smaller ones in most applications.

With regard to the control of applications, pneumatic systems are deemed to be simpler and easier to handle than hydraulic systems. Most operators say that using pneumatics is just like the light switch that makes you choose between two simple choices of 'on' or 'off.' This is true because most pneumatics is designed with simple cylinders and standard components only. An exception in the simplicity of either a hydraulic or pneumatic device would come if the entire system is automated.

Summary:

1. By definition, hydraulics is used in controlling or harnessing power with the use of pressurized fluids whereas pneumatics studies how pressurized gases influences mechanical motion or movement.

2. Hydraulics uses an incompressible fluid medium like oil whereas pneumatics uses a compressible gas like air.

3. Hydraulic applications demand greater pressures during operations that reach thousands of pounds per square inch whereas pneumatic applications only require 100 psi pressures more or less.

4. Most hydraulic applications generally use bigger components that pneumatic applications.

5. Hydraulic systems are generally more difficult to operate compared to pneumatic applications.

PNEUMATIC SYSTEMS

Most pneumatic circuits run at low power -- usually around 2 to 3 horsepower. Two main advantages of air-operated circuits are their low initial cost and design simplicity. Because air systems operate at relatively low pressure, the components can be made of relatively inexpensive material -- often by mass production processes such as plastic injection molding, or zinc or aluminum die-casting. Either process cuts secondary machining operations and cost.

First cost of an air circuit may be less than a hydraulic circuit but operating cost can be five to ten times higher. Compressing atmospheric air to a nominal working pressure requires a lot of horsepower. Air motors are one of the most costly components to operate. It takes approximately one horsepower to compress 4 cfm of atmospheric air to 100 psi. A 1-hp air motor can take up to 60 cfm to operate, so the 1-hp air motor requires (60/4) or 15 compressor horsepower when it runs. Fortunately, an air motor does not have to run continuously but can be cycled as often as needed.

Air-driven machines are usually quieter than their hydraulic counterparts. This is mainly because the power source (the air compressor) is installed remotely from the machine in an enclosure that helps contain its noise.

Because air is compressible, an air-driven actuator cannot hold a load rigidly in place like a hydraulic actuator does. An air-driven device can use a combination of air for power and oil as the driving medium to overcome this problem, but the combination adds cost to the circuit.

Air-operated systems are always cleaner than hydraulic systems because atmospheric air is the force transmitter. Leaks in an air circuit do not cause housekeeping problems, but they are very expensive. It takes approximately 5 compressor horsepower to supply air to a standard hand-held blow-off nozzle and maintain 100 psi. Several data books have charts showing cfm loss through different size orifices at varying pressures. Such charts give an idea of the energy losses due to leaks or bypassing.

HYDRAULIC SYSTEMS

A hydraulic system circulates the same fluid repeatedly from a fixed reservoir that is part of the prime mover. The fluid is an almost non-compressible liquid, so the actuators it drives can be controlled to very accurate positions, speeds, or forces. Most hydraulic systems use mineral oil for the operating media but other fluids such as water, ethylene glycol, or synthetic types are not uncommon. Hydraulic systems usually have a dedicated power unit for each machine. Rubber-molding plants depart from this scheme. They usually have a central power unit with pipes running to and from the presses out in the plant. Because these presses require no flow during their long closing times, a single large pump can operate several of them. These hydraulic systems operate more like a compressed-air installation because the power source is in one location.

A few other manufacturers are setting up central power units when the plant has numerous machines that use hydraulics. Some advantages of this arrangement are: greatly reduced noise levels at the machine, the availability of backup pumps to take over if a working pump fails, less total horsepower and flow, and increased uptime of all machines.

Another advantage hydraulic-powered machines have over pneumatic ones is that they operate at higher pressure -- typically 1500 to 2500 psi. Higher pressures generate high force from smaller actuators, which means less clutter at the work area.

The main disadvantage of hydraulics is increased first cost because a power unit is part of the machine. If the machine life is longer than two years, the higher initial cost is often offset by lower operating cost due to the much higher efficiency of hydraulics. Another problem area often cited for hydraulics is housekeeping. Leaks caused by poor plumbing practices and lack of pipe supports can be profuse. This can be exaggerated by overheated low-viscosity fluid that results from poor circuit design. With proper plumbing procedures, correct materials, and preventive maintenance, hydraulic leaks can be virtually eliminated.

Another disadvantage could be that hydraulic systems are usually more complex and require maintenance personnel with higher skills. Many companies do not have fluid power engineers or maintenance personnel to handle hydraulic problems.

HYDRAULIC CONTROL SYSTEMS

Hydraulic control systems include any controls that use fluid-based operation rather than electronics or pneumatic power. When control units within these systems are shifted, the internal fluid moves throughout the machine's inner workings. As it moves, this non-compressible fluid transfers force throughout the system to shift gears or influence motion. Hydraulic control systems rely on Pascal's law, which states that liquid pressure will remain equally distributed within a sealed system. Many modern machines rely on either hydraulic controls or a hybrid electric-hydraulic system.

Many types of equipment rely on some form of hydraulic control system, including aircraft and aerospace vessels. Marine vessels and elevators also use these types of controls, as do hydraulic cranes. Cars and trucks typically contain hydraulic brake systems, and a variety of industrial and manufacturing machines also rely on these controls for safe and effective operation.

Hydraulic control systems can influence the motion or operation of a machine in several ways. The most basic involves manual control, where a human or robotic users flips a switch, pulls a lever or turns a steering wheel. This motion drives hydraulic fluid throughout the system to accomplish the desired action.

Other systems rely on automatic controls rather than manual input. For example, a sensor on a crane may detect heavy loads and automatically send extra fluid towards the crane's lifting system. This fluid in turn creates excess lifting power to safely move the heavy load. Similar systems rely on pressure sensors, electronic eyes, and a variety of additional inputs.

One of the primary advantages to using hydraulic control systems is the ability to handle very large loads or accommodate tremendous forces. Compared to electric or pneumatic control systems, hydraulic controls are better able to handle sudden changes in load while maintaining an even level of power distribution. Hydraulic systems also allow for very precise and accurate handling in more specialized applications. Compared to air-based pneumatic systems, hydraulic controls make it much easier to spot a potential leak due to the visibility of the fluid.

Buyers should also be aware of the potential drawbacks to this system before investing in hydraulic controls. The hydraulic fluid used within these systems can be highly corrosive, and may lead to extended maintenance and repairs over time. Hydraulic control systems also utilize a large number of seals, which could fail or leak. Finally, because this fluid consists primarily of petroleum products, it poses risks to the environment during use and disposal.

PNEUMATIC CIRCUIT

A **pneumatic circuit** is an interconnected set of components that convert compressed gas (usually air) into mechanical work. In the normal sense of the term, the circuit must include a compressor or compressor-fed tank.

Components

The circuit comprises the following components:
- Active components
  - o Compressor
- Transmission lines
  - o Air tank
  - o Pneumatic hoses
  - o Open atmosphere (for returning the spent gas to the compressor)
  - o Valves
- Passive components
  - o Pneumatic cylinders
  - o Service Unit
- FRL - Filter Regulator and Lubricator

Pneumatic Cylinder

In general based on application pneumatic cylinder is selected which are single acting cylinder, it will have single port in cylinder were extension of cylinder is by compressed air and retraction by means of open coiled spring. In double acting cylinder two ports are available both extension and retraction by means compressed air

Direction Control Valve (DCV)

The direction control valve is used to control the direction of flow of compressed air. Usually classified into normally open (NO)and normally closed (NC) valves. The normally open valves will permit flow from inlet port of valve to outlet port normally the flow will be cut by changing the position of the valve. The normally closed valves will not permit flow from inlet port of valve to outlet port normally the flow will be permitted only by changing the position of the valve. In general valves are designated as 2/2 DCV, 3/2DCV, 5/2 DCV,5/3 DCV *etc.* In which the first numerical indicates number of ports and second numerical indicates number of position To change the position, the valves are generally actuated by:

- Pedal Operated
- Push button operated
- Spring operated
- Solenoid operated
- By using Pneumatic source itself *etc.*

The other auxiliary valves are:

- Two pressure valve (And Valve): Usually two valve actuators are used when both the push buttons are pressed at a time the air flow takes place if either any one is pressed at a time air flow will not take place in valve outlet. Generally used in mechanical press and machine tools to ensure operator's hands are safe during operation.
- OR Valve : Usually two valve actuators are used when either one push button is pressed the air flow takes place.
- Check valve: Allows air flow in one direction
- Quick exhaust valve: The valve construction is OR valve with exhaust port,ensures quick return of cylinder therefore cycle time reduces
- Flow control valve:
- Time delay valve:
- Pressure relief valve *etc.*

The following devices operate using compressed gases, but are not normally thought of as being pneumatic circuits:

- Guns
- Rockets
- Refrigerators
- Internal combustion engines
- Scuba sets.

HYDRAULIC CIRCUIT

A **hydraulic circuit** is a system comprising an interconnected set of discrete components that transport liquid. The purpose of this system may be to control where fluid flows (as in a network of tubes of coolant in a thermodynamic system) or to control fluid pressure (as in hydraulic amplifiers). The approach of describing a fluid system in terms of discrete components is inspired by the success of electrical circuit theory. Just as electric circuit theory works when elements are discrete and linear, hydraulic circuit theory works best when the elements (passive component such as pipes or transmission lines or active components such as power packs or pumps) are discrete and linear. This usually means that hydraulic circuit analysis works best for long, thin tubes with discrete pumps, as found in chemical process flow systems or microscale devices.

Components

The circuit comprises the following components:

*   Active components
    *   o   Hydraulic power pack
*   Transmission lines
    *   o   Hydraulic hoses
*   Passive components
    *   o   Hydraulic cylinders

HYDRAULIC FLUIDS

Hydraulic fluid can be the most vital component of a hydraulic system, so you must carefully consider dozens of characteristics before making a final selection.

The demands placed on hydraulic systems constantly change as industry requires greater efficiency and speed at higher operating temperatures and pressures. Selecting the best hydraulic fluid requires a basic understanding of each particular fluid's characteristics in comparison with an ideal fluid. An ideal fluid would have these characteristics:

*   thermal stability
*   hydrolytic stability
*   low chemical corrosiveness
*   high anti-wear characteristics
*   low tendency to cavitate
*   long life
*   total water rejection
*   constant viscosity, regardless of temperature, and
*   low cost.

Although no single fluid has all of these ideal characteristics, it is possible to select one that is the best compromise for a particular hydraulic system. This selection requires knowledge of the system in which a hydraulic fluid will be used. The designer should know such basic characteristics of the system as:

- maximum and minimum operating and ambient temperatures
- type of pump or pumps used
- operating pressures
- operating cycle
- loads encountered by various components, and
- type of control and power valves

Influential Factors

Each of the following factors influences hydraulic fluid performance:

Viscosity - Maximum and minimum operating temperatures, along with the system's load, determine the fluid's viscosity requirements. The fluid must maintain a minimum viscosity at the highest operating temperature. However, the hydraulic fluid must not be so viscous at low temperature that it cannot be pumped.

Wear - Of all hydraulic system problems, wear is most frequently misunderstood because wear and friction usually are considered together. Friction should be considered apart from wear.

Wear is the unavoidable result of *metal-to-metal* contact. The designer's goal is to minimize metal breakdown through an additive that protects the metal. By comparison, friction is reduced by *preventing* metal-to-metal contact through the use of fluids that create a thin protective oil or additive film between moving metal parts.

Note that excessive wear may not be the fault of the fluid. It may be caused by poor system design, such as excessive pressure or inadequate cooling.

Anti-wear - The compound most frequently added to hydraulic fluid to reduce wear is zinc dithiophosphate (ZDP), but today, ashless anti-wear hydraulic fluids have become popular with some companies and in certain states to reduce loads on waste treatment plants. No ZDP or other type heavy metals have been used in the formulation of ashless anti-wear fluids.

The pump is the critical dynamic element in any hydraulic system, and each pump type (vane, gear, piston) has different requirements for wear protection. Vane and gear pumps need anti-wear protection. With piston pumps, rust and oxidation (R & O) protection is more important. This is because gear and vane pumps operate with inherent metal-to-metal contact, while pistons ride on an oil film.

When two or more types of pumps are used in the same system, it is impractical to have a separate fluid for each, even though their operating requirements differ. The common fluid selected, therefore, must bridge the operating require-ments of all pump types.

Foaming - When foam is carried by a fluid, it degrades system performance and therefore should be eliminated. Foam usually can be prevented by eliminat-ing air leaks within the system. However, two general types of foam still occur frequently:

- surface foam, which usually collects on the fluid surface in a reservoir, and

- entrained air.

Surface foam is the easiest to eliminate, with defoaming additives or by proper sump design so that foam enters the sump and has time to dissipate.

Entrained air can cause more serious problems because this foam is drawn into the system. In worst cases, it causes cavitation, a hammering action that can destroy parts. Entrained air is usually prevented by properly selecting the additive and base oils. Caution: certain anti-foam agents, when used at a high concentra-tion to reduce surface foam, will increase entrained air.

Also linked to the foam problem, is fluid viscosity, which determines how easily air bubbles can migrate through the fluid and escape.

R & O - Most fluids need rust and oxidation inhibitors. These additives both protect the metal and contain anti-oxidation chemicals that help prolong fluid life.

Corrosion - Two potential corrosion problems must be considered: *system rusting* and *acidic chemical corrosion*. System rusting occurs when water carried by the fluid attacks ferrous metal parts. Most hydraulic fluids contain rust inhibitors to protect against system rusting. The tests used to measure this capability are ASTM D 665 A and B. To protect against chemical corrosion, other additives must be considered. The additives must also exhibit good stability in the presence of water (hydrolytic stability) to prevent break down and acidic attack on system metals.

Oxidation and thermal stability - Over time, fluids oxidize and form acids, sludge, and varnish. Acids can attack system parts, particularly soft metals. Extended high-temperature operation and thermal cycling also encourage the formation of fluid decomposition products. The system should be designed to minimize these thermal problems, and the fluid should have additives that ex-hibit good thermal stability, inhibit oxidation, and neutralize acids as they form.

Although not always practical or easy to attain, constant moderate tempera-ture and steady-state operation are best for system and fluid life.

Water retention - Large quantities of water in a hydraulic fluid system can be removed by draining the sump periodically. However, small amounts of water can become entrained, particularly if the sump is small. Usually, demulsifiers are added to the fluid to speed the separation of water. Filters can then physically remove any remaining water from the hydraulic fluid. The water should leave the fluid without taking fluid or additives with it.

Temperature - System operating temperature varies with job requirements. Here are a few general rules: the maximum recommended operating temperature usually is 150° F. Operating temperatures of 180° to 200° F are practical, but the fluid will have to be changed two to three times as often. Systems can operate at temperatures as high as 250° F, but the penalty is fairly rapid decomposition of the fluid and especially rapid decomposition of the additives - sometimes within 24 hours!

Fluid Makeup

Most fluids are evaluated based on their ratings for rust and oxidation (R & O), thermal stability, and wear protection, plus other characteristics that must be considered for efficient operation:

Seal compatibility - In most systems, seals are selected so that when they encounter the fluid they will not change size or they will expand only slightly, thus ensuring tight fits. The fluid selected should be checked to be sure that the fluid and seal materials are compatible, so the fluid will not interfere with proper seal operation.

Fluid life, disposability - There are two other important considerations that do not directly relate to fluid performance in the hydraulic system, but have a great influence on total cost. They are *fluid life* and *disposability*.

Fluids that have long operating lives bring added savings through reduced maintenance and replacement-fluid costs. The cost of changing a fluid can be substantial in a large system. Part life should also be longer with the higher-quality, longer-lived fluid.

Longer fluid life also reduces disposal problems. With greater demands to keep the environment clean, and ever-changing definitions of what is toxic, the problem of fluid disposability increases. Fluids and local anti-pollution laws should both be evaluated to determine any potential problems.

Synthesized hydrocarbon (synthetic) hydraulic fluids contain no waxes that congeal at low temperatures nor compounds that readily oxidize at high temperatures which are inevitable in natural mineral oils. Synthetic hydraulic fluids are being used for applications with very low, very high, or a very wide range of temperatures.

Fire-resistant Fluids

The overwhelming majority of hydraulic components and systems are designed to use oil-based hydraulic fluids. No wonder; these fluids rarely present significant operating, safety, or maintenance problems. Unfortunately, there are circumstances where using oil-based fluid should be avoided. One common fluid power application is in an environment with potential ignition sources - an open flame, sparks, or hot metal. In these environments, a leak spraying from a high-pressure hydraulic system could cause a serious fire and result in major property damage, personnel injury, or even death.

Even though most oil-based hydraulic fluids have relatively high flash/fire points (>300° F), small leaks in a high-pressure system can produce a finely atomized spray that can travel significant distances. If an ignition source is encountered, complete ignition of the spray envelope can occur. The alternative is to use a hydraulic fluid that eliminates or significantly reduces this hazard: any of several fire-resistant hydraulic fluids (FRHFs).

How Far We've Come

Apart from isolated segments of basic research, little progress was made in developing suitable FRHFs until the end of World War II. During the war, tragic incidents related to hydraulic fluid fires and major property losses at steel mills and foundries graphically illustrated the urgent need for something to be done. Similar incidents in captive environments such as coal mines during the rapid post-war industrial expansion helped motivate a major joint research effort between government and industry. This work was directed at developing fluids that could replace oil-based hydraulic fluids at a reasonable cost and with no significant reduction in hydraulic system performance. Two basic approaches were undertaken. One involved the introduction of water into the fluid to act as a "snuffer" if the fluid ignited. The other involved synthetic, non-aqueous products whose chemistry resisted burning or generated products of combustion that helped extinguish any flame.

Commercial products in both categories evolved during the 1950s and 1960s and are still in use today. In the early 1970s, an additional synthetic type of fluid was introduced to address many of the drawbacks inherent in the earlier types. Since the introduction of each type, many improvements have been made in fire resistance, anti-wear properties, and overall quality.

Where We Are

Water glycol and *invert emulsion* constitute the major fluid types of water-containing products. Water glycol is a true solution of a glycol (such as ethylene glycol) in water, along with a variety of additives to impart viscosity, corrosion protection, and anti-wear properties. A shear-stable thickener, which has improved over the years, represents the novel technology aspect of the fluid. Water glycol contains approximately 40% water. Despite a number of drawbacks, water glycol is the dominant FRHF on the market today and is used in a wide variety of applications.

An invert emulsion also contains approximately 40% water but is a stable emulsion of water dispersed in oil. The outer phase, oil, represents the wetting surface; the inner phase, water, provides the fire retardant-element. Oil-soluble additives provide anti-wear properties, corrosion protection, and emulsion stability. Inverts, at one time, were commonly used but are losing favor in industry today.

Synthetic fluids initially were represented by a class of chemical compounds known as **phosphate esters**, which are reaction products between phosphoric acid and aromatic ring-structure alcohols. These fluids are extremely fire resistant and have widespread industrial use, as well as military and aircraft service. However, their popularity has declined because of environmental, cost, and compatibility factors.

The other type of synthetic fluids in use are synthetic hydrocarbons, more specifically, **polyol esters**. These fluids are the reaction products between long-chain fatty acids (derived from animal and vegetable fats) and synthesized organic alcohols. These products contain additives to impart anti-wear properties, corrosion protection, and viscosity modification. Fire resistance results from a combination of high thermal properties and physical characteristics. This is the most recent category of FRHFs and has gained widespread and growing use.

What is Fire Resistance?

The term "fire resistant" often is misunderstood or interpreted to be overly inclusive; it seems appropriate to standardize the terminology and review the accepted test methods for judging the fire resistance of a given fluid. First, there is no single property or test of a fluid, such as flash/fire point, auto ignition temperature (AIT), *etc.* that will quantitatively rate its relative fire resistance. This has led to a "simulated incident" approach in which tests are designed to replicate a worst-case scenario in typical applications where fluid power is used near a potential fire hazard. Fluids generally pass or fail these tests, and those that pass are incorporated into an Approval Guide or List of Qualified Fluids.

In the United States, two test protocols have evolved and are generally regarded as benchmarks in the industry. One was developed by Factory Mutual Research Corporation (FMRC). Their original intent was to use the test results in the risk-assessment programs of those insurance companies under the Factory Mutual System umbrella. The test has since become the chief qualification for commercial companies using FRHFs; all fluid suppliers submit products seeking "FMRC Approval." The FMRC Approval Guide lists over 300 FRHFs from approximately 50 suppliers. Factory Mutual's program is now global in scope.

FMRC addresses the definition of FRHF in the following excerpt in their introduction to the hydraulic fluids sections of their Approval Guide: *Less flammable hydraulic fluids approved and listed here have been tested to evaluate fire hazard only. All presently available fluids will burn under certain conditions. In each case the fire hazard has been reduced to an acceptable degree, meeting the Approval Standards of FMRC; other fluid properties are not investigated.*

This paragraph accurately puts the intent of FRHFs into the proper perspective. They are not fireproof but, rather, they significantly reduce the potential hazard associated with oil-based products. In the FMRC tests, the fluid is conditioned to 140°F, pressurized to 1,000 psi in a steel cylinder, and discharged through an oil burner-type nozzle. The spray generated is intended to simulate a high-pressure hydraulic system leak. A gas flame is passed through the spray envelope at two distances downstream of the nozzle. There may be local burning at the point of flame entry, and the pass criteria dictate that any flame must self-extinguish when the ignition source is removed; no flame may propagate back to the nozzle. This process is repeated 20 times, and the burn duration timed. Any burn duration over 5 sec is considered a fail.

A second test uses the same spray directed at an inclined metal channel heated to 1,300°F. In this test, the spray is continuous for 60 sec. The criteria are:

1. The spray in contact with the channel may not burn, or
2. If spray ignition takes place, fluid rolling off the channel cannot continue to burn, and the flame cannot follow the spray if directed away from the channel.

If these conditions are satisfied, the fluid is approved. Statistics are not available, but many products in all of the fluid categories described do not pass this test.

The Mine Safety & Health Administration (MSHA) has had in place for many years an evaluation program for qualifying fluids that are used underground, primarily in coal mines. MSHA testing is similar to FMRC's in the sense that a spray mist of the candidate fluid is generated. However, the ignition mechanism is somewhat different in the MSHA test. Under this procedure, a spray mist is directed continuously at a variety of ignition sources that include an open gas flame, a welding arc, and burning rags. The pass criteria are that localized burning in the spray mist extinguish within 5 sec, and there can be no sustained propagation along the spray axis. They also have an AIT criterion and a wick test to assess the rate of evaporation of water from a candidate product. MSHA tests also have a relatively high rate of product rejections.

Since both of these tests involve fluids submitted by the supplier to the testing agency, both FMRC and MSHA have comprehensive manufacturer auditing programs in which quality-assurance programs are carefully evaluated and monitored by periodic, on-site inspections. This may include retests of approved fluids.

Other Tests

In addition to these "third party" ratings of FRHFs, many companies have developed their own fire-resistance tests that must be considered in addition to a product having FMRC approval. Again, these tests generally follow the simulated incident philosophy and are specific to the type of industry involved. Examples of these include exposing the candidate fluid - in spray or non-spray form - to a hot manifold, molten metal, heated blocks of a representative metal, burning rags, hot sand, *etc.* The evaluation criteria may be no burning, limited burning, no smoke, non-propagation, *etc.* Minimum AIT and flash/fire point temperatures also are used either independently or in combination with a test described above.

In all of these tests, a product is either approved or rejected; there is no ranking or rating of approved products. This aspect, the occasional lack of reproducibility, and the absence of service history of a fluid has led FMRC to develop a new test that will quantify the relative fire resistance of various fluids. The test procedure involves measuring the heat release of a fluid under a fixed-burn condition and combining this value with a separately determined measurement of the energy required to initiate burning. These values are used to establish a Spray Flammability Parameter for each product evaluated. This test and a new approval standard currently are under review by FMRC and have not been formally adopted.

Other Concerns

The major problem facing a designer converting a hydraulic system from an oil-based fluid to FRHF is selecting the particular type that will minimize the cost of conversion and maximize the operating and safety benefits. The choice becomes a trade-off of characteristics associated with each type. Each product group offers advantages and disadvantages for any given application. It is beyond the scope of this article to attempt to make recommendations for certain end-users, but the major attributes and shortfalls of the various fluid types can be addressed.

Where We're Going

Significant improvements continue to be made with both water glycol and polyol ester fluids. The impact of more-stringent environmental regulations will be more strongly felt in the next few years and may even restrict the choice.

The motivation for converting from an oil-based fluid will also strengthen as waste control regulations expand for any product containing oil. In some areas, "hydraulic oil" already is considered a hazardous material. As their prices decrease, fluids having the capability of being non-toxic and readily biodegradable will further expand the motivation to replace oil-based hydraulic fluids.

Environmental Fluids

In some cases, environmental considerations necessitate the selection of a zinc-free ashless petroleum or a biodegradable hydraulic fluid.

The Environmental Protection Agency (EPA) continues to advocate the use of environmentally safe hydraulic fluids in place of conventional petroleum-based hydraulic oils - particularly in applications where fluid leakage could have a negative impact on the environment. Spills of standard, petroleum-based hydraulic fluids are known to kill marine life and contaminate soil. Environmentally safe hydraulic fluids are formulated to avoid those undesirable results.

To be classified as environmentally safe, a fluid must be readily biodegradable (more than 60% of the fluid must break down into innocuous products when tested in standardized laboratory tests a 28-day period) and virtually non-toxic (more than half the rainbow trout fingerlings in a population must survive after four days in an aquatic solution with concentrations of the fluid greater than 1000 ppm). A major benefit of these highly biodegradeable fluids is that spills may have lower cleanup costs, depending on local regulations. Also, they are less likely to harm plant and animal life that comes in contact with a spill.

Hydraulic applications that could be considered environmentally sensitive include mobile equipment in general, with emphasis on forestry and construction machinery, and marine equipment used on fishing boats, off-shore drilling operations, and hydraulically operated bridges, locks, and dams. Other locations are commercial elevators and equipment in amusement parks.

Three Base Oils

Three different base oils have been tried as environmentally safe hydraulic fluids. They are synthetic esters, polyglycols, and vegetable oils. Synthetic esters can be formulated as biodegradable fluids with superior lubrication performance, but their high cost has limited their usage. Polyglycols — attractive because they have excellent lubricity characteristics and are usually less expensive than synthetic esters — have been used more commonly. However, polyglycols lack required biodegradability and are potentially toxic in water when mixed with lubricating additives. Vegetable oils, such as sunflower, soy, or canola oils, have excellent natural biodegradability, are in plentiful supply, and are inexpensive. They have become the most commonly used environmentally safe fluids in hydraulic systems.

The base fluids of biodegradable hydraulic fluids are usually vegetable oils, selected synthetic esters, or a blend of the two. Biodegradable hydraulic fluids typically contain ashless inhibitors with low toxicity and additives to enhance performance. Properly formulated biodegradable hydraulic fluids can provide effective wear resistance similar to petroleum anti-wear hydraulic fluids. However, some biodegradable base oils, especially vegetable oils, may exhibit poor oxidation stability. The use of a synthetic-ester base usually improves the and oxidation resistance of the fluids.

The tradeoff between environmental advantages and potential performance deficiencies of biodegradable hydraulic fluids suggests that these fluids are most suitable for applications in environmentally sensitive areas. Their use should be considered wherever contamination of ground or water by petroleum lubricants could be a problem.

Additives

Like petroleum oils, vegetable oils or synthetic esters rely on specially selected additives to improve their performance as lubricants. The additives contained in biodegradable hydraulic fluids typically exhibit very low toxicity. Unlike petroleum oils, vegetable oils contain unsaturated hydrocarbons and are natural occurring esters. The unsaturation leads to rapid oxidation at elevated temperatures and poor low temperature flow properties. This low-temperature fluidity can be improved by additives, but their oxidation stability remains a performance concern.

International Guidelines

Throughout Europe, the development of guidelines for biodegradable lubricants is typically left to local authorities or non-government organizations. In

Germany, *Blue Angel* labels will be awarded to biodegradable hydraulic fluids. The Blue Angel for biodegradable hydraulic fluids will likely require that the base fluids must be readily biodegradable - greater than 80% biodegradation in 21 days by CEC L-33-A93 Test, or greater than 70% biodegradation in 28 days by the Modified Sturm Test. In addition, all components must be Water Hazard Class 0 or 1, which means the components are not water pollutants. Environmental Choice Program of Canada is currently in the process of reviewing a guideline on biodegradable, non-toxic hydraulic fluids. It will likely include a requirement that base fluids exhibit greater than 90% biodegradation in 21 days by CEC L-33-A93.

In the United States, ASTM D-2.N.3 on eco-evaluated hydraulic fluids has drafted an information guide that addresses the means of assessing the biodegradability of hydraulic fluids. D-2.N.3 is currently developing environmental classifications for hydraulic fluids. A In December 1995, ASTM D-2.12 on Environmental Standards for Lubricants completed a standard test especially designed to determine the aerobic aquatic biodegradability of all lubricants and their components. The test is similar to the Modified Sturm Test, which measures the evolution of carbon dioxide in 28 days. This standard is being published as ASTM D 5864. ASTM D-2.12 is currently developing other environmental standard tests for lubricants, which include an aquatic toxicity test for fish and large invertebrates; a manometric respirometry biodegradation test method; and a Gladhill Shake Flask biodegradation test.

Initially designed to measure the biodegradability of 2-cycle engine oils, CEC L-33-A93 has been the most widely applied biodegradation test for lubricants in Europe since the early 1980s. The test uses infrared spectroscopy to measure the disappearance of certain hydrocarbons over a 21-day period when the lubricant is mixed with an inoculum containing micro-organisms. Thus, the CEC test is a only a measure of primary biodegradation.

Unlike the CEC test, the Modified Sturm Test is a measure of ultimate biodegradation. By measuring the production of CO_2 over 28 days, the test estimates the extent to which the carbon in a lubricant is converted by micro-organisms to the elements found in nature - namely: CO_2, water, inorganic compounds, and biotic mass. Because this test was designed originally for water-soluble, pure compounds, it is difficult to use for testing lubricants, most of which are water-insoluble, complex mixtures.

The new ASTM D 5864 test is similar to the Modified Sturm Test. It is specially designed for testing water insoluble complex lubricants.

The Readily Biodegradable Question

One question that often comes up is whether a fluid is *readily biodegradable* or just *biodegradable*. Most things are biodegradable, given enough time and proper conditions. Readily biodegradable means that a substance exhibits a result equal to or greater than a pre-set requirement in a standard test.

For example, XYZ Standard requires 80% or higher biodegradation by CEC L-33-A93 in 21 days. If a lubricant meets this requirement, it is considered readily biodegradable by the XYZ Standard. Ideally, any claim that a lubricant is readily biodegradable also also specify the test and standard.

Vegetable Oils or Synthetic Esters?

Being naturally occurring esters, vegetable oils are susceptible to hydrolysis, which leads to fluid decomposition and degradation, especially in the presence of heat. Because of their polarity, vegetable oils tend to cause elastomers to swell, though in most cases the degree of swell is insufficient to cause any serious concern in hydraulic applications.

On the other hand, vegetable oils offer excellent lubricity, intrinsic high viscosity index, and good anti-wear and extreme-pressure properties. Well-formulated, biodegradable hydraulic fluids based on vegetable oils can easily pass the demanding Vickers 35VQ25 or Denison T5D-42 vane-pump wear tests. They also can meet the requirements of major OEMs for premium hydraulic fluids, except hydrolytic, thermal, and oxidation stability. Experience has shown that vegetable oil-based biodegradable hydraulic fluids can perform satisfactory for years under mild climate and operation conditions (temperatures below 160° F, and hydraulic systems kept free of water contamination).

The use of synthetic esters - typically polyol esters - provides better hydrolytic, thermal, and oxidative stability, and excellent low-temperature fluidity, while preserving the high biodegradability and low toxicity of the fluids. For nearly 30 years, polyol esters have been used to formulate aviation gas turbine lubricants, which demand high thermal and oxidation stability at extreme temperatures. While a vegetable oil-based hydraulic fluid can perform between 0° to 180° F, a similar fluid based on synthetic esters can be used between 25° and 200° F. Similar to vegetable oils, synthetic esters have the tendency to swell and soften elastomers, although again, the swell should not be a concern for most hydraulic applications.

Fluid Handling

Vegetable oil or synthetic ester-based biodegradable hydraulic fluids are fully miscible with each other and with petroleum hydraulic fluids.

However, when a biodegradable hydraulic fluid is mixed with petroleum lubricants, its biodegradability typically decreases, and its toxicity increases. Because of their susceptibility to hydrolysis, vegetable oil- or synthetic ester-based fluids should be kept free of water contamination, both in storage and in everyday use.

There is no regulation permitting shortcuts in the disposal of biodegradable hydraulic fluids. Such disposal should be handled in the same manner as the disposal of petroleum fluids, in accordance with applicable federal, state, and local laws and regulations.

The Future of Biodegradable Fluids

Government regulations and codes, and the environmental awareness of lubricant users are the driving forces for the growing use of biodegradable hydraulic fluids. However, the lack of definition and standards for biodegradable fluids in the United States impedes the market development for these fluids. Development of new standards and guidelines by ASTM and other industrial and governmental organizations will inevitably influence the growth of biodegradable fluids.

Meanwhile, lubricant suppliers continue to develop and evaluate new additive chemistries that provide greater oxidative, thermal, and hydrolytic stability properties for biodegradable fluids. Vegetable oil suppliers are using genetic engineering to produce new vegetable oils with improved stability. Ester manufacturers are considering improving ester performance by incorporating additive-type functional groups into molecular structures. The improvement in the performance quality of biodegradable hydraulic fluids will eventually lead to more applications and increased popularity of these important fluids.

Environmentally Safe and Fireproof

A drawback of most hydraulic fluids, including some fire-resistant fluids, is their toxicity - either to personnel, the environment, or both. Furthermore, they are only fire *resistant*, and most will burn under certain conditions. Recently introduced synthetic water additives, on the other hand, mix with water (usually in a concentration of 5%) to become fire *proof*; the solution actually could extinguish a fire.

These water-based fluids, in general, also offer a cost advantage over most other fluids because one gallon of concentrate produces 20 gallons of hydraulic fluid. When disposal expenses enter calculations, the cost differential becomes even greater - especially with a solution containing non-toxic, readily biodegradable synthetic water additives that require no treatment. The accompanying table summarizes characteristics of common fire-resistant and fire-proof fluids.

There are, however, important performance and operating characteristics of water-based fluids that cannot be ignored. First, water-based fluids in general have much lower viscosity, film strength, and lubricating qualities than oil-based fluids. This means that system components - especially pumps, valves, and actuators - must be designed specifically for operation with water-based fluid. You can't just drain fluid from a system containing oil-based fluid and expect it to run on water-based fluid.

A perception remains today that components for water-based fluid are much more expensive and larger - especially valves - than their conventional counterparts. While this may have been true 20 years ago, the cost premium for valves and other components designed for water-based fluid has narrowed to about 30%. This investment can easily be recovered in the cost of fluid alone, not to mention disposal and treatment costs. Moreover, valve size has been reduced dramatically: many are available with standard NFPA footprints.

					Fire-resistant fluids				
Vendor	Synthetic water additive	Oil-in-water emulsion	Micro emulsion	Invert emulsion	Ethylene glycol	Diethylene glycol	Propylene glycol	Phosphate ester	Polyol ester
E. F. Houghton	Hydro-Lubric® 120B				Houghto-Safe® 620			Hougto-Safe® 1000 Series	Cosmo-Lubric® HF Series
Quaker Chemical	Quinto-Lubric® 807		Quinto-Lubric® 200						Quinto-Lubric® 822 Series
MBSA	Muzanin®								
Lubrizol			Microzol®						
Nalco	Fyresafe 126®								
Texas Chemical				Hy-Guard®					
UCON						Hydro-Lube ®HP5046			
Properties									
Flame resistance	Very good	Good (avoid evaporation)	Very good	Good (avoid evaporation)	Very good	Very good	Very good	Very good	Very good
Non-toxicity	Excellent	Good	Good	Fair	Very poor	Poor	Very good	Poor	Good
Biodegradability	Very good	Not possible	Not possible	Not possible	Not possible	Not possible	Very good	Not possible	See below
Viscosity at 50°C, cSt	1 - 40	1 - 40	1 - 40	Non-Newtonian	20 - 76	20 - 76	20 - 76	12 - 52	12 - 52

Some newer formulations of polyol ester fluids meet federal EPA requirements for being "readily biodegradable."

Next, any potential for freezing must be considered. Traditionally, ethylene glycol is added to water to lower the solution's freezing point. However, using highly toxic ethylene glycol in a solution containing the synthetic additive would completely negate the purpose of using an environmentally safe additive. Using *propylene* glycol instead as anti-freeze maintains the environmental integrity of the solution because propylene glycol is so non-toxic that it is approved for use in food by the U. S. Food & Drug Administration.

Finally, because the fluid is non-toxic, it naturally tends to support microbial growth. To minimize or prevent consequences associated with this problem, judicious use of bacteriostatic additives and effective sealing and reservoir design should be practiced.

Glossary of Hydraulic Fluid Terminology

Absolute viscosity - the ratio of shear stress to shear rate. It is a fluid's internal resistance to flow. The common unit of absolute viscosity is the poise. Absolute viscosity divided by fluid density equals kinematic viscosity.

Absorption - the assimilation of one material into another.

Additive - chemical substance added to a fluid to impart or improve certain properties.

Adsorption - adhesion of the molecules of gases, liquids, or dissolved substances to a solid surface, resulting in relatively high concentration of the molecules at the place of contact; *e.g.* the plating out of an anti-wear additive on metal surfaces.

Anti-foam agent - one of two types of additives used to reduce foaming in petroleum products: silicone oil to break up large surface bubbles, and various kinds of polymers to decrease the amount of small bubbles entrained in the oils.

Asperities - microscopic projections on metal surfaces resulting from normal surface-finishing processes. Interference between opposing asperities in sliding or rolling applications is a source of friction, and can lead to metal welding and scoring. Ideally, the lubricating film between two moving surfaces should be thicker than the combined height of the opposing asperities.

Bactericide - additive included in the formulations of water-mixed fluids to inhibit the growth of bacteria.

Boundary lubrication - form of lubrication between two rubbing surfaces without development of a full-fluid lubricating film. Boundary lubrication can be made more effective by including additives in the lubricating oil that provide a stronger oil film, thus preventing excessive friction and possible scoring.

Bulk modulus - the measure of a fluid's resistance to compressibility; the reciprocal of compressibility.

Cavitation - formation of a vapor pocket (bubble) due to sudden lowering of pressure in a liquid, and often causing metal erosion and eventual pump destruction.

Corrosion inhibitor - additive for protecting wetted metal surfaces from chemical attack by water or other contaminants. Polar compounds wet the metal surface preferentially, protecting it with a film of oil. Other compounds may absorb water by incorporating it in a water-in-oil emulsion so that only the oil touches the metal surface. Still others combine chemically with the metal to present a non-reactive surface.

Demulsibility - ability of an oil to separate from water.

Dewaxing - removal of paraffin wax from lubricating oils to improve low temperature properties, especially to lower the cloud point and pour point.

Emulsifier - additive that promotes the formation of a stable mixture, or emulsion, of oil and water. Common emulsifiers are: metallic soaps, animal and vegetable oils, and polar compounds.

Emulsion - intimate mixture of oil and water, generally of a milky or cloudy appearance. Emulsions may be of two types: oil-in water (where water is the continuous phase) or water-in-oil (where water is the discontinuous phase).

EP additive - lubricant additive that prevents sliding metal surfaces from seizing under conditions of extreme pressure (EP). At the high local temperatures associated with metal-to-metal contact, an EP additive combines chemically with the metal to form a surface film that prevents scoring that destroys sliding surfaces under high loads.

Fire-resistant fluid - hydraulic oil used especially in high-temperature or hazardous applications. Three common types of fire-resistant fluids are: water-petroleum oil emulsions, in which the water prevents burning of the petroleum constituent; water-glycol fluids; and non-aqueous fluids of low volatility, such as phosphate esters, silicones, polyolesters, and halogenated hydrocarbon-type fluids.

Full-fluid-film lubrication - presence of a continuous lubricating film sufficient to completely separate two surfaces, as distinct from boundary lubrication. Full-fluid-film lubrication is normally hydrodynamic lubrication, whereby the oil adheres to the moving part and is drawn into the area between the sliding surfaces, where it forms a pressure, or hydrodynamic wedge.

Hydraulic fluid - fluid serving as the power transmission medium in a hydraulic system. The principal requirements of a premium hydraulic fluid are proper viscosity, high viscosity index, anti-wear protection (if needed), good oxidation stability, adequate pour point, good demulsibility, rust inhibition, resistance to foaming, and compatibility with seal materials. Anti-wear oils are frequently used in compact, high-pressure, and high-capacity pumps that require extra lubrication protection.

Immiscible - incapable of being mixed without separation of phases. Water and petroleum oil are immiscible under most conditions, although they can be made miscible with the addition of a proper emulsifier.

Inhibitor - additive that improves the performance of a petroleum product through the control of undesirable chemical reactions.

Kinematic viscosity - absolute viscosity of a fluid divided by its density at the same temperature of measurement. It is the measure of a fluid's resistance to flow under gravity.

Lubricity - ability of an oil or grease to lubricate.

Miscible - capable of being mixed in any concentration without separation of phases; *e.g.,* water and ethyl alcohol are miscible.

Newtonian fluid - fluid, such as a straight mineral oil, whose viscosity does not change with rate of flow.

Non-Newtonian fluid - fluid, such as a grease or a polymer containing oil (*e.g.* multi-grade oil), in which shear stress is not proportional to shear rate.

Oxidation inhibitor - substance added in small quantities to petroleum product to increase its oxidation resistance, thereby lengthening its service or storage life; also called anti-oxidant.

Polar compound - a chemical compound whose molecules exhibit electrically positive characteristics at one extremity and negative characteristics at the other. Polar compounds are used as additives in many petroleum products.

Pour point - lowest temperature at which an oil or distillate fuel will flow, when cooled under conditions prescribed by specific test methods. The pour point is 3° C (5° F) above the temperature at which the oil in a test vessel shows no movement when the container is held horizontally for five seconds.

Shear rate - rate at which adjacent layers of fluid move with respect to each other, usually expressed as reciprocal seconds.

Shear stress - frictional force overcome in sliding one layer of fluid along another, as in any fluid flow. The shear stress of a petroleum oil or other Newtonian fluid at a given temperature varies directly with shear rate (velocity). The ratio between shear stress and shear rate is constant; this ratio is termed viscosity.

Surfactant - surface-active agent that reduces interfacial tension of a liquid. A surfactant used in a petroleum oil may increase the oil's affinity for metals and other material.

Vapor pressure - pressure of a confined vapor in equilibrium with its liquid at a specified temperature; thus, a measure of a liquid's volatility.

Viscosity - measurement of a fluid's resistance to flow. The common metric unit of absolute viscosity is the poise, which is defined as the force in dynes required to move a surface one square centimeter in area past a parallel surface at a speed of one centimeter per second, with the surfaces separated by a fluid film one centimeter thick. In addition to kinematic viscosity, there are other methods for determining viscosity, including, Saybolt Universal viscosity, Saybolt Furol viscosity, Engier viscosity, and Redwood viscosity. Since viscosity varies inversely with temperature, its value is meaningless until the temperature at which it is determined is reported.

Viscosity index (VI) - empirical, unitless number indicating the effect of temperature changes on the kinematic viscosity of an oil. Liquids change viscosity with temperature, becoming less viscous when heated; the higher the V.I. of an oil, the lower its tendency to change viscosity with temperature.

HYDRAULIC SYSTEMS SAFETY

Quick Facts

- Hydraulic systems must store fluid under high pressure.
- Three kinds of hazards exist: burns from the hot, high pressure spray of fluid; bruises, cuts or abrasions from flailing hydraulic lines; and injection of fluid into the skin.
- Safe hydraulic system performance requires general maintenance.
- Proper coupling of high and low pressure hydraulic components and pressure relief valves are important safety measures.

Hydraulic systems are popular on many types of agricultural equipment because they reduce the need for complex mechanical linkages and allow remote control of numerous operations. Hydraulic systems are used to lift implements, such as plows; to change the position of implement components, such as a combine header or bulldozer blade; to operate remote hydraulic motors; and to assist steering and braking.

To do their work, hydraulic systems must store fluid under high pressure, typically 2,000 pounds or more per square inch. One hazard comes from removing or adjusting components without releasing the pressure. The fluid, under tremendous pressure, is also hot. The worker then is exposed to three kinds of hazards: burns from hot, high-pressure fluid; bruises, cuts or abrasions from flailing hydraulic lines; and injection of fluid into the skin.

Many systems store hydraulic energy in accumulators. These accumulators are designed to store oil under pressure when the hydraulic pump cannot keep up with demand, when the engine is shut down, or when the hydraulic pump malfunctions. Even though the pump may be stopped or an implement disconnected, the system is still under pressure. To work on the system safely, relieve the pressure first.

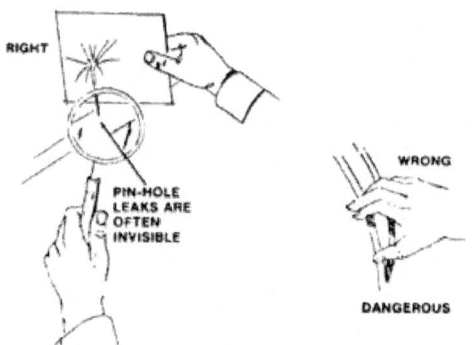

Detecting pinhole leaks in a hydraulic system.

Pinhole Leak Injuries

Probably the most common injury associated with hydraulic systems is the result of pinhole leaks in hoses. These leaks are difficult to locate. A person may notice a damp, oily, dirty place near a hydraulic line. Not seeing the leak, the person runs a hand or finger along the line to find it. When the pinhole is reached, the fluid can be injected into the skin as if from a hypodermic syringe.

Immediately after the injection, the person experiences only a slight stinging sensation and may not think much about it. Several hours later, however, the wound begins to throb and severe pain begins. By the time a doctor is seen, it is often too late, and the individual loses a finger or entire arm.

Unfortunately, this kind of accident is not uncommon. To reduce the chances of this type of injury, run a piece of wood or cardboard along the hose (rather than fingers) to detect the leak.

Improper Coupling

Another hazard is improper coupling of low- and high-pressure hydraulic components. Do not connect a high-pressure pump to a low-pressure system. Do not incorporate a low-pressure component, hose or fitting into a high-pressure system. Component, hose or fitting ruptures are likely to occur.

Pressure relief valves incorporated into the hydraulic system will avoid pressure buildups during use. Keep these valves clean and test them periodically to ensure correct operation.

Maintenance

An improperly maintained hydraulic system can lead to component failures. Safe hydraulic system performance requires general maintenance.

- Periodically check for oil leaks and worn hoses.
- Keep contaminants from hydraulic oil and replace filters periodically.
- Coat cylinder rods with protective lubricants to avoid rusting.

Tips for Safe Operation

Follow these rules for safe hydraulics operation:

- Always lower the hydraulic working units to the ground before leaving the machine.
- Park the machinery where children cannot reach it.
- Block up the working units when you must work on the system while raised; do not rely on the hydraulic lift.
- Never service the hydraulic system while the machine engine is running unless absolutely necessary (bleeding the system).
- Do not remove cylinders until the working units are resting on the ground or securely on safety stands or blocks; shut off the engine.
- When transporting the machine, lock the cylinder stops to hold the working units solidly in place.
- Before disconnecting oil lines, relieve all hydraulic pressure and discharge the accumulator (if used).
- Be sure all line connections are tight and lines are not damaged; escaping oil under pressure is a fire hazard and can cause personal injury.
- Some hydraulic pumps and control valves are heavy. Before removing them, provide a means of support such as a chain hoist, floor jack or blocks.
- When washing parts, use a nonvolatile cleaning solvent.
- To ensure control of the unit, keep the hydraulics in proper adjustment.

Chapter 4

PUMPS AND COMPRESSORS

PUMP

A **pump** is a device that moves fluids (liquids or gases), or sometimes slurries, by mechanical action. Pumps can be classified into three major groups according to the method they use to move the fluid: *direct lift*, *displacement*, and *gravity* pumps.

Pumps operate by some mechanism (typically reciprocating or rotary), and consume energy to perform mechanical work by moving the fluid. Pumps operate *via* many energy sources, including manual operation, electricity, engines, or wind power, come in many sizes, from microscopic for use in medical applications to large industrial pumps.

Mechanical pumps serve in a wide range of applications such as pumping water from wells, aquarium filtering, pond filtering and aeration, in the car industry for water-cooling and fuel injection, in the energy industry for pumping oil and natural gas or for operating cooling towers. In the medical industry, pumps are used for biochemical processes in developing and manufacturing medicine, and as artificial replacements for body parts, in particular the artificial heart and penile prosthesis.

In biology, many different types of chemical and bio-mechanical pumps have evolved, and biomimicry is sometimes used in developing new types of mechanical pumps.

Types

Mechanical pumps may be **submerged** in the fluid they are pumping or be placed **external** to the fluid.

Pumps can be classified by their method of displacement into positive displacement pumps, impulse pumps, velocity pumps, gravity pumps, steam pumps and valveless pumps.

Positive Displacement Pump

Lobe Pump

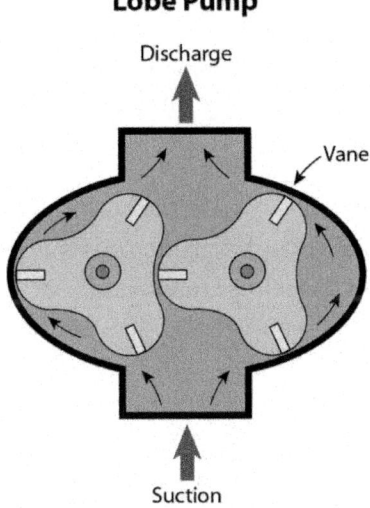

Discharge

Vane

Suction

Lobe pump internals.

A positive displacement pump makes a fluid move by trapping a fixed amount and forcing (displacing) that trapped volume into the discharge pipe.

Some positive displacement pumps use an expanding cavity on the suction side and a decreasing cavity on the discharge side. Liquid flows into the pump as the cavity on the suction side expands and the liquid flows out of the discharge as the cavity collapses. The volume is constant through each cycle of operation.

Positive Displacement Pump Behavior and Safety

Positive displacement pumps, unlike centrifugal or roto-dynamic pumps, theoretically can produce the same flow at a given speed (RPM) no matter what the discharge pressure. Thus, positive displacement pumps are *constant flow machines*. However, a slight increase in internal leakage as the pressure increases prevents a truly constant flow rate.

A positive displacement pump must not operate against a closed valve on the discharge side of the pump, because it has no shutoff head like centrifugal pumps. A positive displacement pump operating against a closed discharge valve continues to produce flow and the pressure in the discharge line increases until the line bursts, the pump is severely damaged, or both.

A relief or safety valve on the discharge side of the positive displacement pump is therefore necessary. The relief valve can be internal or external. The pump manufacturer normally has the option to supply internal relief or safety valves. The internal valve is usually only used as a safety precaution. An external relief valve in the discharge line, with a return line back to the suction line or supply tank provides increased safety.

Positive Displacement Types

A positive displacement pump can be further classified according to the mechanism used to move the fluid:

- Rotary-type positive displacement: internal gear, screw, shuttle block, flexible vane or sliding vane, circumferential piston, flexible impeller, helical twisted roots (*e.g.* the Wendelkolben pump) or liquid ring vacuum pumps
- Reciprocating-type positive displacement: piston or diaphragm pumps
- Linear-type positive displacement: rope pumps and chain pumps

Rotary positive displacement pumps

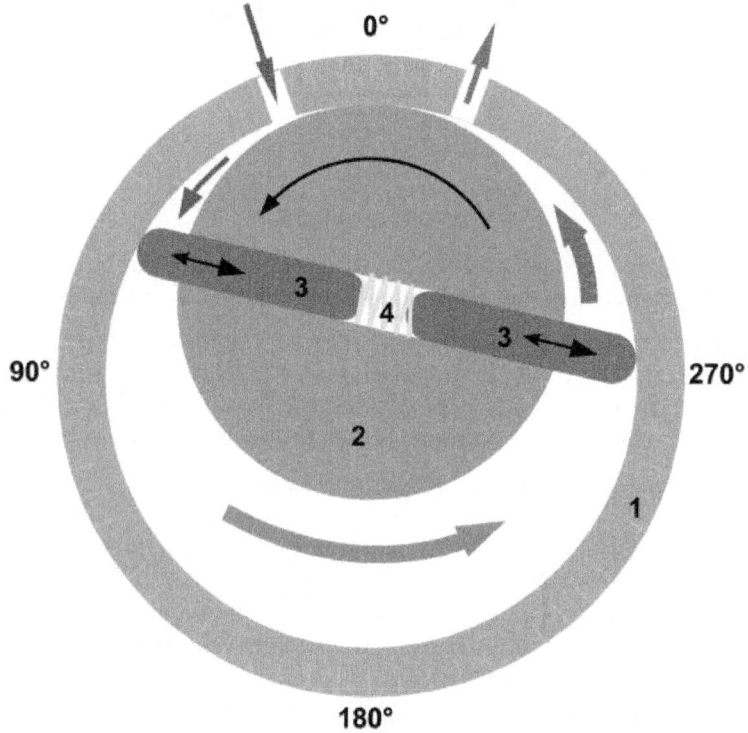

Rotary vane pump

Positive displacement rotary pumps move fluid using a rotating mechanism that creates a vacuum that captures and draws in the liquid.

Advantages: Rotary pumps are very efficient because they naturally remove air from the lines, eliminating the need to bleed the air from the lines manually.

Drawbacks: The nature of the pump demands very close clearances between the rotating pump and the outer edge, making it rotate at a slow, steady speed. If rotary pumps are operated at high speeds, the fluids cause erosion, which eventually causes enlarged clearances that liquid can pass through, which reduces efficiency.

Rotary positive displacement pumps fall into three main types:

- Gear pumps - a simple type of rotary pump where the liquid is pushed between two gears
- Screw pumps - the shape of the internals of this pump is usually two screws turning against each other to pump the liquid
- Rotary vane pumps - similar to scroll compressors, these have a cylindrical rotor encased in a similarly shaped housing. As the rotor orbits, the vanes trap fluid between the rotor and the casing, drawing the fluid through the pump.

Reciprocating Positive Displacement Pumps

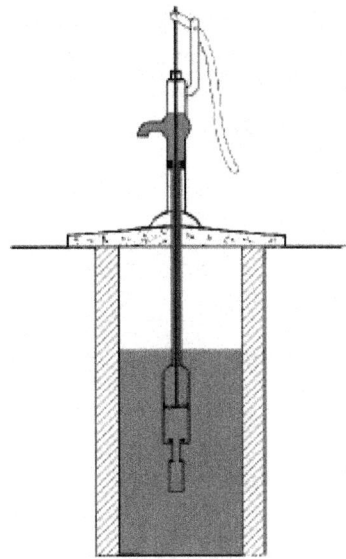

Simple hand pump.

Reciprocating pumps move the fluid using one or more oscillating pistons, plungers, or membranes (diaphragms), while valves restrict fluid motion to the desired direction.

Pumps in this category range from *simplex*, with one cylinder, to in some cases *quad* (four) cylinders, or more. Many reciprocating-type pumps are *duplex* (two) or *triplex* (three) cylinder. They can be either *single-acting* with suction during one direction of piston motion and discharge on the other, or *double-acting* with suction and discharge in both directions. The pumps can be powered manually, by air or steam, or by a belt driven by an engine. This type of pump was used extensively in the 19th century – in the early days of steam propulsion – as boiler feed water pumps. Now reciprocating pumps typically pump highly viscous fluids like concrete and heavy oils, and serve in special applications that demand low flow rates against high resistance. Reciprocating hand pumps were widely used to pump water from wells. Common bicycle pumps and foot pumps for inflation use reciprocating action.

These positive displacement pumps have an expanding cavity on the suction side and a decreasing cavity on the discharge side. Liquid flows into the pumps as the cavity on the suction side expands and the liquid flows out of the discharge as the cavity collapses. The volume is constant given each cycle of operation.

Typical reciprocating pumps are:

- Plunger pumps - a reciprocating plunger pushes the fluid through one or two open valves, closed by suction on the way back.
- Diaphragm pumps - similar to plunger pumps, where the plunger pressurizes hydraulic oil which is used to flex a diaphragm in the pumping cylinder. Diaphragm valves are used to pump hazardous and toxic fluids.
- Piston pumps displacement pumps - usually simple devices for pumping small amounts of liquid or gel manually. The common hand soap dispenser is such a pump.
- Radial piston pumps

Various Positive Displacement Pumps

The positive displacement principle applies in these pumps:
- Rotary lobe pump
- Progressive cavity pump
- Rotary gear pump
- Piston pump
- Diaphragm pump
- Screw pump
- Gear pump
- Hydraulic pump
- Rotary vane pump
- Regenerative (peripheral) pump
- Peristaltic pump
- Rope pump
- Flexible impeller

Gear Pump

This is the simplest of rotary positive displacement pumps. It consists of two meshed gears that rotate in a closely fitted casing. The tooth spaces trap fluid and force it around the outer periphery. The fluid does not travel back on the meshed part, because the teeth mesh closely in the centre. Gear pumps see wide use in car engine oil pumps and in various hydraulic power packs.

Screw Pump

A Screw pump is a more complicated type of rotary pump that uses two or three screws with opposing thread — e.g., one screw turns clockwise and the other counterclockwise. The screws are mounted on parallel shafts that have gears that mesh so the shafts turn together and everything stays in place. The screws turn on the shafts and drive fluid through the pump. As with other forms of rotary pumps, the clearance between moving parts and the pump's casing is minimal.

Progressing Cavity Pump

Widely used for pumping difficult materials, such as sewage sludge contaminated with large particles, this pump consists of a helical rotor, about ten times as long as its width. This can be visualized as a central core of diameter x with, typically, a curved spiral wound around of thickness half x, though in reality it is manufactured in single casting. This shaft fits inside a heavy duty rubber sleeve, of wall thickness also typically x. As the shaft rotates, the rotor gradually forces fluid up the rubber sleeve. Such pumps can develop very high pressure at low volumes.

Roots-type Pumps

Named after the Roots brothers who invented it, this lobe pump displaces the liquid trapped between two long helical rotors, each fitted into the other when perpendicular at 90°, rotating inside a triangular shaped sealing line configuration, both at the point of suction and at the point of discharge. This design produces a continuous flow with equal volume and no vortex. It can work at low pulsation rates, and offers gentle performance that some applications require.

Applications Include:

- High capacity industrial air compressors
- Roots superchargers on internal combustion engines.
- A brand of civil defense siren, the Federal Signal Corporation's Thunderbolt.

Peristaltic Pump

A *peristaltic pump* is a type of positive displacement pump. It contains fluid within a flexible tube fitted inside a circular pump casing (though linear peristaltic pumps have been made). A number of *rollers, shoes,* or *wipers* attached to a rotor compresses the flexible tube. As the rotor turns, the part of the tube under compression closes (or *occludes*), forcing the fluid through the tube. Additionally, when the tube opens to its natural state after the passing of the cam it draws (*restitution*) fluid flow into the pump. This process is called peristalsis and is used in many biological systems such as the gastrointestinal tract.

Plunger Pumps

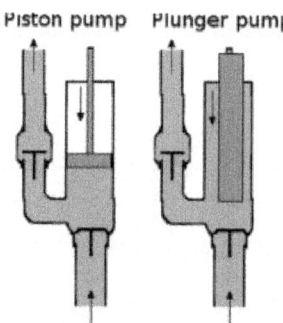

A plunger pump compared to a piston pump.

Plunger pumps are reciprocating positive displacement pumps.

These consist of a cylinder with a reciprocating plunger. The suction and discharge valves are mounted in the head of the cylinder. In the suction stroke the plunger retracts and the suction valves open causing suction of fluid into the cylinder. In the forward stroke the plunger pushes the liquid out of the discharge valve. Efficiency and common problems: With only one cylinder in plunger pumps, the fluid flow varies between maximum flow when the plunger moves through the middle positions, and zero flow when the plunger is at the end positions. A lot of energy is wasted when the fluid is accelerated in the piping system. Vibration and *water hammer* may be a serious problem. In general the problems are compensated for by using two or more cylinders not working in phase with each other.

Triplex-style Plunger Pumps

Triplex plunger pumps use three plungers, which reduces the pulsation of single reciprocating plunger pumps. Adding a pulsation dampener on the pump outlet can further smooth the *pump ripple,* or ripple graph of a pump transducer. The dynamic relationship of the high-pressure fluid and plunger generally requires high-quality plunger seals. Plunger pumps with a larger number of plungers have the benefit of increased flow, or smoother flow without a pulsation dampener. The increase in moving parts and crankshaft load is one drawback.

Car washes often use these triplex-style plunger pumps (perhaps without pulsation dampeners). In 1968, William Bruggeman significantly reduced the size of the triplex pump and increased the lifespan so that car washes could use equipment with smaller footprints. Durable high pressure seals, low pressure seals and oil seals, hardened crankshafts, hardened connecting rods, thick ceramic plungers and heavier duty ball and roller bearings improve reliability in triplex pumps. Triplex pumps now are in a myriad of markets across the world.

Triplex pumps with shorter lifetimes are commonplace to the home user. A person who uses a home pressure washer for 10 hours a year may be satisfied with a pump that lasts 100 hours between rebuilds. Industrial-grade or continuous duty triplex pumps on the other end of the quality spectrum may run for as much as 2,080 hours a year.

The oil and gas drilling industry uses massive semi trailer-transported triplex pumps called mud pumps to pump drilling mud, which cools the drill bit and carries the cuttings back to the surface. Drillers use triplex or even quintuplex pumps to inject water and solvents deep into shale in the extraction process called *fracking*.

Compressed-air-powered Double-diaphragm Pumps

One modern application of positive displacement diaphragm pumps is compressed-air-powered double-diaphragm pumps. Run on compressed air these pumps are intrinsically safe by design, although all manufacturers offer ATEX certified models to comply with industry regulation. These pumps are relatively inexpensive and can perform a wide variety of duties, from pumping water out of bunds, to pumping hydrochloric acid from secure storage (dependent on how the pump is manufactured – elastomers / body construction). Lift is normally limited to roughly 6m although heads can reach almost 200 Psi..

Rope Pumps

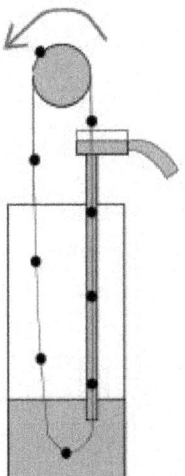

Rope pump schematic.

Devised in China as chain pumps over 1000 years ago, these pumps can be made from very simple materials: A rope, a wheel and a PVC pipe are sufficient to make a simple rope pump. For this reason they have become extremely popular around the world since the 1980s. Rope pump efficiency has been studied by grass roots organizations and the techniques for making and running them have been continuously improved.

Flexible Impeller Pump

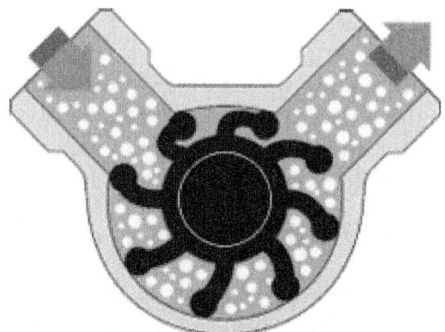

Flexible impeller pump.

The variation of vane volume during the rotation cause the dry selfpriming feature of the pump.

Pump is also reversible.

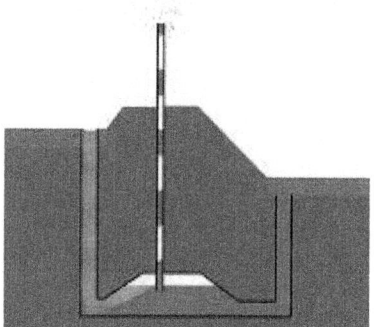

The pulser pump

Impulse Pumps

Impulse pumps use pressure created by gas (usually air). In some impulse pumps the gas trapped in the liquid (usually water), is released and accumulated somewhere in the pump, creating a pressure that can push part of the liquid upwards.

Impulse pumps include:

* Hydraulic ram pumps – kinetic energy of a low-head water supply is stored temporarily in an air-bubble hydraulic accumulator, then used to drive water to a higher head.
* Pulser pumps - run with natural resources, by kinetic energy only.

- Airlift pumps - run on air inserted into pipe, pushing up the water, when bubbles move upward, or on pressure inside pipe pushing water up.

Hydraulic Ram Pumps

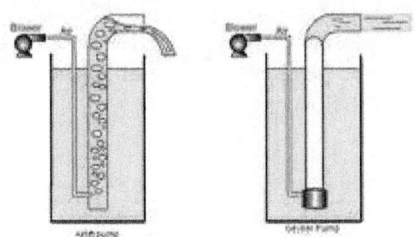

Airlift pump *vs*. Geyser pump.

A hydraulic ram is a water pump powered by hydropower.

It takes in water at relatively low pressure and high flow-rate and outputs water at a higher hydraulic-head and lower flow-rate. The device uses the water hammer effect to develop pressure that lifts a portion of the input water that powers the pump to a point higher than where the water started.

The hydraulic ram is sometimes used in remote areas, where there is both a source of low-head hydropower, and a need for pumping water to a destination higher in elevation than the source. In this situation, the ram is often useful, since it requires no outside source of power other than the kinetic energy of flowing water.

Velocity Pumps

Rotodynamic pumps (or dynamic pumps) are a type of velocity pump in which kinetic energy is added to the fluid by increasing the flow velocity. This increase in energy is converted to a gain in potential energy (pressure) when the velocity is reduced prior to or as the flow exits the pump into the discharge pipe. This conversion of kinetic energy to pressure is explained by the *First law of thermodynamics*, or more specifically by *Bernoulli's principle*.

Dynamic pumps can be further subdivided according to the means in which the velocity gain is achieved.

These types of pumps have a number of characteristics:

1. Continuous energy
2. Conversion of added energy to increase in kinetic energy (increase in velocity)
3. Conversion of increased velocity (kinetic energy) to an increase in pressure head

A practical difference between dynamic and positive displacement pumps is how they operate under closed valve conditions. Positive displacement pumps physically displace fluid, so closing a valve downstream of a positive displacement pump produces a continual pressure build up that can cause mechanical failure of pipeline or pump. Dynamic pumps differ in that they can be safely operated under closed valve conditions.

Radial-flow Pumps

These simply referred to as centripetal design pumps. The fluid enters along the axial plane, is accelerated by the impeller and exits at right angles to the shaft(radially). Radial-flow pumps operate at higher pressures and lower flow rates than axial and mixed-flow pumps.

Axial-flow Pumps

Axial pump (propeller in pipe)

Axial-flow pumps These simply referred to as centrifugal design pumps the fluid is pushed outward from the center or axis. Axial-flow pumps / Centrifugal design pumps operate at much lower pressures and higher flow rates than radial-flow pumps/cepumps.

Mixed-flow Pumps

Mixed-flow pumps function as a compromise between radial and axial-flow pumps. The fluid experiences both radial acceleration and lift and exits the impeller somewhere between 0 and 90 degrees from the axial direction. As a consequence mixed-flow pumps operate at higher pressures than axial-flow pumps while delivering higher discharges than radial-flow pumps. The exit angle of the flow dictates the pressure head-discharge characteristic in relation to radial and mixed-flow.

Eductor-jet Pump

This uses a jet, often of steam, to create a low pressure. This low pressure sucks in fluid and propels it into a higher pressure region.

Gravity Pumps

Gravity pumps include the *syphon* and *Heron's fountain*. The *hydraulic ram* is also sometimes called a gravity pump.

Steam Pumps

Steam pumps have been for a long time mainly of historical interest. They include any type of pump powered by a steam engine and also pistonless pumps such as Thomas Savery's or the Pulsometer steam pump.

Recently there has been a resurgence of interest in low power solar steam pumps for use in smallholder irrigation in developing countries. Previously small steam engines have not been viable because of escalating inefficiencies as vapour engines decrease in size. However the use of modern engineering materials coupled with alternative engine configurations has meant that these types of system are now a cost effective opportunity.

Valveless Pumps

Valveless pumping assists in fluid transport in various biomedical and engineering systems. In a valveless pumping system, no valves (or physical occlusions) are present to regulate the flow direction. The fluid pumping efficiency of a valveless system, however, is not necessarily lower than that having valves. In fact, many fluid-dynamical systems in nature and engineering more or less rely upon valveless pumping to transport the working fluids therein. For instance, blood circulation in the cardiovascular system is maintained to some extent even when the heart's valves fail. Meanwhile, the embryonic vertebrate heart begins pumping blood long before the development of discernible chambers and valves. In microfluidics, valveless impedance pumps have been fabricated, and are expected to be particularly suitable for handling sensitive biofluids. Ink jet printers operating on the Piezoelectric transducer principal also use valveless pumping. The pump chamber is emptied through the printing jet due to reduced flow impedance in that direction and refilled by capillary action.

Pump Repairs

Examining pump repair records and mean time between failures (MTBF) is of great importance to responsible and conscientious pump users.

In early 2005, Gordon Buck, John Crane Inc.'s chief engineer for Field Operations in Baton Rouge, LA, examined the repair records for a number of refinery and chemical plants to obtain meaningful reliability data for centrifugal pumps. A total of 15 operating plants having nearly 15,000 pumps were included in the survey. The smallest of these plants had about 100 pumps; several plants had over 2000. All facilities were located in the United States. In addition, considered as "new", others as "renewed" and still others as "established". Many of these plants — but not all — had an alliance arrangement with John Crane. In some cases, the alliance contract included having a John Crane Inc. technician or engineer on-site to coordinate various aspects of the program.

Not all plants are refineries, however, and different results occur elsewhere. In chemical plants, pumps have traditionally been "throw-away" items as chemical attack limits life. Things have improved in recent years, but the somewhat restricted space available in "old" DIN and ASME-standardized stuffing boxes places limits on the type of seal that fits. Unless the pump user upgrades the seal chamber, the pump only accommodates more compact and simple versions. Without this upgrading, lifetimes in chemical installations are generally around 50 to 60 percent of the refinery values.

Unscheduled maintenance is often one of the most significant costs of ownership, and failures of mechanical seals and bearings are among the major causes. Keep in mind the potential value of selecting pumps that cost more initially, but last much longer between repairs. The MTBF of a better pump may be one to four years longer than that of its non-upgraded counterpart. Consider that published average values of avoided pump failures range from US$2600 to US$12,000. This does not include lost opportunity costs. One pump fire occurs per 1000 failures. Having fewer pump failures means having fewer destructive pump fires.

As has been noted, a typical pump failure based on actual year 2002 reports, costs US$5,000 on average. This includes costs for material, parts, labor and overhead. Extending a pump's MTBF from 12 to 18 months would save US$1,667 per year — which might be greater than the cost to upgrade the centrifugal pump's reliability.

Applications

Pumps are used throughout society for a variety of purposes. Early applications includes the use of the windmill or watermill to pump water. Today, the pump is used for irrigation, water supply, gasoline supply, air conditioning systems, refrigeration (usually called a compressor), chemical movement, sewage movement, flood control, marine services, *etc.*

Because of the wide variety of applications, pumps have a plethora of shapes and sizes: from very large to very small, from handling gas to handling liquid, from high pressure to low pressure, and from high volume to low volume.

Priming a Pump

Typically, a liquid pump can't simply draw air until the feed line and pump fill with the liquid that requires pumping. An operator must introduce liquid into the system to initiate the pumping. This is called *priming* the pump. Loss of prime is usually due to ingestion of air into the pump. The clearances and displacement ratios in pumps for liquids, whether thin or more viscous, usually cannot displace air due to its lower density.

Pumps as Public Water Supplies

One sort of pump once common worldwide was a hand-powered water pump, or 'pitcher pump'. It was commonly installed over community water wells in the days before piped water supplies.

In parts of the British Isles, it was often called *the parish pump*. Though such community pumps are no longer common, people still used the expression *parish pump* to describe a place or forum where matters of local interest are discussed.

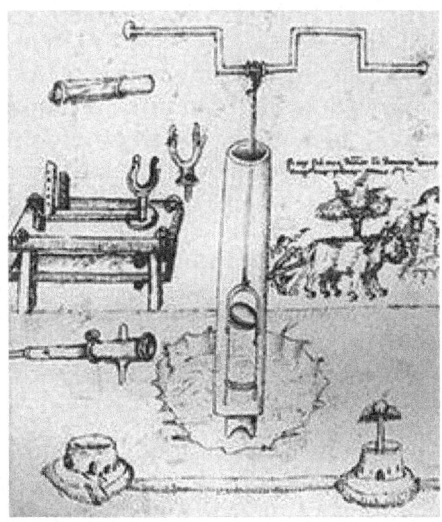

First European depiction of a piston pump, by Taccola, c.1450.

Because water from pitcher pumps is drawn directly from the soil, it is more prone to contamination. If such water is not filtered and purified, consumption of it might lead to gastrointestinal or other water-borne diseases. A notorious case is the 1854 Broad Street cholera outbreak. At the time it was not known how cholera was transmitted, but physician John Snow suspected contaminated water and had the handle of the public pump he suspected removed; the outbreak then subsided.

Modern hand-operated community pumps are considered the most sustainable low-cost option for safe water supply in resource-poor settings, often in rural areas in developing countries. A hand pump opens access to deeper groundwater that is often not polluted and also improves the safety of a well by protecting the water source from contaminated buckets. Pumps such as the Afridev pump are designed to be cheap to build and install, and easy to maintain with simple parts. However, scarcity of spare parts for these type of pumps in some regions of Africa has diminished their utility for these areas.

Sealing Multiphase Pumping Applications

Multiphase pumping applications, also referred to as tri-phase, have grown due to increased oil drilling activity. In addition, the economics of multiphase production is attractive to upstream operations as it leads to simpler, smaller in-field installations, reduced equipment costs and improved production rates. In essence, the multiphase pump can accommodate all fluid stream properties with one piece of equipment, which has a smaller footprint. Often, two smaller multiphase pumps are installed in series rather than having just one massive pump.

For midstream and upstream operations, multiphase pumps can be located onshore or offshore and can be connected to single or multiple wellheads. Basically, multiphase pumps are used to transport the untreated flow stream produced from oil wells to downstream processes or gathering facilities. This means that the pump may handle a flow stream (well stream) from 100 percent gas to 100 percent liquid and every imaginable combination in between. The flow stream can also contain abrasives such as sand and dirt. Multiphase pumps are designed to operate under changing/fluctuating process conditions. Multiphase pumping also helps eliminate emissions of greenhouse gases as operators strive to minimize the flaring of gas and the venting of tanks where possible.

Types and Features of Multiphase Pumps

Helico-Axial Pumps (Centrifugal) A rotodynamic pump with one single shaft that requires two mechanical seals, this pump uses an open-type axial impeller. It's often called a *Poseidon pump*, and can be described as a cross between an axial compressor and a centrifugal pump.

Twin Screw (Positive Displacement) The twin screw pump is constructed of two inter-meshing screws that move the pumped fluid. Twin screw pumps are often used when pumping conditions contain high gas volume fractions and fluctuating inlet conditions. Four mechanical seals are required to seal the two shafts.

Progressive Cavity Pumps (Positive Displacement) Progressive cavity pumps are single-screw types typically used in shallow wells or at the surface. This pump is mainly used on surface applications where the pumped fluid may contain a considerable amount of solids such as sand and dirt.

Electric Submersible Pumps (Centrifugal) These pumps are basically multistage centrifugal pumps and are widely used in oil well applications as a method for artificial lift. These pumps are usually specified when the pumped fluid is mainly liquid.

Buffer Tank A buffer tank is often installed upstream of the pump suction nozzle in case of a slug flow. The buffer tank breaks the energy of the liquid slug, smoothes any fluctuations in the incoming flow and acts as a sand trap.

As the name indicates, multiphase pumps and their mechanical seals can encounter a large variation in service conditions such as changing process fluid composition, temperature variations, high and low operating pressures and exposure to abrasive/erosive media. The challenge is selecting the appropriate mechanical seal arrangement and support system to ensure maximized seal life and its overall effectiveness.

Specifications

Pumps are commonly rated by horsepower, flow rate, outlet pressure in metres (or feet) of head, inlet suction in suction feet (or metres) of head. The head can be simplified as the number of feet or metres the pump can raise or lower a column of water at atmospheric pressure.

From an initial design point of view, engineers often use a quantity termed the specific speed to identify the most suitable pump type for a particular combination of flow rate and head.

Pump Material

The pump material can be Stainless steel, cast iron *etc.* It depends on the application of the pump. In the water industry and for pharma applications SS 316 is normally used, as stainless steel gives better results at high temperatures.

Pumping Power

The power imparted into a fluid increases the energy of the fluid per unit volume. Thus the power relationship is between the conversion of the mechanical energy of the pump mechanism and the fluid elements within the pump. In general, this is governed by a series of simultaneous differential equations, known as the Navier–Stokes equations. However a more simple equation relating only the different energies in the fluid, known as Bernoulli's equation can be used. Hence the power, P, required by the pump:

$$P = \frac{\Delta p Q}{\eta}$$

where Δp is the change in total pressure between the inlet and outlet (in Pa), and Q, the volume flow-rate of the fluid is given in m³/s. The total pressure may have gravitational, static pressure and kinetic energy components; *i.e.* energy is distributed between change in the fluid's gravitational potential energy (going up or down hill), change in velocity, or change in static pressure. η is the pump efficiency, and may be given by the manufacturer's information, such as in the form of a pump curve, and is typically derived from either fluid dynamics simulation (*i.e.* solutions to the Navier–Stokes for the particular pump geometry), or by testing. The efficiency of the pump depends upon the pump's configuration and operating conditions (such as rotational speed, fluid density and viscosity *etc.*)

$$\Delta P = \frac{(v_2^2 - v_1^2)}{2} + \Delta z g + \frac{\Delta p_{static}}{\rho}$$

For a typical "pumping" configuration, the work is imparted on the fluid, and is thus positive. For the fluid imparting the work on the pump (*i.e.* a turbine), the work is negative power required to drive the pump is determined by dividing the output power by the pump efficiency. Furthermore, this definition encompasses pumps with no moving parts, such as a siphon.

Pump Efficiency

Pump efficiency is defined as the ratio of the power imparted on the fluid by the pump in relation to the power supplied to drive the pump. Its value is not fixed for a given pump, efficiency is a function of the discharge and therefore also operating head. For centrifugal pumps, the efficiency tends to increase with flow rate up to a point midway through the operating range (peak efficiency) and then declines as flow rates rise further. Pump performance data such as this is usually supplied by the manufacturer before pump selection. Pump efficiencies tend to decline over time due to wear(*e.g.* increasing clearances as impellers reduce in size).

When a system design includes a centrifugal pump, an important issue it its design is matching the *head loss-flow characteristic* with the pump so that it operates at or close to the point of its maximum efficiency.

Pump efficiency is an important aspect and pumps should be regularly tested. Thermodynamic pump testing is one method.

GAS COMPRESSOR

A **gas compressor** is a mechanical device that increases the pressure of a gas by reducing its volume. An air compressor is a specific type of gas compressor.

Compressors are similar to pumps: both increase the pressure on a fluid and both can transport the fluid through a pipe. As gases are compressible, the compressor also reduces the volume of a gas. Liquids are relatively incompressible; while some can be compressed, the main action of a pump is to pressurize and transport liquids.

Types of Compressors

The main types of gas compressors are illustrated and discussed below:

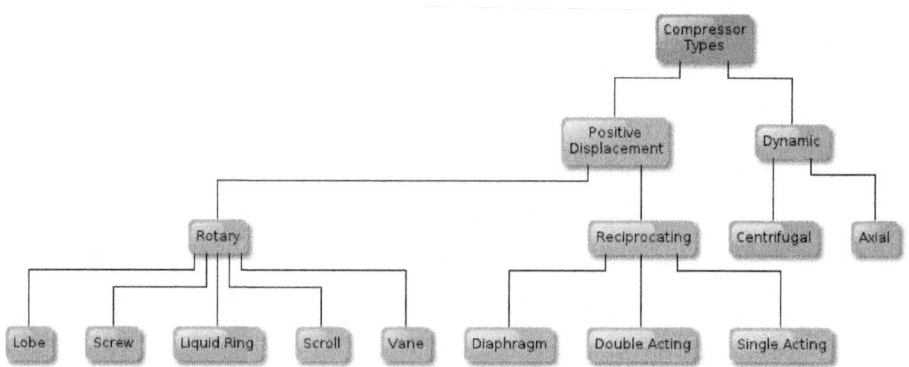

Centrifugal Compressors

Centrifugal compressors use a rotating disk or impeller in a shaped housing to force the gas to the rim of the impeller, increasing the velocity of the gas. A diffuser (divergent duct) section converts the velocity energy to pressure energy. They are primarily used for continuous, stationary service in industries such as oil refineries, chemical and petrochemical plants and natural gas processing plants. Their application can be from 100 horsepower (75 kW) to thousands of horsepower. With multiple staging, they can achieve high output pressures greater than 10,000 psi (69 MPa).

Many large snowmaking operations (like ski resorts) use this type of compressor. They are also used in internal combustion engines as superchargers and turbochargers. Centrifugal compressors are used in small gas turbine engines or as the final compression stage of medium sized gas turbines.

Diagonal or Mixed-flow Compressors

Diagonal or **mixed-flow compressors** are similar to centrifugal compressors, but have a radial and axial velocity component at the exit from the rotor. The diffuser is often used to turn diagonal flow to an axial rather than radial direction.

Axial-flow Compressors

Axial-flow compressors are dynamic rotating compressors that use arrays of fan-like airfoils to progressively compress the working fluid. They are used where there is a requirement for a high flow rate or a compact design.

The arrays of airfoils are set in rows, usually as pairs: one rotating and one stationary. The rotating airfoils, also known as blades or *rotors*, accelerate the fluid. The stationary airfoils, also known as *stators* or vanes, decelerate and redirect the flow direction of the fluid, preparing it for the rotor blades of the next stage. Axial compressors are almost always multi-staged, with the cross-sectional area of the gas passage diminishing along the compressor to maintain an optimum axial Mach number. Beyond about 5 stages or a 4:1 design pressure ratio, variable geometry is normally used to improve operation.

Axial compressors can have high efficiencies; around 90% polytropic at their design conditions. However, they are relatively expensive, requiring a large number of components, tight tolerances and high quality materials. Axial-flow compressors can be found in medium to large gas turbine engines, in natural gas pumping stations, and within certain chemical plants.

Reciprocating Compressors

Reciprocating compressors use pistons driven by a crankshaft. They can be either stationary or portable, can be single or multi-staged, and can be driven by electric motors or internal combustion engines. Small reciprocating compressors from 5 to 30 horsepower (hp) are commonly seen in automotive applications and are typically for intermittent duty. Larger reciprocating compressors well over 1,000 hp (750 kW) are commonly found in large industrial and petroleum applications. Discharge pressures can range from low pressure to very high pressure (>18000 psi or 180 MPa). In certain applications, such as air compression, multi-stage double-acting compressors are said to be the most efficient compressors available, and are typically larger, and more costly than comparable rotary units.

Another type of reciprocating compressor is the swash plate compressor, which uses pistons moved by a swash plate mounted on a shaft.

Household, home workshop, and smaller job site compressors are typically reciprocating compressors 1½ hp or less with an attached receiver tank.

Ionic Liquid Piston Compressor

An ionic liquid piston compressor, *ionic compressor* or *ionic liquid piston pump* is a hydrogen compressor based on an ionic liquid piston instead of a metal piston as in a piston-metal diaphragm compressor.

Rotary Screw Compressors

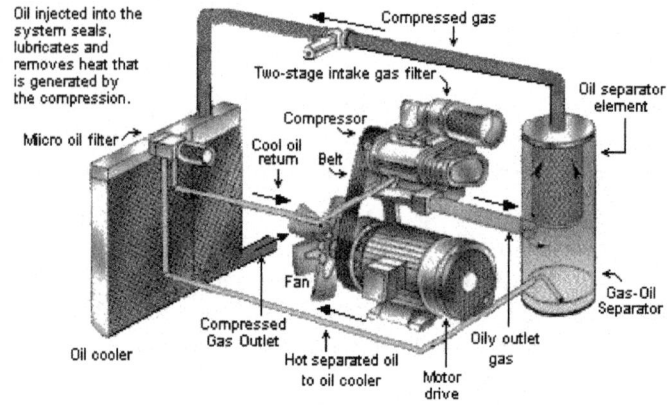

Diagram of a rotary screw compressor.

Rotary screw compressors use two meshed rotating positive-displacement helical screws to force the gas into a smaller space. These are usually used for continuous operation in commercial and industrial applications and may be either stationary or portable. Their application can be from 3 horsepower (2.2 kW) to over 1,200 horsepower (890 kW) and from low pressure to moderately high pressure (>1,200 psi or 8.3 MPa).

Rotary screw compressors are commercially produced in Oil Flooded, Water Flooded and Dry type. The efficiency of rotary compressors depends on the air drier, and the selection of air drier is always 1.5 times volumetric delivery of the compressor.

Rotary Vane Compressors

Rotary vane compressors consist of a rotor with a number of blades inserted in radial slots in the rotor. The rotor is mounted offset in a larger housing that is either circular or a more complex shape. As the rotor turns, blades slide in and out of the slots keeping contact with the outer wall of the housing. Thus, a series of decreasing volumes is created by the rotating blades. Rotary Vane compressors are, with piston compressors one of the oldest of compressor technologies.

With suitable port connections, the devices may be either a compressor or a vacuum pump. They can be either stationary or portable, can be single or multi-staged, and can be driven by electric motors or internal combustion engines. Dry vane machines are used at relatively low pressures (*e.g.*, 2 bar or 200 kPa or 29 psi) for bulk material movement while oil-injected machines have the necessary volumetric efficiency to achieve pressures up to about 13 bar (1,300 kPa; 190 psi) in a single stage. A rotary vane compressor is well suited to electric motor drive and is significantly quieter in operation than the equivalent piston compressor.

Rotary vane compressors can have mechanical efficiencies of about 90%.

Scroll Compressors

A **scroll compressor**, also known as **scroll pump** and **scroll vacuum pump**, uses two interleaved spiral-like vanes to pump or compress fluids such as liquids and gases. The vane geometry may be involute, archimedean spiral, or hybrid curves. They operate more smoothly, quietly, and reliably than other types of compressors in the lower volume range.

Often, one of the scrolls is fixed, while the other orbits eccentrically without rotating, thereby trapping and pumping or compressing pockets of fluid between the scrolls.

Due to minimum clearance volume between the fixed scroll and the orbiting scroll, these compressors have a very high volumetric efficiency.

This type of compressor was used as the supercharger on Volkswagen G60 and G40 engines in the early 1990s.

Diaphragm Compressors

A **diaphragm compressor** (also known as a **membrane compressor**) is a variant of the conventional reciprocating compressor. The compression of gas occurs by the movement of a flexible membrane, instead of an intake element. The back and forth movement of the membrane is driven by a rod and a crankshaft mechanism. Only the membrane and the compressor box come in contact with the gas being compressed.

The degree of flexing and the material constituting the diaphragm affects the maintenance life of the equipment. Generally stiff metal diaphragms may only displace a few cubic centimeters of volume because the metal can not endure large degrees of flexing without cracking, but the stiffness of a metal diaphragm allows it to pump at high pressures. Rubber or silicone diaphragms are capable of enduring deep pumping strokes of very high flexion, but their low strength limits their use to low-pressure applications, and they need to be replaced as plastic embrittlement occurs.

Diaphragm compressors are used for hydrogen and compressed natural gas (CNG) as well as in a number of other applications.

The photograph included in this section depicts a three-stage diaphragm compressor used to compress hydrogen gas to 6,000 psi (41 MPa) for use in a prototype compressed hydrogen and compressed natural gas (CNG) fueling station built in downtown Phoenix, Arizona by the Arizona Public Service company. Reciprocating compressors were used to compress the natural gas.

The prototype alternative fueling station was built in compliance with all of the prevailing safety, environmental and building codes in Phoenix to demonstrate that such fueling stations could be built in urban areas.

Air Bubble Compressor

Also known as a trompe. A mixture of air and water generated through turbulence is allowed to fall into a subterranean chamber where the air separates from the water. The weight of falling water compresses the air in the top of the chamber. A submerged outlet from the chamber allows water to flow to the surface at a lower height than the intake. An outlet in the roof of the chamber supplies the compressed air to the surface. A facility on this principal was built on the Montreal River at Ragged Shutes near Cobalt, Ontario in 1910 and supplied 5,000 horsepower to nearby mines.

Hermetically Sealed, Open, or Semi-Hermetic

Compressors used in refrigeration systems are often described as being either hermetic, open or semi-hermetic, to describe how the compressor and motor drive are situated in relation to the gas or vapor being compressed. The industry name for a hermetic is **hermetically sealed compressor**, while a semi-hermetic is commonly called a **semi-hermetic compressor**.

In hermetic and most semi-hermetic compressors, the compressor and motor driving the compressor are integrated, and operate within the pressurized gas envelope of the system. The motor is designed to operate in, and be cooled by, the refrigerant gas being compressed.

The difference between the hermetic and semi-hermetic, is that the hermetic uses a one-piece welded steel casing that cannot be opened for repair; if the hermetic fails it is simply replaced with an entire new unit. A semi-hermetic uses a large cast metal shell with gasketed covers that can be opened to replace motor and pump components.

The primary advantage of a hermetic and semi-hermetic is that there is no route for the gas to leak out of the system. Open compressors rely on either natural leather or synthetic rubber seals to retain the internal pressure, and these seals require a lubricant such as oil to retain their sealing properties.

An open pressurized system such as an automobile air conditioner can leak its operating gases, if it is not operated frequently enough. Open systems rely on lubricant in the system to splash on pump components and seals. If it is not operated frequently enough, the lubricant on the seals slowly evaporates, and then the seals begin to leak until the system is no longer functional and must be recharged.

By comparison, a hermetic system can sit unused for years, and can usually be started up again at any time without requiring maintenance or experiencing any loss of system pressure.

The disadvantage of hermetic compressors is that the motor drive cannot be repaired or maintained, and the entire compressor must be removed if a motor fails. A further disadvantage is that burnt-out windings can contaminate whole systems, thereby requiring the system to be entirely pumped down and the gas replaced. Typically, hermetic compressors are used in low-cost factory-assembled consumer goods where the cost of repair is high compared to the value of the device, and it would be more economical to just purchase a new device.

An advantage of open compressors is that they can be driven by non-electric power sources, such as an internal combustion engine or turbine. However, open compressors that drive refrigeration systems are generally not totally *maintenance-free* throughout the life of the system, since some gas leakage will occur over time.

Thermodynamics of Gas Compression

Isentropic Compressor

A compressor can be idealized as internally reversible and adiabatic, thus an isentropic steady state device, meaning the change in entropy is 0. By defining the compression cycle as isentropic, an ideal efficiency for the process can be attained, and the ideal compressor performance can be compared to the actual performance of the machine.

Isentropic efficiency of Compressors:

$$\eta_C = \frac{Isentropic\ Compressor\ Work}{Actual\ Compressor\ Work} = \frac{W_s}{W_a} \cong \frac{h_{2s} - h_1}{h_{2a} - h_1}$$

h_1 is the enthalpy at the initial state

h_{2a} is the enthalpy at the final state for the actual process

h_{2s} is the enthalpy at the final state for the isentropic process

Minimizing Work Required by a Compressor

Comparing Reversible to Irreversible Compressors

Comparison of the differential form of the energy balance for each device-
Actual Compressor:

$$\delta q_{act} - \delta w_{act} = dh + dke + dpe$$

Reversible Compressor:

$$\delta q_{rev} - \delta w_{rev} = dh + dke + dpe$$

The right hand side of each compressor type is equivalent, thus:

$$\delta q_{act} - \delta w_{act} = \delta q_{rev} - \delta w_{rev}$$

re-arranging:

$$\delta w_{rev} - \delta w_{act} = \delta q_{rev} - \delta q_{act}$$

By substituting the know equation $\delta q_{rev} = Tds$ into the last equation and dividing both terms by T:

$$\frac{\delta w_{rev} - \delta w_{act}}{T} = ds - \frac{\delta q_{act}}{T} \geq 0$$

Furthermore, $ds \geq \dfrac{\delta q_{act}}{T}$ and T is [asolute temperature] $(T \geq 0)$ which produces: $\delta w_{rev} \geq \delta w_{act}$ or $w_{rev} \geq w_{act}$

Therefore, work-consuming devices such as pumps and compressors require less work when they operate reversibly.

Effect of Cooling During the Compression Process

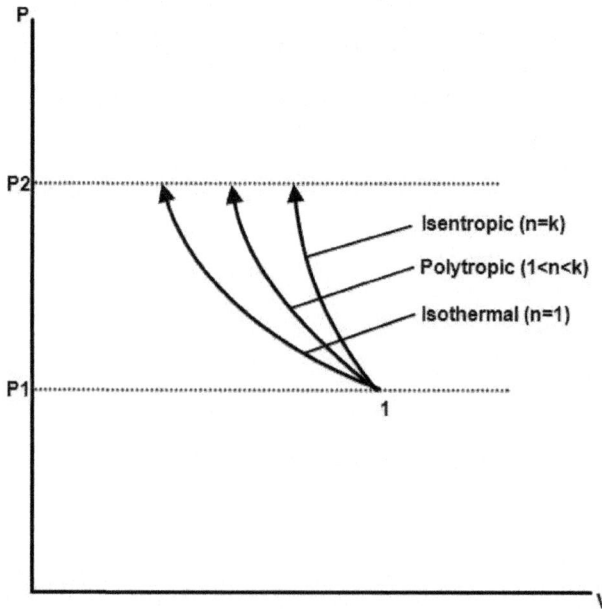

P-v (Specific volume vs. Pressure) diagram comparing isentropic, polytropic, and isothermal processes between the same pressure limits.

isentropic process: involves no cooling

polytropic process: involves some cooling

isothermal process: involves maximum cooling

By making the following assumptions the required work for the compressor to compress a gas from P_1 to P_2 is the following for each process: Assumptions:

P_1 and P_2 are the same for each process

All processes are internally reversible

The gas behaves like an ideal gas with constant specific heats

Isentropic ($Pv^k = constant$):

$$W_{comp,in} = \frac{kR(T_2 - T_1)}{k-1} = \frac{kRT_1}{k-1}\left[\left(\frac{P_2}{P_1}\right)^{(k-1)/k} - 1\right]$$

Polytropic ($Pv^n = constant$):

$$W_{comp,in} = \frac{nR(T_2 - T_1)}{n-1} = \frac{kRT_1}{n-1}\left[\left(\frac{P_2}{P_1}\right)^{(n-1)/n} - 1\right]$$

Isothermal ($T = constant$ or $Pv = constant$):

$$W_{comp,in} = RT\ln\left(\frac{P_2}{P_1}\right)$$

By comparing the three internally reversible processes compressing an ideal gas from P_1 to P_2, the results show that isentropic compression ($Pv^k = constant$) requires the most work in and the isothermal compression($T = constant$ or $Pv = constant$) requires the least amount of work in. For the polytropic process ($Pv^n = constant$) work in decreases as the exponent, n, decreases, by increasing the heat rejection during the compression process. One common way of cooling the gas during compression is to use cooling jackets around the casing of the compressor.

Compressors in Ideal Thermodynamic Cycles

Ideal Rankine Cycle 1->2 Isentropic compression in a pump

Ideal Carnot Cycle 4->1 Isentropic compression

Ideal Otto Cycle 1->2 Isentropic compression

Ideal Diesel Cycle 1->2 Isentropic compression

Ideal Brayton Cycle 1->2 Isentropic compression in a **compressor**

Ideal Vapor-compression refrigeration Cycle 1->2 Isentropic compression in a **compressor**

NOTE: The isentropic assumptions are only applicable with ideal cycles. Real world cycles have inherent losses due to inefficient compressors and turbines. The real world system are not truly isentropic but are rather idealized as isentropic for calculation purposes.

Temperature

Compression of a gas increases its temperature, often referred to as the *heat of compression.*

$$W = \int_{V_1}^{V_2} p\,dV = p_1 V_1^n = \int_{V_1}^{V_2} V^{-n} dV$$

where

$$\frac{p_2}{p_1} = \left(\frac{V_1}{V_2}\right)^n$$

or

$$p_1 V_1^n = p_2 V_2^n = p V^n$$

and

$$p = \frac{p_1 V_1^n}{V^n}$$

so

$$W = \frac{p_1 V_1^n}{1-n}(V_2^{1-n} - V_1^{1-n})$$

in which p is pressure, V is volume, n takes different values for different compression processes, and 1 & 2 refer to initial and final states.

- Adiabatic - This model assumes that no energy (heat) is transferred to or from the gas during the compression, and all supplied work is added to the internal energy of the gas, resulting in increases of temperature and pressure. Theoretical temperature rise is:

$$T_2 = T_1 \left(\frac{p_2}{p_1}\right)^{(k-1)/k}$$

with T_1 and T_2 in degrees Rankine or kelvins, p_2 and p_1 being absolute pressures and k = ratio of specific heats. The rise in air and temperature ratio means compression does not follow a simple pressure to volume ratio. This is less efficient, but quick. Adiabatic compression or expansion more closely model real life when a compressor has good insulation, a large gas volume, or a short time scale (*i.e.,* a high power level). In practice there will always be a certain amount of heat flow out of the compressed gas. Thus, making a perfect adiabatic compressor would require perfect heat insulation of all parts of the machine. For example, even a bicycle tire pump's metal tube becomes hot as you compress the air to fill a tire. The relation between temperature and compression ratio described above means that the value of n for an adiabatic process is k (the ratio of specific heats).

- Isothermal - This model assumes that the compressed gas remains at a constant temperature throughout the compression or expansion process. In this cycle, internal energy is removed from the system as heat at the same rate that it is added by the mechanical work of compression. Isothermal compression or expansion more closely models real life when the compressor has a large heat exchanging surface, a small gas volume, or a long time scale (*i.e.*, a small power level). Compressors that utilize inter-stage cooling between compression stages come closest to achieving perfect isothermal compression. However, with practical devices perfect isothermal compression is not attainable. For example, unless you have an infinite number of compression stages with corresponding intercoolers, you will never achieve perfect isothermal compression.

For an isothermal process, n is 1, so the value of the work integral for an isothermal process is:

$$W = -p_1 V_1 \ln\left(\frac{p_2}{p_1}\right)$$

When evaluated, the isothermal work is found to be lower than the adiabatic work.

- Polytropic - This model takes into account both a rise in temperature in the gas as well as some loss of energy (heat) to the compressor's components. This assumes that heat may enter or leave the system, and that input shaft work can appear as both increased pressure (usually useful work) and increased temperature above adiabatic (usually losses due to cycle efficiency). Compression efficiency is then the ratio of temperature rise at theoretical 100 percent (adiabatic) vs. actual (polytropic). Polytropic compression will use a value of n between 0 (a constant-pressure process) and infinity (a constant volume process). For the typical case where an effort is made to cool the gas compressed by an approximately adiabatic process, the value of n will be between 1 and k.

Staged Compression

In the case of centrifugal compressors, commercial designs currently do not exceed a compression ratio of more than a 3.5 to 1 in any one stage (for a typical gas). Since compression generates heat, the compressed gas is to be cooled between stages making the compression less adiabatic and more isothermal. The inter-stage coolers typically result in some partial condensation that is removed in vapor-liquid separators.

In the case of small reciprocating compressors, the compressor flywheel may drive a cooling fan that directs ambient air across the intercooler of a two or more stage compressor.

Because rotary screw compressors can make use of cooling lubricant to remove the heat of compression, they very often exceed a 9 to 1 compression ratio.

For instance, in a typical diving compressor the air is compressed in three stages. If each stage has a compression ratio of 7 to 1, the compressor can output 343 times atmospheric pressure (7 × 7 × 7 = 343 atmospheres). (343 atm or 34.8 MPa or 5.04 ksi)

Prime Movers

There are many options for the "prime mover" or motor that powers the compressor:

- Gas turbines power the axial and centrifugal flow compressors that are part of jet engines.
- Steam turbines or water turbines are possible for large compressors.
- Electric motors are cheap and quiet for static compressors. Small motors suitable for domestic electrical supplies use single-phase alternating current. Larger motors can only be used where an industrial electrical three phase alternating current supply is available.
- Diesel engines or petrol engines are suitable for portable compressors and support compressors.
- In automobiles and other types of vehicles (including piston-powered airplanes, boats, trucks, *etc.*), diesel or gasoline engines power output can be increased by compressing the intake air, so that more fuel can be burned per cycle. These engines can power compressors using their own crankshaft power (this setup known as a supercharger), or, use their exhaust gas to drive a turbine connected to the compressor (this setup known as a turbocharger).

Applications

Gas compressors are used in various applications where either higher pressures or lower volumes of gas are needed:

- In pipeline transport of purified natural gas from the production site to the consumer, a compressor is driven by a gas turbine fueled by gas bled from the pipeline. Thus, no external power source is necessary.
- Petroleum refineries, natural gas processing plants, petrochemical and chemical plants, and similar large industrial plants require compressing for intermediate and end-product gases.
- Refrigeration and air conditioner equipment use compressors to move heat in refrigerant cycles.
- Gas turbine systems compress the intake combustion air.
- Small-volume purified or manufactured gases require compression to fill high pressure cylinders for medical, welding, and other uses.
- Various industrial, manufacturing, and building processes require compressed air to power pneumatic tools.

- In the manufacturing and blow moulding of PET plastic bottles and containers.
- Some aircraft require compressors to maintain cabin pressurization at altitude.
- Some types of jet engines — such as turbojets and turbofans) — compress the air required for fuel combustion. The jet engine's turbines power the combustion air compressor.
- In SCUBA diving, hyperbaric oxygen therapy, and other life support devices, compressors put breathing gas into small volume containers, such as diving cylinders.
- In surface supplied diving, an air compressor frequently supplies low pressure air (10 to 20 bar) for breathing.
- Submarines use compressors to store air for later use in displacing water from buoyancy chambers to adjust depth.
- Turbochargers and superchargers are compressors that increase internal combustion engine performance by increasing the mass flow of air inside the cylinder, so the engine can burn more fuel and hence produce more power.
- Rail and heavy road transport vehicles use compressed air to operate rail vehicle or road vehicle brakes — and various other systems (doors, windscreen wipers, engine, gearbox control, *etc.*).
- Service stations and auto repair shops use compressed air to fill pneumatic tires and power pneumatic tools.
- Fire pistons and heat pumps exist to heat air or other gasses, and compressing the gas is only a means to that end.

In the United States, there were 300 gas compressor manufacturers in 2011 producing compressors for all of these uses. Although these factories were classified as small business, the total 2011 sales for gas and air compressors was over $9 billion.

AIR COMPRESSOR

An **air compressor** is a device that converts power (usually from an electric motor, a diesel engine or a gasoline engine) into kinetic energy by compressing and pressurizing air, which, on command, can be released in quick bursts. There are numerous methods of air compression, divided into either positive-displacement or negative-displacement types.

Types of Air Compressor

According to the Design and Principle of Operation

1. Reciprocating compressor

2. Rotary screw compressor
3. Turbo compressor

Positive Displacement

Positive-displacement air compressors work by forcing air into a chamber whose volume is decreased to compress the air. Piston-type air compressors use this principle by pumping air into an air chamber through the use of the constant motion of pistons. They use one-way valves to guide air into a chamber, where the air is compressed. Rotary screw compressors also use positive-displacement compression by matching two helical screws that, when turned, guide air into a chamber, whose volume is decreased as the screws turn. Vane compressors use a slotted rotor with varied blade placement to guide air into a chamber and compress the volume. A type of compressor that delivers a fixed volume of air at high pressures. Common types of positive displacement compressors include piston compressors and rotary screw compressors.

Negative Displacement

Negative-displacement air compressors include centrifugal compressors. These use centrifugal force generated by a spinning impeller to accelerate and then decelerate captured air, which pressurizes it.

Cooling

Due to adiabatic heating, air compressors require some method of disposing of waste heat. Generally this is some form of air- or water-cooling, although some (particularly rotary type) compressors may be cooled by oil (that is then in turn air- or water-cooled) and the atmospheric changes also considered during cooling of compressors.

Applications

- To supply high-pressure clean air to fill gas cylinders
- To supply moderate-pressure clean air to a submerged surface supplied diver
- To supply moderate-pressure clean air for driving some office and school building pneumatic HVAC control system valves
- To supply a large amount of moderate-pressure air to power pneumatic tools, such as jackhammers
- For filling tires
- To produce large volumes of moderate-pressure air for large-scale industrial processes (such as oxidation for petroleum coking or cement plant bag house purge systems).

Most air compressors either are reciprocating piston type, rotary vane or rotary screw. Centrifugal compressors are common in very large applications. There are two main types of air compressor's pumps: oil-lubed and oil-less. The oil-less system has more technical development, but is more expensive, louder and lasts for less time than oil-lubed pumps. The oil-less system also delivers air of better quality.

Chapter 5

CYLINDERS AND MOTORS

CYLINDER

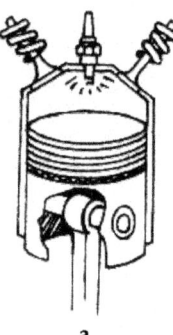

a

Illustration of an engine cylinder with a cross-section view of the piston, connecting rod, valves and spark plug.

A **cylinder** is the central working part of a reciprocating engine or pump, the space in which a piston travels. Multiple cylinders are commonly arranged side by side in a bank, or engine block, which is typically cast from aluminum or cast iron before receiving precision machine work. Cylinders may be **sleeved** (*lined* with a harder metal) or **sleeveless** (with a wear-resistant coating such as Nikasil). A sleeveless engine may also be referred to as a "patent-bore engine".

A cylinder's displacement, or swept volume, can be calculated by multiplying its cross-sectional area (the square of half the bore by pi) and again by the distance the piston travels within the cylinder (the stroke). The engine displacement can be calculated by multiplying the swept volume of one cylinder by the number of cylinders.

Presented mathematically,

$$Cylinder\ Volume = \pi * (\frac{bore}{2})^2 * Stroke$$

*Engine Displacement = Cylinder Volume * Number of Cylinders*

A piston is seated inside each cylinder by several metal piston rings fitted around its outside surface in machined grooves; typically two for compressional sealing and one to seal the oil. The rings make near contact with the cylinder walls (sleeved or sleeveless), riding on a thin layer of lubricating oil; essential to keep the engine from seizing and necessitating a cylinder wall's durable surface.

During the earliest stage of an engine's life, its initial *breaking-in* or *running-in* period, small irregularities in the metals are encouraged to gradually form congruent grooves by avoiding extreme operating conditions. Later in its life, after mechanical wear has increased the spacing between the piston and the cylinder (with a consequent decrease in power output) the cylinders may be machined to a slightly larger diameter to receive new sleeves (where applicable) and piston rings, a process sometimes known as *reboring*.

Heat Engines

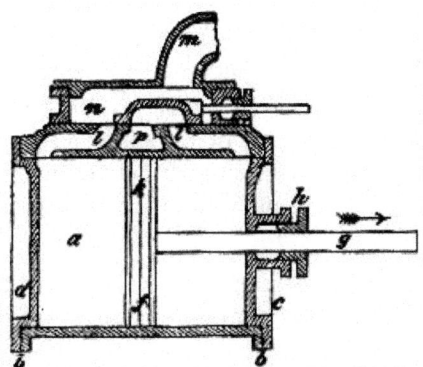

Cylinder with piston in a double acting steam engine.

Heat engines, including Stirling engines, are sealed machines using pistons within cylinders to transfer energy from a heat source to a colder reservoir, often using steam or another gas as the working substance. The first illustration depicts a longitudinal section of a cylinder in a steam engine. The sliding part at the bottom is the piston, and the upper sliding part is a distribution valve (in this case of the D slide valve type) that directs steam alternately into either end of the cylinder. Refrigerator and air conditioner compressors are heat engines driven in reverse cycle as pumps.

Internal Combustion Engines

Internal combustion engines operate on the inherent volume change accompanying oxidation of gasoline (petrol), diesel fuel (or some other hydrocarbon) or ethanol, an expansion which is greatly enhanced by the heat produced. They are not classical heat engines since they expel the working substance, which is also the combustion product, into the surroundings.

The reciprocating motion of the pistons is translated into crankshaft rotation *via* connecting rods. As a piston moves back and forth, a connecting rod changes its angle; its distal end has a rotating link to the crankshaft. A typical four-cylinder automobile engine has a single row of water-cooled cylinders. V engines (V_6 or V_8) use two angled cylinder banks. The "V" configuration is utilized to create a more compact configuration relative to the number of cylinders. Many other engine configurations exist.

For example, there are also rotary turbines. The Wankel engine is a rotary adaptation of the cylinder-piston concept which has been used by Mazda and NSU in automobiles. Rotary engines are relatively quiet because they lack the clatter of reciprocating motion.

Air-cooled engines generally use individual cases for the cylinders to facilitate cooling. Inline motorcycle engines are an exception, having two-, three-, four-, or even six-cylinder air-cooled units in a common block. Water-cooled engines with only a few cylinders may also use individual cylinder cases, though this makes the cooling system more complex. The Ducati motorcycle company, which for years used air-cooled motors with individual cylinder cases, retained the basic design of their V-twin engine while adapting it to water-cooling.

In some engines, especially French designs, the cylinders have "wet liners". They are formed separately from the main casting so that liquid coolant is free to flow around their outsides. Wet-lined cylinders have better cooling and a more even temperature distribution, but this design makes the engine as a whole somewhat less rigid.

During use, the cylinder is subject to wear from the rubbing action of the piston rings and piston skirt. This is minimized by the thin oil film which coats the cylinder walls and also by a layer of glaze which naturally forms as the engine is run-in, but eventually the cylinder becomes worn and slightly oval in shape, usually necessitating a rebore to an oversize diameter and the fitting of new, oversize pistons. The cylinder does not wear above the highest point reached by the top compression ring of the piston, which can result in a detectable ridge. If an engine is only operated at low rpm for its early life (*e.g.* in a gently driven automobile) then abruptly used in the higher rpm range (*e.g.* by a new owner), the slight stretching of the connecting rods at high speed can enable the top compression ring to contact the wear ridge, breaking the ring. For this reason it is important that all engines, once initially run-in, are occasionally "exercised" through their full speed range to develop a tapered wear profile rather than a sharp ridge.

Cylinder Sleeving

Cylinder walls can become very worn or damaged from use. If the engine is not equipped with replaceable sleeves there is a limit to how far the cylinder walls can be bored or worn before the block must be sleeved or replaced. In such cases where the use of a sleeve or liner can restore proper clearances to an engine.

Sleeves are made out of iron alloys and are very reliable. A sleeve is installed by a machinist at a machine shop. The engine block is mounted on a precision boring machine where the cylinder is then bored to a size much larger than normal and a new cast-iron sleeve can be inserted with an interference fit. The sleeves can be pressed into place, or they can be held in by a shrink fit. This is done by boring the cylinder (between 3 to 6 thousandths of an inch) smaller than the sleeve being installed, then heating the engine block and while hot, the cold sleeve can be inserted easily. When the engine block cools down it shrink fits around the sleeve holding it into place. Cylinder wall thickness is important to efficient thermal conductivity in the engine. When choosing sleeves, engines have specifications to how thick the cylinder walls should be to prevent overworking the coolant system. Each engine's needs are different, dependent on designed work load duty cycle and energy produced. After selecting and installing the sleeve, the cylinder needs to be finish bored and honed to match the piston. Care needs to be given to the finish of the cylinder walls to prevent improper ring seating at break in.

CYLINDER CUSHIONING

Cushioning of some sort normally is required to decelerate a cylinder's piston before it strikes the end cap. Reducing the piston velocity as it approaches the end cap lowers the stresses on cylinder components and reduces vibration transmitted to the machine structure.

End-of-stroke impact can be dealt with in three ways: by simple impact cushioning, pneumatic cushioning, or by installing shock absorbers. This discussion deals with pneumatic cushioning and is intended to serve as an instruction on how to optimize cushioning for a given mass.

The Concept of Ideal Cushioning

Ideal pneumatic cushioning occurs when all kinetic energy is dissipated to decelerate the piston to exactly zero velocity when it reaches the end of its travel. Any contact between the piston and end cap would be negligible, so the piston would not rebound off the end cap. Ideal pneumatic cushioning produces minimal noise from end cover contact, and minimum piston deceleration time. Thus, properly adjusted pneumatic cushioning can improve the work environment and increase machine throughput in rapid-cycle applications.

Knowing operating pressure, cylinder characteristics, and the specified load mass, the first step is to ensure that the piston velocity is within that specified in cushioning charts in the manufacturer's catalog. The best results in adjusting piston velocity are obtained by installing throttling non-return valves directly in the cylinder end ports. This permits free inlet flow while allowing the outlet pressure to be adjusted simply by altering the area of the exhaust port with an adjusting screw. Directional-control valves with integral restrictors may be used as an alternative.

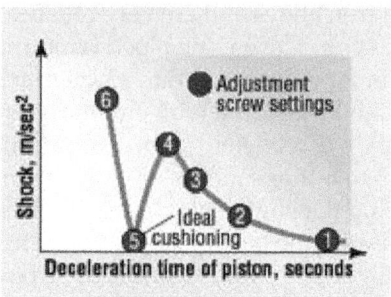

A critical aid in achieving ideal conditioning is an electronic instrument to measure piston velocity. Such an instrument also allows measuring the time for all sequences in a cylinder cycle.

Understanding the Dynamics

Although the cushioning adjustment from the factory may prevent the piston from striking the end cap on its first stroke, the cushioning is far from ideal. Furthermore, the factory adjustment may also provide too much damping, making it difficult to achieve rapid cycling.

Figure illustrates what can occur when the adjusting screw is opened, with all other parameters being constant. Starting at point 1, which represents the initial factory setting, opening the adjusting screw a turn at a time moves the setting to the left. During the first two to three adjustments, the deceleration time becomes progressively shorter, but the end impact becomes correspondingly higher. The normal reaction at this point is to stop, and return the adjusting screw toward the original setting to reduce the increasing shock.

However, ideal cushioning is achieved by continuing to open the screw an additional one to two turns. At some point, the end impact reaches its minimum, point 5 in Figure. Continuing to open the adjusting screw will cause end impact to suddenly increase substantially without a significant reduction in deceleration time. So at the ideal pneumatic cushioning point, not only is end shock minimized, but deceleration time is also reduced, which translates to shorter cycle times. A 20% to 40% reduction in total cycle time is not unusual when ideal cushioning is achieved.

Effects of Pressure Variations

So far, all parameters have been assumed to be constant, with changes made only to the cushioning setting. In reality, though, operating pressure often fluctuates, especially in large circuits having dozens of actuators and end effectors. The effects of pressure variations demonstrate the importance of providing consistent pressure control. This is best accomplished by specifying properly sized flow passages to minimize pressure drop and strategically placing pressure regulators where ever they are needed.

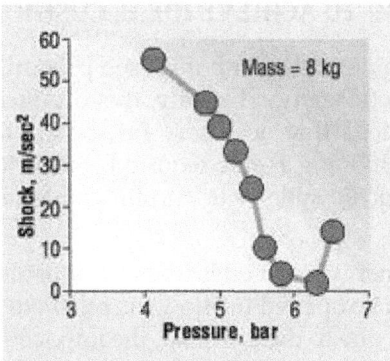

As shown, ideal cushioning occurs at 6.3 bar for a given mass. Any variation in pressure — whether higher or lower — will increase end-of-stroke impact and piston deceleration time.

Operating pressure can have a dramatic effect on end-of-stroke shock, Figure. As shown, ideal pneumatic cushioning for a given mass occurs at 6.3 bar. But end impact shock increases sharply if pressure decreases. At 5 bar, the cylinder is subjected to a shock equivalent to 30 to 40 times the mass it was intended to cushion. Apart from loud noise, severe vibration, and a longer cycle time, this shortens the life of the cylinder appreciably. Unfortunately, higher pressure produces the same result.

Over and Underdamping

Ideal cushioning cannot be achieved with an overdamped cylinder. This is because the effects of damping will become progressively worse, regardless of the adjustment. Any of three actions may be taken to solve this problem.

Increase piston velocity — Piston velocity can be raised by increasing the area of the cylinder outlet port, adjusting throttling non-return valves, or altering the restrictors in the outlet ports of directional control valves. Sufficient kinetic energy for ideal cushioning can be developed by this means.

Reduce operating pressure — This may be accomplished simply by installing a pressure regulator in the feed line to the cylinder.

Increase the moving mass — A higher kinetic energy can be achieved by increasing the moving mass, although, in practice, this can be difficult to achieve.

If severe shock occurs regardless of the cushioning adjustment setting, the cylinder is underdamped. In this case, the possible courses of action are the opposite of those described for overdamped conditions. To correct underdamping:

- reduce the piston velocity,
- increase operating pressure,
- reduce the mass, or
- incorporate external shock absorbers into the cylinder assembly.

HOW TO ACHIEVE IDEAL CUSHIONING

Most people determine cylinder bore based primarily on the thrust required, with no regard for kinetic energy. Usually, this produces a cylinder cushioning capacity that well exceeds that necessary for the application. This will suffice if ideal pneumatic cushioning is not required. But if it is required, pneumatic cushioning will be unachievable with certain combinations of bore, load, and operating pressure.

If the cushioning energy required is within 10% of the cylinder's initial setting, the adjusting screw can be opened to allow the piston to strike the end cover. The shock will be moderate due to the low load, the impact time will be short, and the braking effect of the cushioning will be low because of the short time available for backpressure to develop.

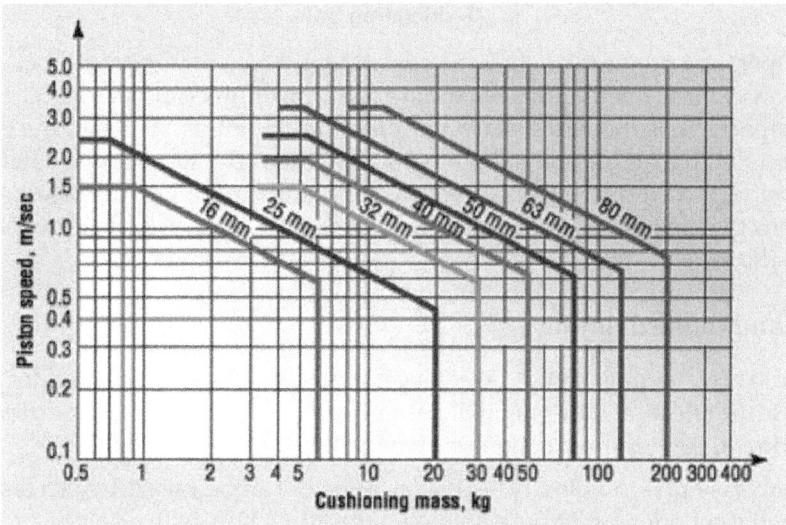

Manufacturers' charts make it easy to select a cylinder with a bore size that provides optimum cushioning for application paramters.

However, if the required cushioning energy exceeds 10% of the initial setting, but is less than 80%, a different scenario applies. If required cushioning is 10% to 80% below the initial setting, the piston can be cushioned partially with air, with impact cushioning accounting for the remainder of the energy absorption. Some impact shock will occur, and the adjusting screw will be almost completely open.

If required cushioning is 10% to 80% higher than the initial setting, the adjusting screw will have to be closed almost completely. The kinetic energy will be damped, and the piston will travel toward the end cover at a certain velocity. However, the shock will be less than in the case described above. The condition is referred to as *rebound cushioning*. With rebound cushioning, the piston changes direction two or three times as deceleration progresses, which makes the total cycle time somewhat longer.

A Better Approach

To this point, scenarios have dealt with cases in which the cylinder was already selected, so the pneumatic cushioning had to be adjusted. Achieving ideal pneumatic cushioning is much less complicated if certain factors are taken into account when specifying the cylinders.

First, cylinders should be specified according to cushioning charts contained in manufacturers' catalogs. A typical example is shown in Figure.

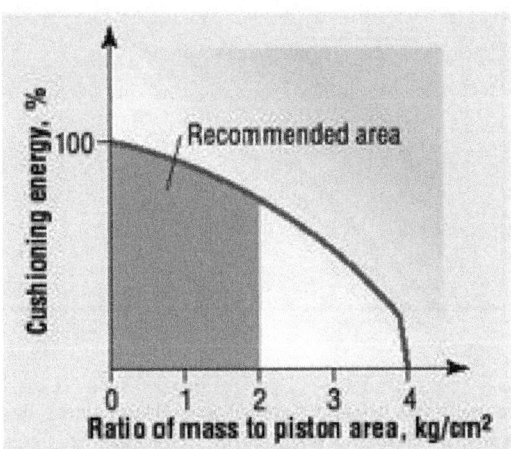

When the load acts downward, the cylinder should be selected so that cushioning falls within the shaded area.

For this discussion, we will assume a horizontal installation, with a minimum stroke of 200 mm, and an operating pressure of 6.3 bar. A good rule of thumb for choosing the correct cylinder is that the ratio of the load mass (in kg) to the piston area (in cm²) should not exceed four. This limit is represented by the vertical line for each bore size shown in Figure. Ideal cushioning can be achieved, without changing the operating pressure, if the intersection of mass and velocity-is on or just below the inclined line in the chart.

Effect of Mounting Orientation

The horizontal cushioning charts also can be used when the piston operates vertically downward. If the cylinder is installed with piston traveling upward, the cushioning capacity will be less, due to the reduction in cushioning pressure. The force of gravity, which acts downward, reduces the cushioning capacity. Figure serves as a guide, as does this rule of thumb:

$m/A < 2$

where m = mass, kg and

A = piston area, cm²

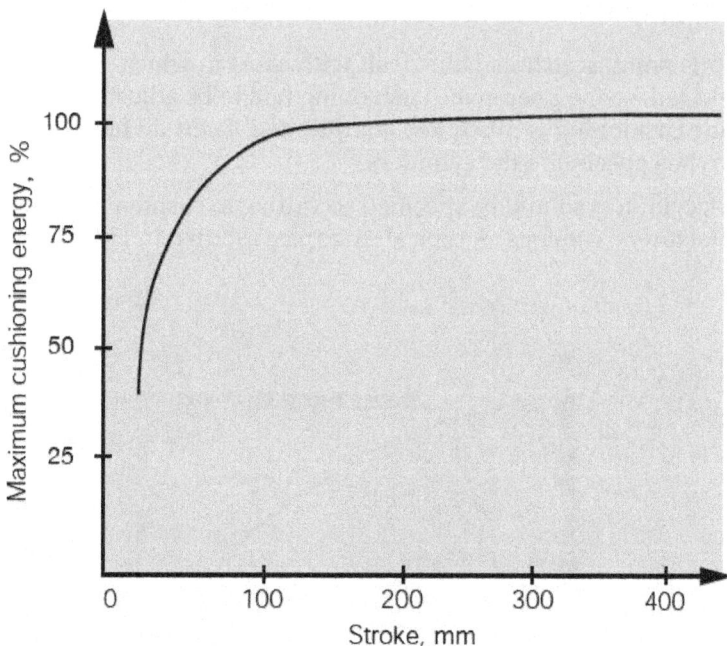

Full cushioning capacity is achieved only when the cylinder has a stroke of about 200 mm for our example. It occurs with stable operating pressure and backpressure.

Influence of Cylinder Stroke

Figure shows that full cushioning capacity is achieved only when our specified cylinder has a stroke of at least 200 mm. Note that the cushioning energy decreases sharply when stroke is less than the recommended minimum.

Again, the operating pressure, mass, and velocity govern the characteristics of cushioning. Once ideal cushioning has been achieved, the application parameters must remain unchanged — otherwise, any variation would require cushioning energy to be readjusted.

Seizing

Failed lubrication can cause the pistons or piston rings to seize to the cylinder walls. Seizing can occur during engine use, *via* overheating and lack of oil, or during storage *via* condensation and corrosion.

ELECTRIC MOTOR

An **electric motor** is an electric machine that converts electrical energy into mechanical energy.

In normal motoring mode, most electric motors operate through the interaction between an electric motor's magnetic field and winding currents to generate force within the motor. In certain applications, such as in the transportation industry with traction motors, electric motors can operate in both motoring and generating or braking modes to also produce electrical energy from mechanical energy.

Found in applications as diverse as industrial fans, blowers and pumps, machine tools, household appliances, power tools, and disk drives, electric motors can be powered by direct current (DC) sources, such as from batteries, motor vehicles or rectifiers, or by alternating current (AC) sources, such as from the power grid, inverters or generators. Small motors may be found in electric watches. General-purpose motors with highly standardized dimensions and characteristics provide convenient mechanical power for industrial use. The largest of electric motors are used for ship propulsion, pipeline compression and pumped-storage applications with ratings reaching 100 megawatts. Electric motors may be classified by electric power source type, internal construction, application, type of motion output, and so on.

Devices such as magnetic solenoids and loudspeakers that convert electricity into motion but do not generate usable mechanical power are respectively referred to as actuators and transducers. Electric motors are used to produce linear force or torque (rotary).

History

Early Motors

Perhaps the first electric motors were simple electrostatic devices created by the Scottish monk Andrew Gordon in the 1740s. The theoretical principle behind production of mechanical force by the interactions of an electric current and a magnetic field, Ampère's force law, was discovered later by André-Marie Ampère in 1820. The conversion of electrical energy into mechanical energy by electromagnetic means was demonstrated by the British scientist Michael Faraday in 1821. A free-hanging wire was dipped into a pool of mercury, on which a permanent magnet (PM) was placed. When a current was passed through the wire, the wire rotated around the magnet, showing that the current gave rise to a close circular magnetic field around the wire. This motor is often demonstrated in physics experiments, brine substituting for toxic mercury. Though Barlow's wheel was an early refinement to this Faraday demonstration, these and similar homopolar motors were to remain unsuited to practical application until late in the century.

In 1827, Hungarian physicist Ányos Jedlik started experimenting with electromagnetic coils. After Jedlik solved the technical problems of the continuous rotation with the invention of commutator, he called his early devices "electromagnetic self-rotors". Although they were used only for instructional purposes, in 1828 Jedlik demonstrated the first device to contain the three main components of practical DC motors: the stator, rotor and commutator. The device employed no permanent magnets, as the magnetic fields of both the stationary and revolving components were produced solely by the currents flowing through their windings.

Success with DC Motors

The first rotating device driven by electromagnetism was built by the Englishman Peter Barlow in 1822. After many other more or less successful attempts with relatively weak rotating and reciprocating apparatus the German-speaking Prussian Moritz Jacobi created the first real rotating electric motor in May 1834 that actually developed a remarkable mechanical output power. His motor set a world record which was improved only four years later in September 1838 by Jacobi himself. His second motor was powerful enough to drive a boat with 14 people across a wide river. It was not until 1839/40 that other developers worldwide managed to build motors of similar and later also of higher performance.

The first commutator DC electric motor capable of turning machinery was invented by the British scientist William Sturgeon in 1832. Following Sturgeon's work, a commutator-type direct-current electric motor made with the intention of commercial use was built by the American inventor Thomas Davenport, which he patented in 1837. The motors ran at up to 600 revolutions per minute, and powered machine tools and a printing press. Due to the high cost of primary battery power, the motors were commercially unsuccessful and Davenport went bankrupt. Several inventors followed Sturgeon in the development of DC motors but all encountered the same battery power cost issues. No electricity distribution had been developed at the time. Like Sturgeon's motor, there was no practical commercial market for these motors.

In 1855, Jedlik built a device using similar principles to those used in his electromagnetic self-rotors that was capable of useful work. He built a model electric vehicle that same year.

The first commercially successful DC motors followed the invention by Zénobe Gramme who had in 1871 developed the anchor ring dynamo which solved the double-T armature pulsating DC problem. In 1873, Gramme found that this dynamo could be used as a motor, which he demonstrated to great effect at exhibitions in Vienna and Philadelphia by connecting two such DC motors at a distance of up to 2 km away from each other, one as a generator.

In 1886, Frank Julian Sprague invented the first practical DC motor, a non-sparking motor that maintained relatively constant speed under variable loads. Other Sprague electric inventions about this time greatly improved grid electric distribution (prior work done while employed by Thomas Edison), allowed power from electric motors to be returned to the electric grid, provided for electric distribution to trolleys *via* overhead wires and the trolley pole, and provided controls systems for electric operations. This allowed Sprague to use electric motors to invent the first electric trolley system in 1887–88 in Richmond VA, the electric elevator and control system in 1892, and the electric subway with independently powered centrally controlled cars, which were first installed in 1892 in Chicago by the South Side Elevated Railway where it became popularly known as the "L". Sprague's motor and related inventions led to an explosion of interest and use in electric motors for industry, while almost simultaneously another great inventor was developing its primary competitor, which would become much more wide-spread. The development of electric motors of acceptable efficiency was delayed for several decades by failure to recognize the extreme importance of a relatively small air gap between rotor and stator. Efficient designs have a comparatively small air gap. The St. Louis motor, long used in classrooms to illustrate motor principles, is extremely inefficient for the same reason, as well as appearing nothing like a modern motor.

Application of electric motors revolutionized industry. Industrial processes were no longer limited by power transmission using line shafts, belts, compressed air or hydraulic pressure. Instead every machine could be equipped with its own electric motor, providing easy control at the point of use, and improving power transmission efficiency. Electric motors applied in agriculture eliminated human and animal muscle power from such tasks as handling grain or pumping water. Household uses of electric motors reduced heavy labor in the home and made higher standards of convenience, comfort and safety possible. Today, electric motors stand for more than half of the electric energy consumption in the US.

Emergence of AC Motors

In 1824, the French physicist François Arago formulated the existence of rotating magnetic fields, termed Arago's rotations, which, by manually turning switches on and off, Walter Baily demonstrated in 1879 as in effect the first primitive induction motor. In the 1880s, many inventors were trying to develop workable AC motors because AC's advantages in long distance high voltage transmission were counterbalanced by the inability to operate motors on AC. Practical rotating AC induction motors were independently invented by Galileo Ferraris and Nikola Tesla, a working motor model having been demonstrated by the former in 1885 and by the latter in 1887. In 1888, the *Royal Academy of Science of Turin* published Ferraris's research detailing the foundations of motor operation while however concluding that "the apparatus based on that principle could not be of any commercial importance as motor." In 1888, Tesla presented his paper *A New System for Alternating Current Motors and Transformers* to the AIEE that described three patented two-phase four-stator-pole motor types: one with a four-pole rotor forming a non-self-starting reluctance motor, another with a wound rotor forming a self-starting induction motor, and the third a true synchronous motor with separately excited DC supply to rotor winding. One of the patents Tesla filed in 1887, however, also described a shorted-winding-rotor induction motor. George Westinghouse promptly bought Tesla's patents, employed Tesla to develop them, and assigned C. F. Scott to help Tesla, Tesla leaving for other pursuits in 1889. The constant speed AC induction motor was found not to be suitable for street cars but Westinghouse engineers successfully adapted it to power a mining operation in Telluride, Colorado in 1891. Steadfast in his promotion of three-phase development, Mikhail Dolivo-Dobrovolsky invented the three-phase cage-rotor induction motor in 1889 and the three-limb transformer in 1890. This type of motor is now used for the vast majority of commercial applications. However, he claimed that Tesla's motor was not practical because of two-phase pulsations, which prompted him to persist in his three-phase work. Although Westinghouse achieved its first practical induction motor in 1892 and developed a line of polyphase 60 hertz induction motors in 1893, these early Westinghouse motors were two-phase motors with wound rotors until B. G. Lamme developed a rotating bar winding rotor. The General Electric Company began developing three-phase induction motors in 1891. By 1896, General Electric and Westinghouse signed a cross-licensing agreement for the bar-winding-rotor design, later called the squirrel-cage rotor. Induction motor improvements flowing from these inventions and innovations were such that a 100 horsepower (HP) induction motor currently has the same mounting dimensions as a 7.5 HP motor in 1897.

Motor Construction

Rotor

In an electric motor the moving part is the rotor which turns the shaft to deliver the mechanical power. The rotor usually has conductors laid into it which carry currents that interact with the magnetic field of the stator to generate the forces that turn the shaft. However, some rotors carry permanent magnets, and the stator holds the conductors.

Stator

The stationary part is the stator, usually has either windings or permanent magnets. The stator is the stationary part of the motor's electromagnetic circuit. The stator core is made up of many thin metal sheets, called laminations. Laminations are used to reduce energy losses that would result if a solid core were used.

Air Gap

In between the rotor and stator is the air gap. The air gap has important effects, and is generally as small as possible, as a large gap has a strong negative effect on the performance of an electric motor.

Windings

Windings are wires that are laid in coils, usually wrapped around a laminated soft iron magnetic core so as to form magnetic poles when energized with current.

Electric machines come in two basic magnet field pole configurations: *salient-pole* machine and *nonsalient-pole* machine. In the salient-pole machine the pole's magnetic field is produced by a winding wound around the pole below the pole face. In the *nonsalient-pole*, or distributed field, or round-rotor, machine, the winding is distributed in pole face slots. A shaded-pole motor has a winding around part of the pole that delays the phase of the magnetic field for that pole.

Some motors have conductors which consist of thicker metal, such as bars or sheets of metal, usually copper, although sometimes aluminum is used. These are usually powered by electromagnetic induction.

Commutator

A commutator is a mechanism used to switch the input of certain AC and DC machines consisting of slip ring segments insulated from each other and from the electric motor's shaft. The motor's armature current is supplied through the stationary brushes in contact with the revolving commutator, which causes required current reversal and applies power to the machine in an optimal manner as the rotor rotates from pole to pole. In absence of such current reversal, the motor would brake to a stop. In light of significant advances in the past few decades due to improved technologies in electronic controller, sensorless control, induction motor, and permanent magnet motor fields, electromechanically commutated motors are increasingly being displaced by externally commutated induction and permanent magnet motors.

Motor Supply and Control

Motor Supply

A DC motor is usually supplied through slip ring commutator as described above. AC motors' commutation can be either slip ring commutator or externally commutated type, can be fixed-speed or variable-speed control type, and can be synchronous or asynchronous type. Universal motors can run on either AC or DC.

Motor Control

Fixed-speed controlled AC motors are provided with direct-on-line or soft-start starters.

Variable speed controlled AC motors are provided with a range of different power inverter, variable-frequency drive or electronic commutator technologies.

The term electronic commutator is usually associated with self-commutated brushless DC motor and switched reluctance motor applications.

Major Categories

Electric motors operate on three different physical principles: magnetic, electrostatic and piezoelectric. By far the most common is magnetic.

In magnetic motors, magnetic fields are formed in both the rotor and the stator. The product between these two fields gives rise to a force, and thus a torque on the motor shaft. One, or both, of these fields must be made to change with the rotation of the motor. This is done by switching the poles on and off at the right time, or varying the strength of the pole.

The main types are DC motors and AC motors, the former increasingly being displaced by the latter.

AC electric motors are either asynchronous or synchronous.

Once started, a synchronous motor requires synchronism with the moving magnetic field's synchronous speed for all normal torque conditions.

In synchronous machines, the magnetic field must be provided by means other than induction such as from separately excited windings or permanent magnets.

Self-commutated Motor

Brushed DC Motor

All self-commutated DC motors are by definition run on DC electric power. Most DC motors are small PM types. They contain a brushed internal mechanical commutation to reverse motor windings' current in synchronism with rotation.

Electrically Excited DC Motor

A commutated DC motor has a set of rotating windings wound on an armature mounted on a rotating shaft. The shaft also carries the commutator, a long-lasting rotary electrical switch that periodically reverses the flow of current in the rotor windings as the shaft rotates. Thus, every brushed DC motor has AC flowing through its rotating windings. Current flows through one or more pairs of brushes that bear on the commutator; the brushes connect an external source of electric power to the rotating armature.

The rotating armature consists of one or more coils of wire wound around a laminated, magnetically "soft" ferromagnetic core. Current from the brushes flows through the commutator and one winding of the armature, making it a temporary magnet (an electromagnet). The magnets field produced by the armature interacts with a stationary magnetic field produced by either PMs or another winding a field coil, as part of the motor frame. The force between the two magnetic fields tends to rotate the motor shaft. The commutator switches power to the coils as the rotor turns, keeping the magnetic poles of the rotor from ever fully aligning with the magnetic poles of the stator field, so that the rotor never stops, but rather keeps rotating as long as power is applied.

Many of the limitations of the classic commutator DC motor are due to the need for brushes to press against the commutator. This creates friction. Sparks are created by the brushes making and breaking circuits through the rotor coils as the brushes cross the insulating gaps between commutator sections. Depending on the commutator design, this may include the brushes shorting together adjacent sections – and hence coil ends – momentarily while crossing the gaps. Furthermore, the inductance of the rotor coils causes the voltage across each to rise when its circuit is opened, increasing the sparking of the brushes. This sparking limits the maximum speed of the machine, as too-rapid sparking will overheat, erode, or even melt the commutator. The current density per unit area of the brushes, in combination with their resistivity, limits the output of the motor. The making and breaking of electric contact also generates electrical noise; sparking generates RFI. Brushes eventually wear out and require replacement, and the commutator itself is subject to wear and maintenance (on larger motors) or replacement (on small motors). The commutator assembly on a large motor is a costly element, requiring precision assembly of many parts. On small motors, the commutator is usually permanently integrated into the rotor, so replacing it usually requires replacing the whole rotor.

While most commutators are cylindrical, some are flat discs consisting of several segments (typically, at least three) mounted on an insulator.

Large brushes are desired for a larger brush contact area to maximize motor output, but small brushes are desired for low mass to maximize the speed at which the motor can run without the brushes excessively bouncing and sparking. (Small brushes are also desirable for lower cost.) Stiffer brush springs can also be used to make brushes of a given mass work at a higher speed, but at the cost of greater friction losses (lower efficiency) and accelerated brush and commutator wear. Therefore, DC motor brush design entails a trade-off between output power, speed, and efficiency/wear.

DC machines are defined as follows:

- Armature circuit - A winding where the load current is carried, such that can be either stationary or rotating part of motor or generator.
- Field circuit - A set of windings that produces a magnetic field so that the electromagnetic induction can take place in electric machines.
- Commutation: A mechanical technique in which rectification can be achieved, or from which DC can be derived, in DC machines.

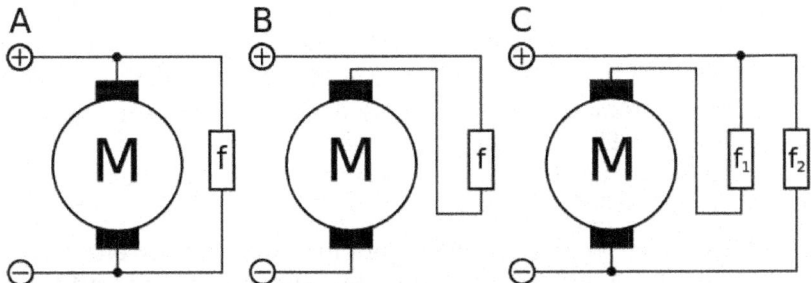

A: shunt B: series C: compound f = field coil

There are five types of brushed DC motor:

- DC shunt-wound motor
- DC series-wound motor
- DC compound motor (two configurations):
    o Cumulative compound
    o Differentially compounded
- PM DC motor
- Separately excited.

Permanent Magnet DC Motor

A PM motor does not have a field winding on the stator frame, instead relying on PMs to provide the magnetic field against which the rotor field interacts to produce torque. Compensating windings in series with the armature may be used on large motors to improve commutation under load. Because this field is fixed, it cannot be adjusted for speed control. PM fields (stators) are convenient in miniature motors to eliminate the power consumption of the field winding. Most larger DC motors are of the "dynamo" type, which have stator windings. Historically, PMs could not be made to retain high flux if they were disassembled; field windings were more practical to obtain the needed amount of flux. However, large PMs are costly, as well as dangerous and difficult to assemble; this favors wound fields for large machines.

To minimize overall weight and size, miniature PM motors may use high energy magnets made with neodymium or other strategic elements; most such are neodymium-iron-boron alloy. With their higher flux density, electric machines with high-energy PMs are at least competitive with all optimally designed singly fed synchronous and induction electric machines. Miniature motors resemble the structure in the illustration, except that they have at least three rotor poles (to ensure starting, regardless of rotor position) and their outer housing is a steel tube that magnetically links the exteriors of the curved field magnets.

Electronic Commutator (EC) Motor

Brushless DC Motor

Some of the problems of the brushed DC motor are eliminated in the BLDC design. In this motor, the mechanical "rotating switch" or commutator is replaced by an external electronic switch synchronised to the rotor's position. BLDC motors are typically 85–90% efficient or more. Efficiency for a BLDC motor of up to 96.5% have been reported, whereas DC motors with brushgear are typically 75–80% efficient.

The BLDC motor's characteristic trapezoidal back-emf waveform is derived partly from the stator windings being evenly distributed, and partly from the placement of the rotor's PMs. Also known as electronically commutated DC or inside out DC motors, the stator windings of trapezoidal BLDC motors can be with single-phase, two-phase or three-phase and use Hall effect sensors mounted on their windings for rotor position sensing and low cost closed-loop control of the electronic commutator.

BLDC motors are commonly used where precise speed control is necessary, as in computer disk drives or in video cassette recorders, the spindles within CD, CD-ROM (*etc.*) drives, and mechanisms within office products such as fans, laser printers and photocopiers. They have several advantages over conventional motors:

- Compared to AC fans using shaded-pole motors, they are very efficient, running much cooler than the equivalent AC motors. This cool operation leads to much-improved life of the fan's bearings.

- Without a commutator to wear out, the life of a BLDC motor can be significantly longer compared to a DC motor using brushes and a commutator. Commutation also tends to cause a great deal of electrical and RF noise; without a commutator or brushes, a BLDC motor may be used in electrically sensitive devices like audio equipment or computers.

- The same Hall effect sensors that provide the commutation can also provide a convenient tachometer signal for closed-loop control (servo-controlled) applications. In fans, the tachometer signal can be used to derive a "fan OK" signal as well as provide running speed feedback.

- The motor can be easily synchronized to an internal or external clock, leading to precise speed control.

- BLDC motors have no chance of sparking, unlike brushed motors, making them better suited to environments with volatile chemicals and fuels. Also, sparking generates ozone which can accumulate in poorly ventilated buildings risking harm to occupants' health.

- BLDC motors are usually used in small equipment such as computers and are generally used in fans to get rid of unwanted heat.

- They are also acoustically very quiet motors which is an advantage if being used in equipment that is affected by vibrations.

Modern BLDC motors range in power from a fraction of a watt to many kilowatts. Larger BLDC motors up to about 100 kW rating are used in electric vehicles. They also find significant use in high-performance electric model aircraft.

Switched Reluctance Motor

The SRM has no brushes or PMs, and the rotor has no electric currents. Instead, torque comes from a slight misalignment of poles on the rotor with poles on the stator. The rotor aligns itself with the magnetic field of the stator, while the stator field stator windings are sequentially energized to rotate the stator field.

The magnetic flux created by the field windings follows the path of least magnetic reluctance, meaning the flux will flow through poles of the rotor that are closest to the energized poles of the stator, thereby magnetizing those poles of the rotor and creating torque. As the rotor turns, different windings will be energized, keeping the rotor turning.

SRMs are now being used in some appliances.

Universal AC-DC Motor

A commutated electrically excited series or parallel wound motor is referred to as a universal motor because it can be designed to operate on both AC and DC power. A universal motor can operate well on AC because the current in both the field and the armature coils (and hence the resultant magnetic fields) will alternate (reverse polarity) in synchronism, and hence the resulting mechanical force will occur in a constant direction of rotation.

Operating at normal power line frequencies, universal motors are often found in a range less than 1000 watts. Universal motors also formed the basis of the traditional railway traction motor in electric railways. In this application, the use of AC to power a motor originally designed to run on DC would lead to efficiency losses due to eddy current heating of their magnetic components, particularly the motor field pole-pieces that, for DC, would have used solid (un-laminated) iron and they are now rarely used.

An advantage of the universal motor is that AC supplies may be used on motors which have some characteristics more common in DC motors, specifically high starting torque and very compact design if high running speeds are used. The negative aspect is the maintenance and short life problems caused by the commutator. Such motors are used in devices such as food mixers and power tools which are used only intermittently, and often have high starting-torque demands. Multiple taps on the field coil provide (imprecise) stepped speed control. Household blenders that advertise many speeds frequently combine a field coil with several taps and a diode that can be inserted in series with the motor (causing the motor to run on half-wave rectified AC). Universal motors also lend themselves to electronic speed control and, as such, are an ideal choice for devices like domestic washing machines. The motor can be used to agitate the drum by switching the field winding with respect to the armature.

Whereas SCIMs cannot turn a shaft faster than allowed by the power line frequency, universal motors can run at much higher speeds. This makes them useful for appliances such as blenders, vacuum cleaners, and hair dryers where high speed and light weight are desirable. They are also commonly used in portable power tools, such as drills, sanders, circular and jig saws, where the motor's characteristics work well. Many vacuum cleaner and weed trimmer motors exceed 10,000 rpm, while many similar miniature grinders exceed 30,000 rpm.

Externally Commutated AC Machine

The design of AC induction and synchronous motors is optimized for operation on single-phase or polyphase sinusoidal or quasi-sinusoidal waveform power such as supplied for fixed-speed application from the AC power grid or for variable-speed application from VFD controllers. An AC motor has two parts: a stationary stator having coils supplied with AC to produce a rotating magnetic field, and a rotor attached to the output shaft that is given a torque by the rotating field.

Induction Motor

Cage and Wound Rotor Induction Motor

An induction motor is an asynchronous AC motor where power is transferred to the rotor by electromagnetic induction, much like transformer action. An induction motor resembles a rotating transformer, because the stator is essentially the primary side of the transformer and the rotor (rotating part) is the secondary side. Polyphase induction motors are widely used in industry.

Induction motors may be further divided into Squirrel Cage Induction Motors and Wound Rotor Induction Motors. SCIMs have a heavy winding made up of solid bars, usually aluminum or copper, joined by rings at the ends of the rotor. When one considers only the bars and rings as a whole, they are much like an animal's rotating exercise cage, hence the name.

Currents induced into this winding provide the rotor magnetic field. The shape of the rotor bars determines the speed-torque characteristics. At low speeds, the current induced in the squirrel cage is nearly at line frequency and tends to be in the outer parts of the rotor cage. As the motor accelerates, the slip frequency becomes lower, and more current is in the interior of the winding. By shaping the bars to change the resistance of the winding portions in the interior and outer parts of the cage, effectively a variable resistance is inserted in the rotor circuit. However, the majority of such motors have uniform bars.

In a WRIM, the rotor winding is made of many turns of insulated wire and is connected to slip rings on the motor shaft. An external resistor or other control devices can be connected in the rotor circuit. Resistors allow control of the motor speed, although significant power is dissipated in the external resistance. A converter can be fed from the rotor circuit and return the slip-frequency power that would otherwise be wasted back into the power system through an inverter or separate motor-generator.

The WRIM is used primarily to start a high inertia load or a load that requires a very high starting torque across the full speed range. By correctly selecting the resistors used in the secondary resistance or slip ring starter, the motor is able to produce maximum torque at a relatively low supply current from zero speed to full speed. This type of motor also offers controllable speed.

Motor speed can be changed because the torque curve of the motor is effectively modified by the amount of resistance connected to the rotor circuit. Increasing the value of resistance will move the speed of maximum torque down. If the resistance connected to the rotor is increased beyond the point where the maximum torque occurs at zero speed, the torque will be further reduced.

When used with a load that has a torque curve that increases with speed, the motor will operate at the speed where the torque developed by the motor is equal to the load torque. Reducing the load will cause the motor to speed up, and increasing the load will cause the motor to slow down until the load and motor torque are equal. Operated in this manner, the slip losses are dissipated in the secondary resistors and can be very significant. The speed regulation and net efficiency is also very poor.

Torque Motor

A torque motor is a specialized form of electric motor which can operate indefinitely while stalled, that is, with the rotor blocked from turning, without incurring damage. In this mode of operation, the motor will apply a steady torque to the load.

A common application of a torque motor would be the supply- and take-up reel motors in a tape drive. In this application, driven from a low voltage, the characteristics of these motors allow a relatively constant light tension to be applied to the tape whether or not the capstan is feeding tape past the tape heads. Driven from a higher voltage, (and so delivering a higher torque), the torque motors can also achieve fast-forward and rewind operation without requiring any additional mechanics such as gears or clutches. In the computer gaming world, torque motors are used in force feedback steering wheels.

Another common application is the control of the throttle of an internal combustion engine in conjunction with an electronic governor. In this usage, the motor works against a return spring to move the throttle in accordance with the output of the governor. The latter monitors engine speed by counting electrical pulses from the ignition system or from a magnetic pickup and, depending on the speed, makes small adjustments to the amount of current applied to the motor. If the engine starts to slow down relative to the desired speed, the current will be increased, the motor will develop more torque, pulling against the return spring and opening the throttle. Should the engine run too fast, the governor will reduce the current being applied to the motor, causing the return spring to pull back and close the throttle.

Synchronous Motor

A synchronous electric motor is an AC motor distinguished by a rotor spinning with coils passing magnets at the same rate as the AC and resulting magnetic field which drives it. Another way of saying this is that it has zero slip under usual operating conditions. Contrast this with an induction motor, which must slip to produce torque. One type of synchronous motor is like an induction motor except the rotor is excited by a DC field. Slip rings and brushes are used to conduct current to the rotor. The rotor poles connect to each other and move at the same speed hence the name synchronous motor. Another type, for low load torque, has flats ground onto a conventional squirrel-cage rotor to create discrete poles. Yet another, such as made by Hammond for its pre-World War II clocks, and in the older Hammond organs, has no rotor windings and discrete poles. It is not self-starting. The clock requires manual starting by a small knob on the back, while the older Hammond organs had an auxiliary starting motor connected by a spring-loaded manually operated switch.

Finally, hysteresis synchronous motors typically are (essentially) two-phase motors with a phase-shifting capacitor for one phase. They start like induction motors, but when slip rate decreases sufficiently, the rotor (a smooth cylinder) becomes temporarily magnetized. Its distributed poles make it act like a PMSM. The rotor material, like that of a common nail, will stay magnetized, but can also be demagnetized with little difficulty. Once running, the rotor poles stay in place; they do not drift.

Low-power synchronous timing motors (such as those for traditional electric clocks) may have multi-pole PM external cup rotors, and use shading coils to provide starting torque. *Telechron* clock motors have shaded poles for starting torque, and a two-spoke ring rotor that performs like a discrete two-pole rotor.

Doubly Fed Electric Machine

Doubly fed electric motors have two independent multiphase winding sets, which contribute active (*i.e.*, working) power to the energy conversion process, with at least one of the winding sets electronically controlled for variable speed operation. Two independent multiphase winding sets (*i.e.*, dual armature) are the maximum provided in a single package without topology duplication. Doubly fed electric motors are machines with an effective constant torque speed range that is twice synchronous speed for a given frequency of excitation. This is twice the constant torque speed range as singly fed electric machines, which have only one active winding set.

A doubly fed motor allows for a smaller electronic converter but the cost of the rotor winding and slip rings may offset the saving in the power electronics components. Difficulties with controlling speed near synchronous speed limit applications.

SPECIAL MAGNETIC MOTORS

Rotary

Hydraulic Cylinder Displacement

Electric motors are replacing hydraulic cylinders in airplanes and military equipment.

Ironless or Coreless Rotor Motor

Nothing in the principle of any of the motors described above requires that the iron (steel) portions of the rotor actually rotate. If the soft magnetic material of the rotor is made in the form of a cylinder, then (except for the effect of hysteresis) torque is exerted only on the windings of the electromagnets. Taking advantage of this fact is the *coreless or ironless DC motor*, a specialized form of a PM DC motor. Optimized for rapid acceleration, these motors have a rotor that is constructed without any iron core. The rotor can take the form of a winding-filled cylinder, or a self-supporting structure comprising only the magnet wire and the bonding material. The rotor can fit inside the stator magnets; a magnetically soft stationary cylinder inside the rotor provides a return path for the stator magnetic flux. A second arrangement has the rotor winding basket surrounding the stator magnets. In that design, the rotor fits inside a magnetically soft cylinder that can serve as the housing for the motor, and likewise provides a return path for the flux.

Because the rotor is much lighter in weight (mass) than a conventional rotor formed from copper windings on steel laminations, the rotor can accelerate much more rapidly, often achieving a mechanical time constant under one ms. This is especially true if the windings use aluminum rather than the heavier copper. But because there is no metal mass in the rotor to act as a heat sink, even small core-less motors must often be cooled by forced air. Overheating might be an issue for coreless DC motor designs.

Among these types are the disc-rotor types, described in more detail in the next section.

Vibrator motors for cellular phones are sometimes tiny cylindrical PM field types, but there are also disc-shaped types which have a thin multipolar disc field magnet, and an intentionally unbalanced molded-plastic rotor structure with two bonded coreless coils. Metal brushes and a flat commutator switch power to the rotor coils.

Related limited-travel actuators have no core and a bonded coil placed between the poles of high-flux thin PMs. These are the fast head positioners for rigid-disk ("hard disk") drives. Although the contemporary design differs considerably from that of loudspeakers, it is still loosely (and incorrectly) referred to as a "voice coil" structure, because some earlier rigid-disk-drive heads moved in straight lines, and had a drive structure much like that of a loudspeaker.

Pancake or Axial Rotor Motor

A rather unusual motor design, the printed armature or pancake motor has the windings shaped as a disc running between arrays of high-flux magnets. The magnets are arranged in a circle facing the rotor with space in between to form an axial air gap. This design is commonly known as the pancake motor because of its extremely flat profile, although the technology has had many brand names since its inception, such as ServoDisc.

The printed armature (originally formed on a printed circuit board) in a printed armature motor is made from punched copper sheets that are laminated together using advanced composites to form a thin rigid disc. The printed armature has a unique construction in the brushed motor world in that it does not have a separate ring commutator. The brushes run directly on the armature surface making the whole design very compact.

An alternative manufacturing method is to use wound copper wire laid flat with a central conventional commutator, in a flower and petal shape. The windings are typically stabilized by being impregnated with electrical epoxy potting systems. These are filled epoxies that have moderate mixed viscosity and a long gel time. They are highlighted by low shrinkage and low exotherm, and are typically UL 1446 recognized as a potting compound insulated with 180 °C, Class H rating.

The unique advantage of ironless DC motors is that there is no cogging (torque variations caused by changing attraction between the iron and the magnets). Parasitic eddy currents cannot form in the rotor as it is totally ironless, although iron rotors are laminated. This can greatly improve efficiency, but variable-speed controllers must use a higher switching rate (>40 kHz) or DC because of the decreased electromagnetic induction.

These motors were originally invented to drive the capstan(s) of magnetic tape drives in the burgeoning computer industry, where minimal time to reach operating speed and minimal stopping distance were critical. Pancake motors are still widely used in high-performance servo-controlled systems, robotic systems, industrial automation and medical devices. Due to the variety of constructions now available, the technology is used in applications from high temperature military to low cost pump and basic servos.

Servo Motor

A servomotor is a motor, very often sold as a complete module, which is used within a position-control or speed-control feedback control system mainly control valves, such as motor operated control valves. Servomotors are used in applications such as machine tools, pen plotters, and other process systems. Motors intended for use in a servomechanism must have well-documented characteristics for speed, torque, and power. The speed vs. torque curve is quite important and is high ratio for a servo motor. Dynamic response characteristics such as winding inductance and rotor inertia are also important; these factors limit the overall performance of the servomechanism loop. Large, powerful, but slow-responding servo loops may use conventional AC or DC motors and drive systems with position or speed feedback on the motor. As dynamic response requirements increase, more specialized motor designs such as coreless motors are used. AC motors' superior power density and acceleration characteristics compared to that of DC motors tends to favor PM synchronous, BLDC, induction, and SRM drive applications.

A servo system differs from some stepper motor applications in that the position feedback is continuous while the motor is running; a stepper system relies on the motor not to "miss steps" for short term accuracy, although a stepper system may include a "home" switch or other element to provide long-term stability of control. For instance, when a typical dot matrix computer printer starts up, its controller makes the print head stepper motor drive to its left-hand limit, where a position sensor defines home position and stops stepping. As long as power is on, a bidirectional counter in the printer's microprocessor keeps track of printhead position.

Stepper Motor

Stepper motors are a type of motor frequently used when precise rotations are required. In a stepper motor an internal rotor containing PMs or a magnetically soft rotor with salient poles is controlled by a set of external magnets that are switched electronically. A stepper motor may also be thought of as a cross between a DC electric motor and a rotary solenoid. As each coil is energized in turn, the rotor aligns itself with the magnetic field produced by the energized field winding. Unlike a synchronous motor, in its application, the stepper motor may not rotate continuously; instead, it "steps" — starts and then quickly stops again — from one position to the next as field windings are energized and de-energized in sequence. Depending on the sequence, the rotor may turn forwards or backwards, and it may change direction, stop, speed up or slow down arbitrarily at any time.

A   B

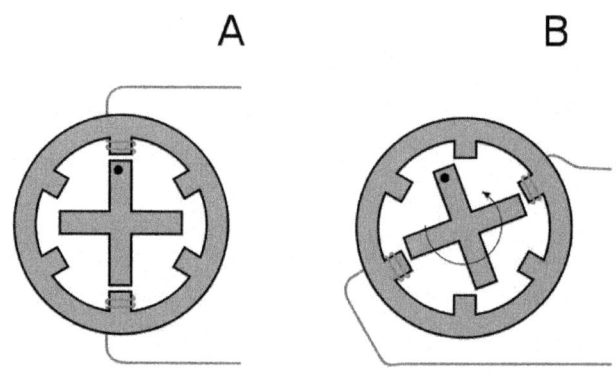

A stepper motor with a soft iron rotor, with active windings shown. In 'A' the active windings tend to hold the rotor in position. In 'B' a different set of windings are carrying a current, which generates torque and rotation.

Simple stepper motor drivers entirely energize or entirely de-energize the field windings, leading the rotor to "cog" to a limited number of positions; more sophisticated drivers can proportionally control the power to the field windings, allowing the rotors to position between the cog points and thereby rotate extremely smoothly. This mode of operation is often called microstepping. Computer controlled stepper motors are one of the most versatile forms of positioning systems, particularly when part of a digital servo-controlled system.

Stepper motors can be rotated to a specific angle in discrete steps with ease, and hence stepper motors are used for read/write head positioning in computer floppy diskette drives. They were used for the same purpose in pre-gigabyte era computer disk drives, where the precision and speed they offered was adequate for the correct positioning of the read/write head of a hard disk drive. As drive density increased, the precision and speed limitations of stepper motors made them obsolete for hard drives — the precision limitation made them unusable, and the speed limitation made them uncompetitive — thus newer hard disk drives use voice coil-based head actuator systems. (The term "voice coil" in this connection is historic; it refers to the structure in a typical (cone type) loudspeaker. This structure was used for a while to position the heads. Modern drives have a pivoted coil mount; the coil swings back and forth, something like a blade of a rotating fan. Nevertheless, like a voice coil, modern actuator coil conductors move perpendicular to the magnetic lines of force.)

Stepper motors were and still are often used in computer printers, optical scanners, and digital photocopiers to move the optical scanning element, the print head carriage (of dot matrix and inkjet printers), and the platen or feed rollers. Likewise, many computer plotters (which since the early 1990s have been replaced with large-format inkjet and laser printers) used rotary stepper motors for pen and platen movement; the typical alternatives here were either linear stepper motors or servomotors with closed-loop analog control systems.

So-called quartz analog wristwatches contain the smallest commonplace stepping motors; they have one coil, draw very little power, and have a PM rotor. The same kind of motor drives battery-powered quartz clocks. Some of these watches, such as chronographs, contain more than one stepping motor.

Closely related in design to three-phase AC synchronous motors, stepper motors and SRMs are classified as variable reluctance motor type. Stepper motors were and still are often used in computer printers, optical scanners, and computer numerical control (CNC) machines such as routers, plasma cutters and CNC lathes.

Linear Motor

A linear motor is essentially any electric motor that has been "unrolled" so that, instead of producing a torque (rotation), it produces a straight-line force along its length.

Linear motors are most commonly induction motors or stepper motors. Linear motors are commonly found in many roller-coasters where the rapid motion of the motorless railcar is controlled by the rail. They are also used in maglev trains, where the train "flies" over the ground. On a smaller scale, the 1978 era HP 7225A pen plotter used two linear stepper motors to move the pen along the X and Y axes.

Comparison by Major Categories

Comparison of motor types				
Type	Advantages	Disadvantages	Typical application	Typical drive, output
Self-commutated motors				
Brushed DC	Simple speed control Low initial cost	Maintenance (brushes) Medium lifespan Costly commutator and brushes	Steel mills Paper making machines Treadmill exercisers Automotive accessories	Rectifier, linear transistor(s) or DC chopper controller.
Brushless DC motor (BLDC) or (BLDM)	Long lifespan Low maintenance High efficiency	Higher initial cost Requires EC controller with closed-loop control	Rigid ("hard") disk drives CD/DVD players Electric vehicles RC Vehicles UAVs	Synchronous; single-phase or three-phase with PM rotor and trapezoidal stator winding; VFD typically VS PWM inverter type.

(Contd...)

Switched reluctance motor (SRM)	Long lifespan Low maintenance High efficiency No permanent magnets Low cost Simple construction	Mechanical resonance possible High iron losses Not possible: * Open or vector control * Parallel operation Requires EC controller	Appliances Electric Vehicles Textile mills Aircraft applications	PWM and various other drive types, which tend to be used in very specialized / OEM applications.
Universal motor	High starting torque, compact, high speed.	Maintenance (brushes) Shorter lifespan Usually acoustically noisy Only small ratings are economical	Handheld power tools, blenders, vacuum cleaners, insulation blowers	Variable single phase AC, half-wave or full-wave phase-angle control with triac(s); closed-loop control optional.
AC asynchronous motors				
AC poly-phase squirrel-cage or wound-rotor induction motor (SCIM) or (WRIM)	Self-starting Low cost Robust Reliable Ratings to 1+ MW Standardized types.	High starting current Lower efficiency due to need for magnetization.	Fixed-speed, traditionally, SCIM the world's workhorse especially in low performance applications of all types Variable-speed, traditionally, low-performance variable-torque pumps, fans, blowers and compressors. Variable-speed, increasingly, other high-performance constant-torque and constant-power or dynamic loads.	Fixed-speed, low performance applications of all types. Variable-speed, traditionally, WRIM drives or fixed-speed V/Hz-controlled VSDs. Variable-speed, increasingly, vector-controlled VSDs displacing DC, WRIM and single-phase AC induction motor drives.
AC SCIM split-phase capacitor-start	High power high starting torque	Speed slightly below synchronous Starting switch or relay required	Appliances Stationary Power Tools	Fixed or variable single-phase AC, variable speed being derived, typically, by full-wave phase-angle control with triac(s); closed-loop control optional.
AC SCIM split-phase capacitor-run	Moderate power High starting torque No starting switch Comparatively long life	Speed slightly below synchronous Slightly more costly	Industrial blowers Industrial machinery	
AC SCIM split-phase, auxiliary start winding	Moderate power Low starting torque	Speed slightly below synchronous Starting switch or relay required	Appliances Stationary power tools	
AC induction shaded-pole motor	Low cost Long life	Speed slightly below synchronous Low starting torque Small ratings low efficiency	Fans, appliances, record players	

AC synchronous motors				
Wound-rotor synchronous motor (WRSM)	Synchronous speed Inherently more efficient induction motor, low power factor	More costly	Industrial motors	Fixed or variable speed, three-phase; VFD typically six-step CS load-commutated inverter type or VS PWM inverter type.
Hysteresis motor	Accurate speed control Low noise No vibration High starting torque	Very low efficiency	Clocks, timers, sound producing or recording equipment, hard drive, capstan drive	Single-phase AC, two-phase capacitor-start, capacitor run motor
Synchronous reluctance motor (SyRM)	Equivalent to SCIM except more robust, more efficient, runs cooler, smaller footprint Competes with PM synchronous motor without demagnetization issues	Requires a controller Not widely available High cost	Appliances Electric vehicles Textile mills Aircraft applications	VFD can be standard DTC type or VS inverter PWM type.
Speciality motors				
Pancake or axial rotor motors	Compact design Simple speed control	Medium cost Medium lifespan	Office Equip Fans/Pumps, fast industrial and military servos	Drives can typically be brushed or brushless DC type.
Stepper motor	Precision positioning High holding torque	Some can be costly Require a controller	Positioning in printers and floppy disc drives; industrial machine tools	Not a VFD. Stepper position is determined by pulse counting.

Electromagnetism

Force and Torque

The fundamental purpose of the vast majority of the world's electric motors is to electromagnetically induce relative movement in an air gap between a stator and rotor to produce useful torque or linear force.

According Lorentz force law the force of a winding conductor can be given simply by:

$$F = I\ell \times B$$

or more generally, to handle conductors with any geometry:

$$F = J \times B$$

The most general approaches to calculating the forces in motors use tensors.

Power

Where rpm is shaft speed and T is torque, a motor's mechanical power output P_{em} is given by,

in British units with T expressed in foot-pounds,

$$P_{em} = \frac{rpm \times T}{5252} \text{ (horsepower), and,}$$

in SI units with shaft speed expressed in radians per second, and T expressed in newton-meters,

$$P_{em} = speed \times T \text{ (watts).}$$

For a linear motor, with force F and velocity v expressed in newtons and meters per second,

$$P_{em} = F \times v \text{ (watts).}$$

In an asynchronous or induction motor, the relationship between motor speed and air gap power is, neglecting skin effect, given by the following:

$$P_{airgap} = \frac{Rr}{s} * I_r^2 \text{, where}$$

R_r - rotor resistance

I_r^2 - square of current induced in the rotor

s - motor slip; i.e, difference between synchronous speed and slip speed, which provides the relative movement needed for current induction in the rotor.

Back Emf

Since the armature windings of a direct-current motor are moving through a magnetic field, they have a voltage induced in them. This voltage tends to oppose the motor supply voltage and so is called "back electromotive force (emf)". The voltage is proportional to the running speed of the motor. The back emf of the motor, plus the voltage drop across the winding internal resistance and brushes, must equal the voltage at the brushes. This provides the fundamental mechanism of speed regulation in a DC motor. If the mechanical load increases, the motor slows down; a lower back emf results, and more current is drawn from the supply. This increased current provides the additional torque to balance the new load.

In AC machines, it is sometimes useful to consider a back emf source within the machine; this is of particular concern for close speed regulation of induction motors on VFDs, for example.

Losses

Motor losses are mainly due to resistive losses in windings, core losses and mechanical losses in bearings, and aerodynamic losses, particularly where cooling fans are present, also occur.

Losses also occur in commutation, mechanical commutators spark, and electronic commutators and also dissipate heat.

Efficiency

To calculate a motor's efficiency, the mechanical output power is divided by the electrical input power:

$$\eta = \frac{P_m}{P_e},$$

where η is energy conversion efficiency, P_e is electrical input power, and P_m is mechanical output power:

$$P_e = IV$$
$$P_m = T\omega$$

where V is input voltage, I is input current, T is output torque, and ω is output angular velocity. It is possible to derive analytically the point of maximum efficiency. It is typically at less than 1/2 the stall torque.

Various regulatory authorities in many countries have introduced and implemented legislation to encourage the manufacture and use of higher efficiency electric motors.

Goodness Factor

Professor Eric Laithwaite proposed a metric to determine the 'goodness' of an electric motor:

$$G = \frac{\grave{u}}{\text{resistance} \times \text{reluctance}} = \frac{\omega\mu\sigma A_m A_e}{l_m l_e}$$

Where:

Gis the goodness factor (factors above 1 are likely to be efficient)

A_m, A_e are the cross sections of the magnetic and electric circuit

l_m, $l_e$are the lengths of the magnetic and electric circuits

μ is the permeability of the core

ω is the angular frequency the motor is driven at

From this, he showed that the most efficient motors are likely to have relatively large magnetic poles. However, the equation only directly relates to non PM motors.

Performance Parameters

Torque Capability of Motor Types

All the electromagnetic motors, and that includes the types mentioned here derive the torque from the vector product of the interacting fields. For calculating the torque it is necessary to know the fields in the air gap. Once these have been established by mathematical anylysis using FEA or other tools the torque may be calculated as the integral of all the vectors of force multiplied by the radius of each vector. The current flowing in the windings is producing the fields and for a motor using a magnetic material the field is not linearilly proprtional to the current. This makes the calculation difficult but a computer can do the many calculations needed. Once this is done a figure relating the current to the torque can be used as a useful parameter for motor selection. The maximum torque for a motor will depend on the maximum current although this will usually be.only usable until thermal considerations take precedence. When optimally designed within a given core saturation constraint and for a given active current (*i.e.*, torque current), voltage, pole-pair number, excitation frequency (*i.e.*, synchronous speed), and air-gap flux density, all categories of electric motors or generators will exhibit virtually the same maximum continuous shaft torque (*i.e.*, operating torque) within a given air-gap area with winding slots and back-iron depth, which determines the physical size of electromagnetic core. Some applications require bursts of torque beyond the maximum operating torque, such as short bursts of torque to accelerate an electric vehicle from standstill. Always limited by magnetic core saturation or safe operating temperature rise and voltage, the capacity for torque bursts beyond the maximum operating torque differs significantly between categories of electric motors or generators.

Capacity for bursts of torque should not be confused with field weakening capability. Field weakening allows an electric machine to operate beyond the designed frequency of excitation. Field weakening is done when the maximum speed cannot be reached by increasing the applied voltage. This applies to only motors with current controlled fields and therefore cannot be achieved with PM motors.

Electric machines without a transformer circuit topology, such as that of WRSMs or PMSMs, cannot realize bursts of torque higher than the maximum designed torque without saturating the magnetic core and rendering any increase in current as useless. Furthermore, the PM assembly of PMSMs can be irreparably damaged, if bursts of torque exceeding the maximum operating torque rating are attempted.

Electric machines with a transformer circuit topology, such as induction machines, induction doubly fed electric machines, and induction or synchronous wound-rotor doubly fed (WRDF) machines, exhibit very high bursts of torque because the emf-induced active current on either side of the transformer oppose each other and thus contribute nothing to the transformer coupled magnetic core flux density, which would otherwise lead to core saturation.

Electric machines that rely on induction or asynchronous principles short-circuit one port of the transformer circuit and as a result, the reactive impedance of the transformer circuit becomes dominant as slip increases, which limits the magnitude of active (*i.e.*, real) current. Still, bursts of torque that are two to three times higher than the maximum design torque are realizable.

The brushless wound-rotor synchronous doubly fed (BWRSDF) machine is the only electric machine with a truly dual ported transformer circuit topology (*i.e.*, both ports independently excited with no short-circuited port). The dual ported transformer circuit topology is known to be unstable and requires a multiphase slip-ring-brush assembly to propagate limited power to the rotor winding set. If a precision means were available to instantaneously control torque angle and slip for synchronous operation during motoring or generating while simultaneously providing brushless power to the rotor winding set, the active current of the BWRSDF machine would be independent of the reactive impedance of the transformer circuit and bursts of torque significantly higher than the maximum operating torque and far beyond the practical capability of any other type of electric machine would be realizable. Torque bursts greater than eight times operating torque have been calculated.

Continuous Torque Density

The continuous torque density of conventional electric machines is determined by the size of the air-gap area and the back-iron depth, which are determined by the power rating of the armature winding set, the speed of the machine, and the achievable air-gap flux density before core saturation. Despite the high coercivity of neodymium or samarium-cobalt PMs, continuous torque density is virtually the same amongst electric machines with optimally designed armature winding sets. Continuous torque density relates to method of cooling and permissible period of operation before destruction by overheating of windings or PM damage.

Continuous Power Density

The continuous power density is determined by the product of the continuous torque density and the constant torque speed range of the electric machine.

Standards

The following are major design and manufacturing standards covering electric motors:

- International Electrotechnical Commission: IEC 60034 Rotating Electrical Machines
- National Electrical Manufacturers Association: MG-1 Motors and Generators
- Underwriters Laboratories: UL 1004 - Standard for Electric Motors

Non-magnetic Motors

An electrostatic motor is based on the attraction and repulsion of electric charge. Usually, electrostatic motors are the dual of conventional coil-based motors. They typically require a high voltage power supply, although very small motors employ lower voltages. Conventional electric motors instead employ magnetic attraction and repulsion, and require high current at low voltages. In the 1750s, the first electrostatic motors were developed by Benjamin Franklin and Andrew Gordon. Today the electrostatic motor finds frequent use in micro-electro-mechanical systems (MEMS) where their drive voltages are below 100 volts, and where moving, charged plates are far easier to fabricate than coils and iron cores. Also, the molecular machinery which runs living cells is often based on linear and rotary electrostatic motors.

A piezoelectric motor or piezo motor is a type of electric motor based upon the change in shape of a piezoelectric material when an electric field is applied. Piezoelectric motors make use of the converse piezoelectric effect whereby the material produces acoustic or ultrasonic vibrations in order to produce a linear or rotary motion. In one mechanism, the elongation in a single plane is used to make a series stretches and position holds, similar to the way a caterpillar moves.

An electrically powered spacecraft propulsion system uses electric motor technology to propel spacecraft in outer space, most systems being based on electrically powering propellant to high speed, with some systems being based on electrodynamic tethers principles of propulsion to the magnetosphere.

HYDRAULIC MOTOR

A **hydraulic motor** is a mechanical actuator that converts hydraulic pressure and flow into torque and angular displacement (rotation). The hydraulic motor is the rotary counterpart of the hydraulic cylinder.

Conceptually, a hydraulic motor should be interchangeable with a hydraulic pump because it performs the opposite function - similar to the way a DC electric motor is theoretically interchangeable with a DC electrical generator. However, most hydraulic pumps cannot be used as hydraulic motors because they cannot be backdriven. Also, a hydraulic motor is usually designed for working pressure at both sides of the motor.

Hydraulic pumps, motors, and cylinders can be combined into hydraulic drive systems. One or more hydraulic pumps, coupled to one or more hydraulic motors, constitute a hydraulic transmission.

One of the first rotary hydraulic motors to be developed was that constructed by William Armstrong for his Swing Bridge over the River Tyne. Two motors were provided, for reliability. Each one was a three-cylinder single-acting oscil-lating engine. Armstrong developed a wide range of hydraulic motors, linear and rotary, that were used for a wide range of industrial and civil engineering tasks, particularly for docks and moving bridges.

Hydraulic Motor Types

Many designs are possible. The following types of hydraulic motors are available:

Symbol hydraulic motor.

Gear and Vane Motors

Hydraulic motor.

Gear and vane motors are used in simple rotating systems. Their benefits include low initial cost and high rpm.

A gear motor (external gear) consists of two gears, the driven gear (attached to the output shaft by way of a key, *etc.*) and the idler gear. High pressure oil is ported into one side of the gears, where it flows around the periphery of the gears, between the gear tips and the wall housings in which it resides, to the outlet port. The gears then mesh, not allowing the oil from the outlet side to flow back to the inlet side. For lubrication, the gear motor uses a small amount of oil from the pressurized side of the gears, bleeds this through the (typically) hydrodynamic bearings, and vents the same oil either to the low pressure side of the gears, or through a dedicated drain port on the motor housing. An especially positive attribute of the gear motor is that catastrophic breakdown is a lot less common than in most other types of hydraulic motors. This is because the gears gradually wear down the housing and/or main bushings, reducing the volumetric efficiency of the motor gradually until it is all but useless. This often happens long before wear causes the unit to seize or break down.

A vane motor consists of a housing with an eccentric bore, in which runs a rotor with vanes in it that slide in and out. The force differential created by the unbalanced force of the pressurized fluid on the vanes causes the rotor to spin in one direction. A critical element in vane motor design is how the vane tips are machined at the contact point between vane tip and motor housing. Several types of "lip" designs are used, and the main objective is to provide a tight seal between the inside of the motor housing and the vane, and at the same time to minimize wear and metal-to-metal contact.

GEROTOR MOTORS

The gerotor motor is in essence a rotor with N-1 teeth, rotating off center in a rotor/stator with N teeth. Pressurized fluid is guided into the assembly using a (usually) axially placed plate-type distributor valve. Several different designs exist, such as the Geroller and Nichols motors. Typically, the Gerotor motors are low-to-medium speed and medium-to-high torque.

Axial Plunger Motors

For high quality rotating drive systems plunger motors are generally used. Whereas the speed of hydraulic pumps range from 1200 to 1800 rpm, the machinery to be driven by the motor often requires a much lower speed. This means that when an axial plunger motor is used, a gearbox is usually needed. For a continuously adjustable swept volume, axial piston motors are used. **PISTON TYPE.** – Like piston (reciprocating) type pumps, the most common design of the piston type of motor is the axial. This type of motor is the most commonly used in hydraulic systems. These motors are, like their pump counterparts, available in both variable and fixed displacement designs. Typical usable rotational speeds range from below 50 rpm to above 14000 rpm. Efficiencies and minimum/maximum rotational speeds are highly dependent on the design of the rotating group, and many different types are in use.

Radial Piston Motors

Hydraulic motor Calzoni.

Radial piston motors are available in two basic types.

- The crankshaft type (*e.g.* Staffa or Sai hydraulic motors) with a single cam and the pistons pushing inwards is basically an old design but is one which has extremely high starting torque characteristics. They are available in displacements from 40 cc/rev up to about 50 litres/rev but can sometimes be limited in power output. Crankshaft type radial piston motors are capable of running at "creep" speeds and some can run seamlessly up to 1500 rpm whilst offering virtually constant output torque characteristics. This makes them still the most versatile design.

- The single-cam-type radial piston motor exists in many different designs itself. Usually the difference lies in the way the fluid is distributed to the different pistons or cylinders, and also the design of the cylinders themselves. Some motors have pistons attached to the cam using rods (much like in an internal combustion engine), while others employ floating "shoes", and even spherical contact telescopic cylinders like the Parker Denison Calzoni type. Each design has its own set of pros and cons, such as freewheeling ability, high volumetric efficiency, high reliability and so on.

- Multi-lobe cam ring types have a cam ring with multiple lobes and the piston rollers push outwardly against the cam ring. This produces a very smooth output with high starting torque but they are often limited in the upper speed range. This type of motor is available in a very wide range from about 1 litre/rev to 250 litres/rev. These motors are particularly good on low speed applications and can develop very high power.

Braking

Hydraulic motors usually have a drain connection for the internal leakage, which means that when the power unit is turned off the hydraulic motor in the drive system will move if an external load is acting on it, such as a crane or winch with suspended load. In these cases there is always a need for a brake or a locking device.

Uses

Hydraulic motors are used for many applications now such as winches and crane drives, wheel motors for military vehicles, self-driven cranes, and excavators. Conveyor and feeder drives, mixer and agitator drives, roll mills, drum drives for digesters, trommels and kilns, shredders for cars, tyres, cable and general garbage, drilling rigs, trench cutters, high-powered lawn trimmers, plastic injection machines

PNEUMATIC MOTOR

A **pneumatic motor** or **compressed air engine** is a type of motor which does mechanical work by expanding compressed air. Pneumatic motors generally convert the compressed air energy to mechanical work through either linear or rotary motion. Linear motion can come from either a diaphragm or piston actuator, while rotary motion is supplied by either a vane type air motor or piston air motor.

Pneumatic motors have existed in many forms over the past two centuries, ranging in size from hand-held turbines to engines of up to several hundred horsepower. Some types rely on pistons and cylinders; others use turbines. Many compressed air engines improve their performance by heating the incoming air or the engine itself. Pneumatic motors have found widespread success in the hand-held tool industry, and continual attempts are being made to expand their use to the transportation industry. However, pneumatic motors must overcome inefficiencies before being seen as a viable option in the transportation industry.

A **pneumatic motor** or **compressed air engine** is a type of motor which does mechanical work by expanding compressed air. Pneumatic motors generally convert the compressed air energy to mechanical work through either linear or rotary motion. Linear motion can come from either a diaphragm or piston actuator, while rotary motion is supplied by either a vane type air motor or piston air motor.

Pneumatic motors have existed in many forms over the past two centuries, ranging in size from hand-held turbines to engines of up to several hundred horsepower. Some types rely on pistons and cylinders; others use turbines. Many compressed air engines improve their performance by heating the incoming air or the engine itself. Pneumatic motors have found widespread success in the hand-held tool industry, and continual attempts are being made to expand their use to the transportation industry. However, pneumatic motors must overcome inefficiencies before being seen as a viable option in the transportation industry.

Flight

Transport category airplanes, such as commercial airliners, use compressed air starters to start the main engines. The air is supplied by the load compressor of the aircraft's auxiliary power unit, or by ground equipment.

Water rockets use compressed air to power their water jet and generate thrust, they are used as toys.

Air Hogs, a toy brand, also uses compressed air to power piston engines in toy airplanes (and some other toy vehicles).

Automotive

There is currently some interest in developing air cars. Several engines have been proposed for these, although none have demonstrated the performance and long life needed for personal transport.

Energine

The Energine Corporation was a South Korean company that claimed to deliver fully assembled cars running on a hybrid compressed air and electric engine. The compressed-air engine is used to activate an alternator, which extends the autonomous operating capacity of the car. The CEO was arrested for fraudulently promoting air motors with false claims.

Engine Air

EngineAir, an Australian company, is making a rotary engine powered by compressed air, called The Di Pietro motor. The Di Pietro motor concept is based on a rotary piston. Different from existing rotary engines, the Di Pietro motor uses a simple cylindrical rotary piston (shaft driver) which rolls, with little friction, inside the cylindrical stator.

It can be used in boat, cars, burden carriers and other vehicles. Only 1 psi (≈ 6,8 kPa) of pressure is needed to overcome the friction. The engine was also featured on the ABC's New Inventors programme in Australia on 24 March 2004.

K'Airmobiles

K'Airmobiles vehicles were intended to be commercialized from a project developed in France in 2006-2007 by a small group of researchers. However, the project has not been able to gather the necessary funds.

People should note that, meantime, the team has recognized the physical impossibility to use on-board stored compressed air due to its poor energy capacity and the thermal losses resulting from the expansion of the gas.

These days, using the patent pending 'K'Air Generator', converted to work as a compressed-gas motor, the project should be launched in 2010, thanks to a North American group of investors, but for the purpose of developing first a green energy power system.

MDI

In the original Nègre air engine, one piston compresses air from the atmosphere to mix with the stored compressed air (which will cool drastically as it expands). This mixture drives the second piston, providing the actual engine power. MDI's engine works with constant torque, and the only way to change the torque to the wheels is to use a pulley transmission of constant variation, losing some efficiency. When vehicle is stopped, MDI's engine had to be on and working, losing energy. In 2001-2004 MDI switched to a design similar to that described in Regusci's patents, which date back to 1990.

It has been reported in 2008 that Indian car manufacturer Tata was looking at an MDI compressed air engine as an option on its low priced Nano automobiles. Tata announced in 2009 that the compressed air car was proving difficult to develop due to its low range and problems with low engine temperatures.

Quasiturbine

The **Pneumatic Quasiturbine** engine is a compressed air pistonless rotary engine using a rhomboidal-shaped rotor whose sides are hinged at the vertices.

The Quasiturbine has demonstrated as a pneumatic engine using stored compressed air

It can also take advantage of the energy amplification possible from using available external heat, such as solar energy.

The Quasiturbine rotates from pressure as low as 0.1 atm (1.47psi).

Since the Quasiturbine is a pure expansion engine, while the Wankel and most other rotary engines are not, it is well-suited as a compressed fluid engine, air engine or air motor.

Regusci

Armando Regusci's version of the air engine couples the transmission system directly to the wheel, and has variable torque from zero to the maximum, enhancing efficiency. Regusci's patents date from 1990.

Team Psycho-Active

Psycho-Active is developing a multi-fuel/air-hybrid chassis which is intended to serve as the foundation for a line of automobiles. Claimed performance is 50 hp/litre. The compressed air motor they use is called the DBRE or Ducted Blade Rotary Engine.

Defunct Air Engine Designs

Conger Motor

Milton M. Conger in 1881 patented and supposedly built a motor that ran off compressed air or steam that using a **flexible tubing** which will form a wedge-shaped or inclined wall or abutment in the rear of the tangential bearing of the wheel, and propel it with greater or less speed according to the pressure of the propelling medium.

Chapter 6

CONTROL VALVES

Control valves are valves used to control conditions such as flow, pressure, temperature, and liquid level by fully or partially opening or closing in response to signals received from controllers that compare a "setpoint" to a "process variable" whose value is provided by sensors that monitor changes in such conditions. Control Valve is also termed as the **Final Control Element**.

The opening or closing of control valves is usually done automatically by electrical, hydraulic or pneumatic actuators. Positioners are used to control the opening or closing of the actuator based on electric, or pneumatic signals. These control signals, traditionally based on 3-15psi (0.2-1.0bar), more common now are 4-20mA signals for industry, 0-10V for HVAC systems, and the introduction of "Smart" systems, HART, Fieldbus Foundation, and Profibus being the more common protocols. Some of the control valve available are Reverse Double-Ported Globe-Style Valve Body,Three-Way Valve with Balanced Valve Plug, Flanged Angle-style control valve body and. Valve Body with Cage-Style Trim, Balanced Valve Plug, and Soft Seat.

A control valve consists of three main parts in which each part exist in several types and designs:

- Valve's actuator
- Valve's positioner
- Valve's body

TYPES OF CONTROL VALVE BODIES

- The most common and versatile types of control valves are sliding-stem globe and angle valves. Their popularity derives from rugged construction and the many options available that make them suitable for a variety of process applications, including severe service. Control valve bodies may be categorized as below:

Angle Valves
- o Cage-style valve bodies
- o DiskStack style valve bodies
- Angle seat piston valves
- Globe valves
  - o Single-port valve bodies
  - o Balanced-plug cage-style valve bodies
  - o High capacity, cage-guided valve bodies
  - o Port-guided single-port valve bodies
  - o Double-ported valve bodies
  - o Three-way valve bodies
- Diaphragm Valves
- Rotary valves
  - o Butterfly valve bodies
  - o V-notch ball control valve bodies
  - o Eccentric-disk control valve bodies
  - o Eccentric-plug control valve bodies
- Sliding cylinder valves
  - o Directional control valve
  - o Spool valve
  - o Piston valve
- Air-operated valves
  - o Air-operated valve
  - o Relay valve
  - o Air-operated pinch valve

Directional Control Valve

Directional control valves are one of the most fundamental parts in hydraulic machinery as well and pneumatic machinery. They allow fluid flow into different paths from one or more sources. They usually consist of a spool inside a cylinder which is mechanically or electrically controlled. The movement of the spool restricts or permits the flow, thus it controls the fluid flow.

Nomenclature

The spool (sliding type) consists of lands and grooves. The lands block oil flow through the valve body. The grooves allow oil or gas to flow around the spool and through the valve body. There are two fundamental positions of directional control valve namely **normal position** where valve returns on removal of actuating force and other is **working position** which is position of a valve when actuating force is applied. There is another class of valves with 3 or more position that can be spring centered with 2 **working position** and a **normal position**.

Classification

Directional control valves can be classified according to-

- number of ports
- number of positions
- actuating methods
- type of spool.

Example: A 5/2 directional control valve would have five ports and two spool positions.

Number of Ports

According to total number of entries or exits connected to the valve through which fluid can enter the valve or leave the valve there are types like two way, three way, four way valves.

Number of Positions

Including the normal and working positions which a valve spool can take there are types like two position, three position and proportional valves.

ACTUATING METHODS

Manually Operated

Manually operated valves work with simple levers or paddles where the operator applies force to operate the valve. Spring force is sometimes used to recover the position of valve. Some manual valves utilize either a lever or an external pneumatic or hydraulic signal to return the spool.

Mechanically Operated

Mechanically operated valves apply forces by using cams, wheels, rollers, *etc.*, hence these valves are subjected to wear.

Hydraulically Operated

A hydraulically operated DCV works at much higher pressures than its pneumatic equivalent. They must therefore be far more robust in nature so are precision machined from higher quality and strength materials.

Solenoid Operated

They are widely used in the hydraulics industry. These valves make use of electromechanical solenoids for sliding of the spool. Because simple application of electrical power provides control, these valves are used extensively. However, electrical solenoids cannot generate large forces unless supplied with large amounts of electrical power. Heat generation poses a threat to extended use of these valves when energized over time. Many have a limited duty cycle. This makes their direct acting use commonly limited to low actuating forces.

Often a low power solenoid valve is used to operate a small hydraulic valve (called the pilot) that starts a flow of fluid that drives a larger hydraulic valve that requires more force.

A bi-stable pneumatic valve is typically a pilot valve that is a 3 ported 2 position detented valve. The valve retains its position during loss of power, hence the bi-stable name.

Bi-stability can be accomplished with a mechanical detent and 2 opposing solenoids or a "magna-latch" magnetic latch with a polarity sensitive coil. Positive opens and negative closes or vice -versa. The coil is held in position magnetically when actuated.

Type of Spool

Spool is of two types namely sliding and rotary. Sliding spool is cylindrical in cross section, and the lands and grooves are also cylindrical. Rotary valves have sphere-like lands and grooves in the form of holes drilled through the spheres.

Specification

They are generally specified using the number of ports and the number of switching positions. It can be represented in general form as n_p/n_s, where n_p is the number of ports connected to the direction control valve and n_s the number of switching positions.

In addition, the method of actuation and the return method can also be specified. A hypothetical valve could be specified as 4-way, 3-position direction control valve or 4/3 DCV since there are four ports and three switching positions for the valve. In this example, one port is called the pressure port which is connected to the pump; one port is the tank port and is connected to the tank (or reservoir); and the two remaining ports are called working ports and are connected to the actuator. Apart from characteristics of valve the fluid suitable for valve,working temperature and viscosity also thought upon before selecting a particular type of valve.

Symbolic Representation

While working with layouts of hydraulic machinery it is cumbersome to draw actual picture of every valve and other components.instead of pictures symbols are used for variety of components in the hydraulic system to highlight the functional aspects. symbol for directional control valve is made of number of square boxes adjacent to each other depending on the number of positions.connections to the valve are shown on these squares by capital letters.usually they are named only in their normal position and not repeated in other positions.actuation system of the valve is also designated in its symbol.

Two Way Two Position Directional Control Valve

Gate valve is example of 2W/2P directional control valve which either turns on or off the flow in normal or working position depending on need of application. Here arrow indicates that fluid flow is taking place whereas other position shows cut-off position.

Four Way Two Position Directional Control Valve

4 way 2 position valve has four connections to it and two valve positions. One of them is normally open.

Flow Control Valve

A **flow control valve** regulates the flow or pressure of a fluid. Control valves normally respond to signals generated by independent devices such as flow meters or temperature gauges.

Control valves are normally fitted with actuators and positioners. Pneumatically-actuated globe valves and Diaphragm Valves are widely used for control purposes in many industries, although quarter-turn types such as (modified) ball, gate and butterfly valves are also used.

Control valves can also work with hydraulic actuators. These types of valves are also known as Automatic Control Valves. The hydraulic actuators will respond to changes of pressure or flow and will open/close the valve. Automatic Control Valves do not require an external power source, meaning that the fluid pressure is enough to open and close the valve. Automatic control valves include: pressure reducing valves, flow control valves, back-pressure sustaining valves, altitude valves, and relief valves. An altitude valve controls the level of a tank. The altitude valve will remain open while the tank is not full and it will close when the tanks reaches its maximum level. The opening and closing of the valve requires no external power source (electric, pneumatic, or man power), it is done automatically, hence its name.

Process plants consist of hundreds, or even thousands, of control loops all networked together to produce a product to be offered for sale. Each of these control loops is designed to keep some important process variable such as pressure, flow, level, temperature, *etc.* within a required operating range to ensure the quality of the end product. Each of these loops receives and internally creates disturbances that detrimentally affect the process variable, and interaction from other loops in the network provides disturbances that influence the process variable.

To reduce the effect of these load disturbances, sensors and transmitters collect information about the process variable and its relationship to some desired set point. A controller will then process this information and decides what must be done to get the process variable back to where it should be after a load disturbance occurs. When all the measuring, comparing, and calculating are done, some type of final control element must implement the strategy selected by the controller. The most common final control element in the process control industries is the control valve. The control valve manipulates a flowing fluid, such as gas, steam, water, or chemical compounds, to compensate for the load disturbance and keep the regulated process variable as close as possible to the desired set point.

Pressure-control Valves

Pressure-control valves are found in virtually every hydraulic system, and they assist in a variety of functions, from keeping system pressures safely below a desired upper limit to maintaining a set pressure in part of a circuit. Types include relief, reducing, sequence, counterbalance, and unloading. All of these are normally closed valves, except for reducing valves, which are normally open. For most of these valves, a restriction is necessary to produce the required pressure control. One exception is the externally piloted unloading valve, which depends on an external signal for its actuation.

Relief Valves

Most fluid power systems are designed to operate within a preset pressure range. This range is a function of the forces the actuators in the system must generate to do the required work. Without controlling or limiting these forces, the fluid power components (and expensive equipment) could be damaged. Relief valves avoid this hazard. They are the safeguards which limit maximum pressure in a system by diverting excess oil when pressures get too high.

Cracking pressure and pressure override — The pressure at which a relief valve first opens to allow fluid to flow through is known as *cracking pressure*. When the valve is bypassing its full rated flow, it is in a state of *full-flow pressure*. The difference between full-flow and cracking pressure is sometimes known as *pressure differential*, also known as *pressure override*.

In some cases, this pressure override is not objectionable. However, it can be a disadvantage if it wastes power (because of the fluid lost through the valve before reaching the maximum setting). This can further permit maximum system pressure to exceed the ratings of other components. (To minimize override, use a pilot-operated relief valve.)

Relief valves are either direct-acting or pilot-operated.

Direct-acting — A direct-acting valve may consist of a poppet or ball, held exposed to system pressure on one side and opposed by a spring of preset force on the other. In a fixed, non-adjustable, normally closed relief valve, Figure, the force exerted by the compression spring exceeds the force exerted by system pressure acting on the ball or poppet. The spring holds the ball or poppet tightly seated. A reservoir port on the spring side of the valve returns leakage fluid to tank.

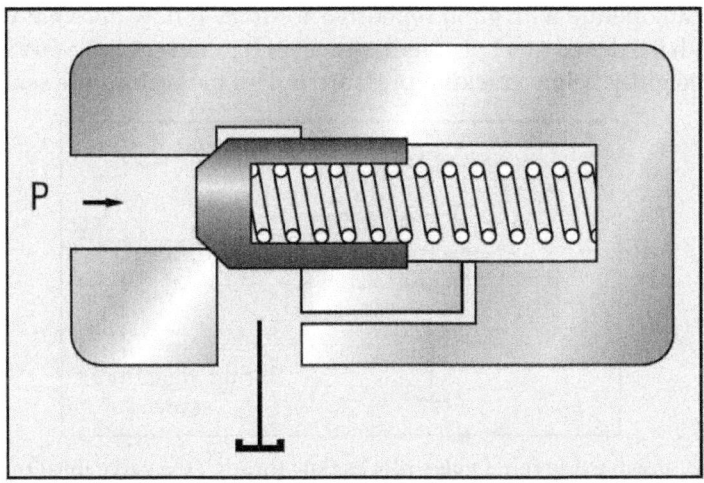

Simple, direct-acting relief valve has no adjusting screw and therefore opens at a fixed, pre-set pressure as controlled by setting of compression spring.

When system pressure begins to exceed the setting of the valve spring, the fluid unseats the ball or poppet, allowing a controlled amount of fluid to bypass to reservoir, maintaining system pressure at the valve setting. The spring re-seats the ball or poppet when enough fluid is released (bypassed) to drop system pressure below the setting of the valve spring.

Because the usefulness of a fixed relief valve is limited to the single setting of its spring, most relief valves are adjustable. This is commonly achieved with an adjusting screw acting on the spring, Figure. By turning the screw in or out, the operator compresses or decompresses the spring respectively. The valve can be set to open at any pressure within a desired range. Aside from the adjustable feature, this valve works just like the fixed valve in Figure.

Spring-loaded *poppet valves* are generally used for small flows. They don't leak below cracking pressure and respond rapidly, making them ideal for relieving shock pressures. They often are used as safety valves to prevent damage to components from high surge pressures, or to relieve pressure caused by thermal expansion in locked cylinders. The differential between cracking and full open pressure on spring-loaded poppet relief valves is high. For this reason they are not recommended for precise pressure control.

Relief valves are also made to relieve flow in either direction. Fluid pressure at the other port acts on a shoulder on the plunger to open the valve. Another type of direct-acting relief valve has a guided piston. In this valve a sliding piston, instead of a poppet, connects the pressure and reservoir ports. System pressure acts on the piston and moves it against a spring force. As the piston moves, it uncovers a reservoir port in the valve body.

These valves have a fast response but may be prone to chatter. They can be damped to eliminate chatter, but this also slows their reaction time. They are reliable and can operate with good repetitive accuracy if flow does not vary widely. Valves with hardened-steel pistons and sleeves have a very long service life. They may leak slightly below cracking pressure unless the pistons are sealed.

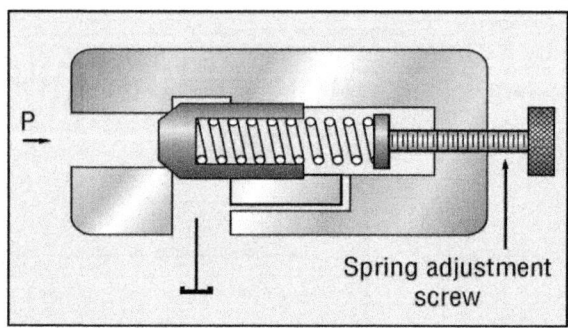

Adjustable, direct-acting relief valve blocks flow through the valve until force of system pressure on the poppet overcomes the adjustable spring force and downstream pressure.

Guided-piston relief valves generally are used for pressures below 800 psi, although they can be made with heavier springs for higher pressures. The heavier springs give the valve a greater differential and consequently increase the size of the valve.

A variation of the guided-piston relief valve is the *differential-piston relief valve*. Here, the pressure acts on an annular area. This annular area is smaller than the valve's seat area. This permits the use of a lighter spring than would be needed if pressure acted on the entire seat area. These valves have a lower pressure differential than poppet or guided-piston relief valves.

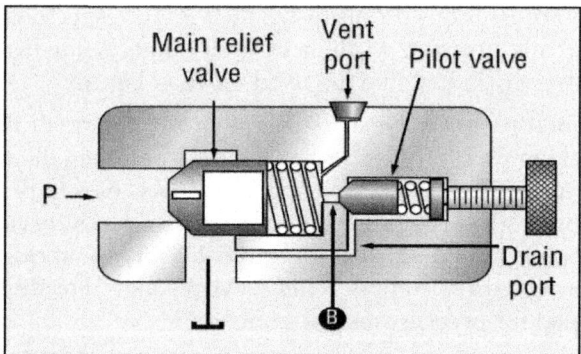

Pilot-operated relief valve has orifice through piston, which is held closed by force of light spring and system pressure acting on larger piston area at spring end.

Pilot-operated Relief Valves

For applications requiring valves that must relieve large flows with small pressure differential, pilot-operated relief valves are often used, Figure. The pilot-operated relief valve operates in two stages. A pilot stage, which consists of a small, spring-biased relief valve (generally built into the main relief valve), acts as a trigger to control the main relief valve. However, the pilot may also be located remotely and connected to the main valve with pipe or tubing.

The main relief valve is normally closed when the pressure of the inlet is below the setting of the main valve spring. Orifice B in the main valve, Figure, permits system pressure fluid to act on a larger area on the spring side of the poppet so that the sum of this force and that of the main spring keep the poppet seated. At this time, the pilot valve is also closed. Pressure in passage B is the same as system pressure and is less than the setting of the pilot valve spring.

As system pressure rises, the pressure in passage Brises as well, and, when it reaches the setting of the pilot valve, the pilot valve opens. Oil is released behind the main valve through passage B through the drain port. The resulting pressure drop across orifice A in the main relief valve opens it and excess oil flows to tank, preventing any further rise in inlet pressure. The valves close again when inlet oil pressure drops below the valve setting. Pilot-operated relief valves have less pressure override than direct-acting relief valves, such as in Figure.

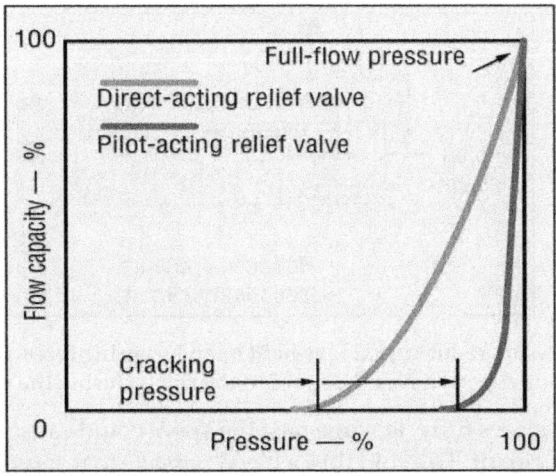

Comparison of action of relief valves at cracking and full flow pressure.

Because these valves do not start opening until the system reaches 90% of full pressure, the efficiency of the system is protected because less oil is released. These valves are best suited for high-pressure, high-volume applications. Although their operation is slower than that of direct-acting relief valves, pilot-operated relief valves maintain a system at a more constant pressure while relieving. Figure plots the operating characteristics of direct-acting and pilot-operated relief valves.

Pressure-reducing Valves

The most practical components for maintaining secondary, lower pressure in a hydraulic system are pressure-reducing valves. Pressure-reducing valves are normally open, 2-way valves that close when subjected to sufficient downstream pressure. There are two types: direct acting and pilot operated.

Direct acting — A pressure-reducing valve limits the maximum pressure available in the secondary circuit regardless of pressure changes in the main circuit. This assumes the work load generates no back flow into the reducing valve port in which case the valve will close, Figure. The pressure-sensing signal comes from the downstream side (secondary circuit). This valve, in effect, operates in reverse fashion from a relief valve (which senses pressure from the inlet and is normally closed).

As pressure rises in the secondary circuit, Figure, hydraulic force acts on area A of the valve, closing it partly. Spring force opposes the hydraulic force, so that only enough oil flows past the valve to supply the secondary circuit at the desired pressure. The spring setting is adjustable.

When outlet pressure reaches that of the valve setting, the valve closes except for a small quantity of oil that bleeds from the low-pressure side of the valve, usually through an orifice in the spool, through the spring chamber, to reservoir.

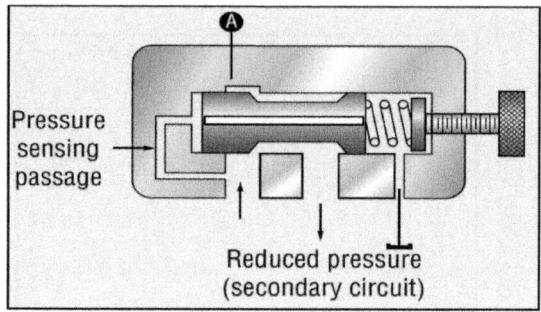

Direct-acting, pressure reducing valve is held open by spring force. Increasing pressure at outlet port moves the spool to the right, closing the valve.

If the valve closes fully, leakage past the spool could cause pressure buildup in the secondary circuit. To avoid this, a bleed passage to reservoir keeps it slightly open, preventing a rise in downstream pressure above the valve setting. The drain passage returns leakage flow to tank.

Constant and Fixed Pressure Reduction

Constant-pressure-reducing valves supply a preset pressure, regardless of main circuit pressure, as long as pressure in the main circuit is *higher* than that in the secondary. These valves balance secondary-circuit pressure against the force exerted by an adjustable spring which tries to open the valve. When pressure in the secondary circuit drops, spring force opens the valve enough to increase pressure and keep a constant reduced pressure in the secondary circuit.

Fixed pressure reducing valves supply a *fixed amount of pressure reduction* regardless of the pressure in the main circuit. For instance, assume a valve is set to provide reduction of 250 psi. If main system pressure is 2750 psi, reduced pressure will be 2500 psi; if main pressure is 2000 psi, reduced pressure will be 1750 psi.

This valve operates by balancing the force exerted by the pressure in the main circuit against the sum of the forces exerted by secondary circuit pressure and the spring. Because the pressurized areas on both sides of the poppet are equal, the fixed reduction is that exerted by the spring.

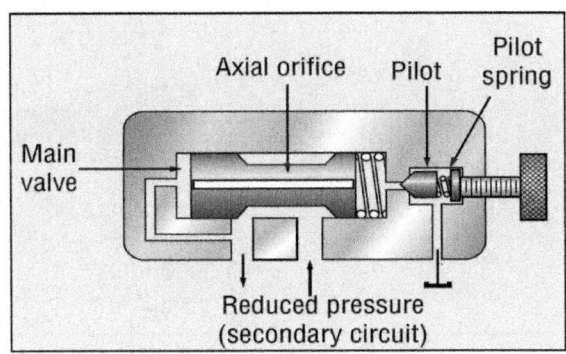

Pilot-operated, pressure reducing valve has reduced pressure on both ends of the spool. A light spring holds the spool open.

Pilot-operated pressure reducing valves — The spool in a pilot-operated, pressure-reducing valve is balanced hydraulically by downstream pressure at both ends, Figure. A light spring holds the valve open. A small pilot relief valve, usually built into the main valve body, relieves fluid to tank when reduced pressure reaches the pilot valve's spring setting. This fluid flow causes a pressure drop across the spool. Pressure differential then shifts the spool toward its closed position against the light spring force.

The pilot valve relieves only enough fluid to position the main valve spool or poppet so that flow through the main valve equals the flow requirements of the reduced pressure circuit. If no flow is required in the low-pressure circuit during a portion of the cycle, the main valve closes. Leakage of high-pressure fluid into the reduced-pressure section of the valve then returns to the reservoir through the pilot operated relief valve.

Pilot-operated pressure reducing valves generally have a wider range of spring adjustment than direct-acting valves and provide better repetitive accuracy. However, oil contamination can block flow to the pilot valve and the main valve will fail to close properly. Pilot-operated valves with built-in reduced pressure system relieving capability also are available.

Sequence Valves

In circuits with more than one actuator, it is often necessary to drive the actuators, such as cylinders, in a definite order or sequence. One way to do this is with limit switches, timers, or other electrical control devices.

Sometimes, this result can also be achieved by sizing cylinders according to the load they must displace. The cylinder requiring the least pressure to move its load extends first. At the end of its stroke, system pressure increases and extends the second cylinder. This continues until all cylinders are actuated.

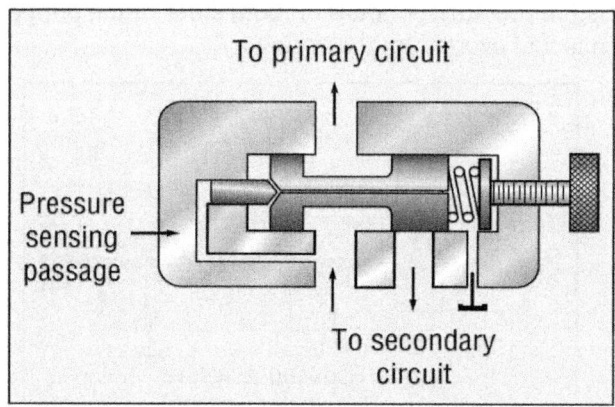

Sequence valve is a 2-way valve held closed by an adjustable spring and opened by pressure at the inlet port acting on the left of the spool.

However, in many installations, space limitations and force requirements determine the cylinder size needed to do the job. In this case, sequence valves can be used to actuate the cylinders in the required order.

Sequence valves are normally closed, 2-way valves. They regulate the sequence in which various functions in a circuit occur, Figure. They resemble direct-acting relief valves except that their spring chambers are generally drained externally to reservoir, instead of internally to the outlet port, as in a relief valve.

A sequence valve usually permits pressurized fluid to flow to a second function only after an earlier, priority function has been completed and satisfied. When normally closed, a sequence valve allows fluid to flow freely to the primary circuit, to perform its first function until the pressure setting of the valve is reached.

When the primary function is satisfied, pressure in the primary circuit rises and is sensed in pressure-sensing passage A. This pressurizes the spool and overcomes the force exerted by the spring. The spring is compressed, the valve spool shifts, and oil flows to the secondary circuit.

Sequence valves sometimes have check valves, which permit reverse flow from the secondary to the primary circuit. However, sequencing action is provided only when the flow is from the primary to the secondary circuit.

In some applications, an interlock can prevent sequencing occurring until the primary actuator reaches a certain position. This is done with remote operations.

Counterbalance Valves

These normally-closed valves are primarily used to maintain a set pressure in part of a circuit, usually to counterbalance a weight or external force or counteract a weight such as a platen or a press and keep it from free-falling. The valve's primary port is connected to the cylinder's rod end, and the secondary port to the directional control valve, Figure. The pressure setting is slightly higher than that required to keep the load from free-falling.

When pressurized fluid flows to the cylinder's cap end, the cylinder extends, increasing pressure in the rod end, and shifting the main spool in the counterbalance valve.

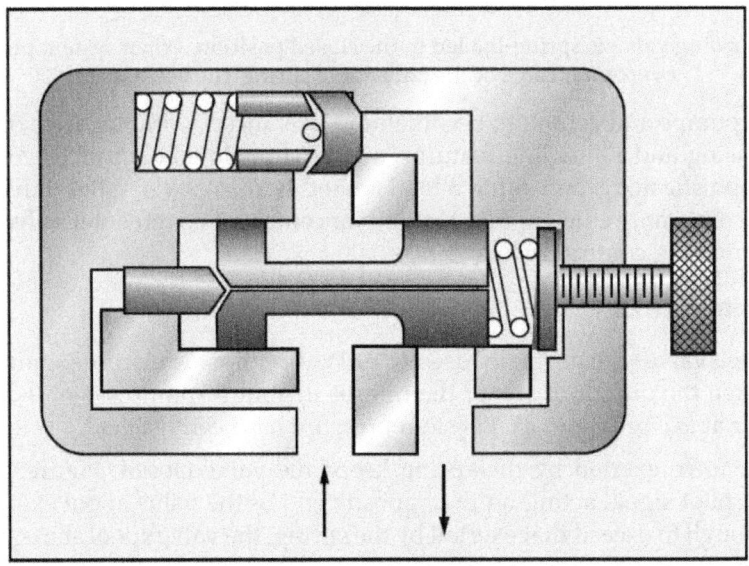

Counterbalance valve stops flow from its inlet port to its outlet port until pressure at the inlet port overcomes adjusting spring force.

This creates a path which permits fluid to flow through the secondary port to the directional control valve and to reservoir. As the load is raised, the integral check valve opens to allow the cylinder to retract freely.

If it is necessary to relieve back pressure at the cylinder, and increase the force at the bottom of the stroke, the counterbalance valve can be operated remotely.

Counterbalance valves are usually drained internally. When the cylinder extends, the valve must open and its secondary port is connected to reservoir. When the cylinder retracts, it matters little that load pressure is felt in the drain passage because the check valve bypasses the valve's spool.

Over Center Valves

Overcenter valves resemble counterbalance valves in that their purpose is to maintain a set pressure opposite a load, to keep it from free-falling. The main difference is that an overcenter valve uses a pilot signal, usually from the inlet of the actuator, to assist in opening the spool. This pilot assist makes the overcenter valve more efficient, and reduces the horsepower requirement and heat generation within the system.

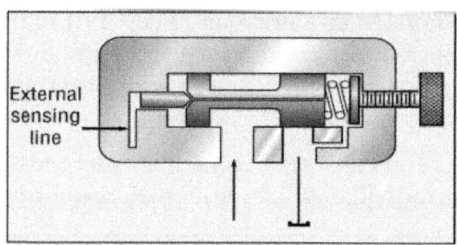

Unloading valve is spring-loaded to the closed position. When system pressure overcomes force of the adjustable spring, the valve opens.

As pumps and actuators become more advanced, with negative or positive load sensing and unloading features, and as directional control valves become more sophisticated, controlling a load smoothly using overcenter valves has, in turn, become more challenging. New advancements in overcenter valve technology are making control easier.

Unloading Valves

These valves are normally used to unload pumps. They direct pump output flow (often the output of one of the pumps in a multi-pump system) directly to reservoir at *low* pressure, *after* system pressure has been reached.

The force exerted by the spring keeps the valve closed, Figure. When an external pilot signal acting on the opposite end of the valve spool exerts a force large enough to exceed that exerted by the spring, the valve spool shifts, diverting pump output to reservoir at low pressure.

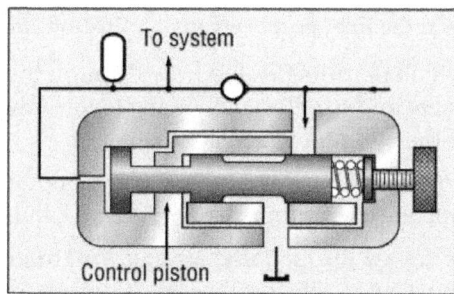

Piloted unloading valve has piston with pump pressure at both ends.

High-low circuits which use two pumps for traverse and speed, or clamping, depend on unloading valves to improve efficiency. Output from both pumps is needed only for fast traverse. During feed or clamping, output from the large pump is unloaded to tank at low pressure.

Piloted unloading valves — Unloading valves are also made with a pilot to control the main valve, Figure. A port through the main valve plunger allows system pressure to act on both ends of the plunger. A light spring plus system pressure acting on the larger area at the spring end of the plunger holds the valve closed. A built-in check valve maintains system pressure. When system pressure drops to a preset value, the pilot valve closes. Pump flow through the port in the main valve spool closes the valve.

Check Valve

A **check valve, clack valve, non-return valve** or **one-way valve** is a valve that normally allows fluid (liquid or gas) to flow through it in only one direction.

Check valves are two-port valves, meaning they have two openings in the body, one for fluid to enter and the other for fluid to leave. There are various types of check valves used in a wide variety of applications. Check valves are often part of common household items. Although they are available in a wide range of sizes and costs, check valves generally are very small, simple, or inexpensive. Check valves work automatically and most are not controlled by a person or any external control; accordingly, most do not have any valve handle or stem. The bodies of most check valves are made of plastic or metal.

An important concept in check valves is the cracking pressure which is the minimum upstream pressure at which the valve will operate. Typically the check valve is designed for and can therefore be specified for a specific cracking pressure.

Heart valves are essentially inlet and outlet check valves for the heart ventricles, since the ventricles act as pumps.

Technical Data

Cracking pressure—the inlet pressure at which the first indication of flow occurs (steady stream of bubbles). Reseal pressure—the pressure at which there is no indication of flow. Back pressure—the differential pressure between the inlet and outlet pressures.

Types of Check Valves

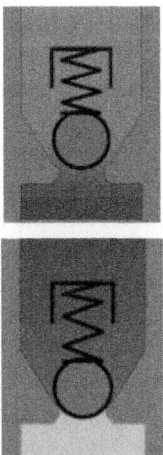

A ball check valve in the open position to allow forward flow and closed
position to block reverse flow

A **ball check valve** is a check valve in which the closing member, the movable part to block the flow, is a spherical ball. In some ball check valves, the ball is spring-loaded to help keep it shut. For those designs without a spring, reverse flow is required to move the ball toward the seat and create a seal. The interior surface of the main seats of ball check valves are more or less conically-tapered to guide the ball into the seat and form a positive seal when stopping reverse flow.

Ball check valves are often very small, simple, and cheap. They are commonly used in liquid or gel minipump dispenser spigots, spray devices, some rubber bulbs for pumping air, *etc.*, manual air pumps and some other pumps, and refillable dispensing syringes. Although the balls are most often made of metal, they can be made of other materials, or in some specialized cases out of artificial ruby. High pressure HPLC pumps and similar applications commonly use small inlet and outlet ball check valves with both balls and seats made of artificial ruby, for both hardness and chemical resistance. After prolonged use, such check valves can eventually wear out or the seat can develop a crack, requiring replacement. Therefore, such valves are made to be replaceable, sometimes placed in a small plastic body tightly-fitted inside a metal fitting which can withstand high pressure and which is screwed into the pump head.

There are similar check valves where the disc is not a ball, but some other shape, such as a poppet energized by a spring. Ball check valves should not be confused with ball valves, which is a different type of valve in which a ball acts as a controllable rotor to stop or direct flow.

A **diaphragm check valve** uses a flexing rubber diaphragm positioned to create a normally-closed valve. Pressure on the upstream side must be greater than the pressure on the downstream side by a certain amount, known as the pressure differential, for the check valve to open allowing flow. Once positive pressure stops, the diaphragm automatically flexes back to its original closed position.

A **swing check valve** or **tilting disc check valve** is check valve in which the disc, the movable part to block the flow, swings on a hinge or trunnion, either onto the seat to block reverse flow or off the seat to allow forward flow. The seat opening cross-section may be perpendicular to the centerline between the two ports or at an angle. Although swing check valves can come in various sizes, large check valves are often swing check valves. The flapper valve in a flush-toilet mechanism is an example of this type of valve. Tank pressure holding it closed is overcome by manual lift of the flapper. It then remains open until the tank drains and the flapper falls due to gravity. Another variation of this mechanism is the clapper valve, used in applications such firefighting and fire life safety systems. A hinged gate only remains open in the inflowing direction. The clapper valve often also has a spring that keeps the gate shut when there is no forward pressure. Another example is the backwater valve that protects against flooding caused by return flow of sewage waters. Such risk occurs most often in sanitary drainage systems connected to combined sewerage systems and in rainwater drainage systems. It may be caused by intense rainfall, thaw or flood.

A **stop-check valve** is a check valve with override control to stop flow regardless of flow direction or pressure. In addition to closing in response to backflow or insufficient forward pressure (normal check-valve behavior), it can also be deliberately shut by an external mechanism, thereby preventing any flow regardless of forward pressure.

A **lift-check valve** is a check valve in which the disc, sometimes called a *lift*, can be lifted up off its seat by higher pressure of inlet or upstream fluid to allow flow to the outlet or downstream side. A guide keeps motion of the disc on a vertical line, so the valve can later reseat properly. When the pressure is no longer higher, gravity or higher downstream pressure will cause the disc to lower onto its seat, shutting the valve to stop reverse flow.

An **in-line check valve** is a check valve similar to the lift check valve. However, this valve generally has a spring that will 'lift' when there is pressure on the upstream side of the valve. The pressure needed on the upstream side of the valve to overcome the spring tension is called the 'cracking pressure'. When the pressure going through the valve goes below the cracking pressure, the spring will close the valve to prevent back-flow in the process.

A **duckbill valve** is a check valve in which flow proceeds through a soft tube that protrudes into the downstream side. Back-pressure collapses this tube, cutting off flow.

Multiple check valves can be connected in series. For example, a double check valve is often used as a backflow prevention device to keep potentially contaminated water from siphoning back into municipal water supply lines. There are also *double ball check valves* in which there are two ball/seat combinations sequentially in the same body to ensure positive leak-tight shutoff when blocking reverse flow; and piston check valves, wafer check valves, and ball-and-cone check valves.

Applications

Pumps

Check valves are often used with some types of pumps. Piston-driven and diaphragm pumps such as metering pumps and pumps for chromatography commonly use inlet and outlet ball check valves. These valves often look like small cylinders attached to the pump head on the inlet and outlet lines. Many similar pump-like mechanisms for moving volumes of fluids around use check valves such as ball check valves. The feed pumps or injectors which supply water to steam boilers are fitted with check valves to prevent back-flow.

Industrial Processes

Check valves are used in many fluid systems such as those in chemical and power plants, and in many other industrial processes.

Check valves are also often used when multiple gases are mixed into one gas stream. A check valve is installed on each of the individual gas streams to prevent mixing of the gases in the original source. For example, if a fuel and an oxidizer are to be mixed, then check valves will normally be used on both the fuel and oxidizer sources to ensure that the original gas cylinders remain pure and therefore nonflammable.

Domestic Use

Some types of irrigation sprinklers and drip irrigation emitters have small check valves built into them to keep the lines from draining when the system is shut off.

Check valves used in domestic heating systems to prevent vertical convection, especially in combination with solar thermal installations, also are called gravity brake.

Rainwater harvesting systems that are plumbed into the main water supply of a utility provider may be required to have one or more check values fitted to prevent contamination of the primary supply by rainwater.

Hydraulic jacks use ball check valves to build pressure on the lifting side of the jack.

History

Frank P. Cotter developed a "simple self sealing check valve, adapted to be connected in the pipe connections without requiring special fittings and which may be readily opened for inspection or repair" 1907 (U.S. patent #865,631).

Nikola Tesla invented a deceptively simple one-way valve for fluids in 1916, called a Tesla valve. It was patented in 1920 (U.S. patent 1,329,559).

Electrohydraulic Servo Valve

An **electrohydraulic servo valve** (**EHSV**) is an electrically operated valve that controls how hydraulic fluid is ported to an actuator. Servo valves and servo-proportional valves are operated by transforming a changing analogue or digital input signal into a smooth set of movements in a hydraulic cylinder. Servo valves can provide precise control of position, velocity, pressure and force with good post movement damping characteristics.

Control

A low voltage is used to control the servo valve. The control voltage is passed into an amplifier which provides the power to alter the valve's position. The valve will then deliver a measured amount of fluid power to an actuator. The use of a feedback transducer on the actuator returns an electrical signal to the amplifier to condition the strength of the voltage to the servo valve.

Examples of Usage

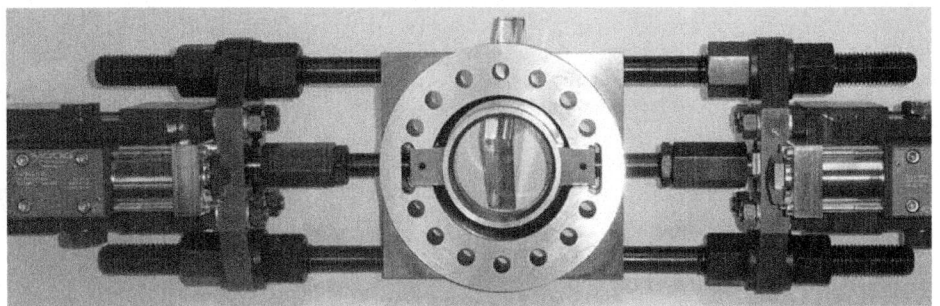

The twin Moog servo valves are used to deform the shape of the die on
this blow molding accessory designed by BMC Limited

One example of servo valve use is in blow molding where the servo valve
controls the wall thickness of extruded plastic making up the bottle or container
by use of a deformable die. The mechanical feedback has been replaced by an
electric feedback with a position transducer. Integrated electronics close the position loop for the spool. These valves are suitable for electrohydraulic position,
velocity, pressure or force control systems with extremely high dynamic response
requirements. Another example is the regulation of fuel flow into a turbofan engine
governed by FADEC. One such example is Honeywell's servo valve which is part
of the fuel control mechanism for the CFM International CFM56 engine powering
the Boeing 737 NG and Airbus A320 passenger aircraft.

Principle of Operation

An electric command signal (flow rate set point) is applied to the integrated
position controller which drives the pilot stage. The thereby deflected nozzle
flapper system produces a pressure difference across the drive areas of the spool
and effects its movement. The position transducer (LVDT) which is excited *via* an
oscillator measures the position of the spool (actual value, position voltage). This
signal is then demodulated and fed back to the controller where it is compared
with the command signal. The controller drives the pilot stage until the error
between command signal and feedback signal will be zero. Thus the position of
the spool is proportional to the electric command signal.

Chapter 7

CIRCUITS

An electrical circuit is a device that uses electricity to perform a task, such as run a vacuum or power a lamp. The circuit is a closed loop formed by a power source, wires, a fuse, a load, and a switch. Electricity flows through the circuit and is delivered to the object it is powering, such as the vacuum motor or lightbulb, after which the electricity is sent back to the original source; this return of electricity enables the circuit to keep the electricity current flowing. Three types of electrical circuits exist: the series circuit, the parallel circuit, and the series-parallel circuit; depending on the circuit type, it may be possible for electricity to continue flowing should a circuit stop working. Two concepts, Ohm's Law and source voltage, can affect the amount of electricity flowing through a circuit, and therefore, how well an electrical circuit functions.

HOW IT WORKS

Most devices that run on electricity contain an electrical circuit; when connected to a source of power, such as being plugged into an electrical outlet, electricity can run through the electrical circuit within the device and then return to the original power source, to continue the flow of electricity. In other words, when a power switch is turned on, the electrical circuit is complete and current flows from the positive terminal of the power source, through the wire to the load, and finally to the negative terminal. Any device that consumes the energy flowing through a circuit and converts that energy into work is called a load. A light bulb is one example of a load; it consumes the electricity from a circuit and converts it into work — heat and light.

TYPES OF CIRCUITS

A series circuit is the simplest because it has only one possible path that the electrical current may flow; if the electrical circuit is broken, none of the load devices will work. The difference with parallel circuits is that they contain more than one path for electricity to flow, so if one of the paths is broken, the other paths will continue to work. A series-parallel circuit, however, is a combination of the first two: it attaches some of the loads to a series circuit and others to parallel circuits. If the series circuit breaks, none of the loads will function, but if one of the parallel circuits breaks, that parallel circuit and the series circuit will stop working, while the other parallel circuits will continue to work.

OHM'S LAW

Many "laws" apply to electrical circuits, but Ohm's Law is probably the most well known. Ohm's Law states that an electrical circuit's current is directly proportional to its voltage, and inversely proportional to its resistance. So, if voltage increases, for example, the current will also increase, and if resistance increases, current decreases; both situations directly influence the efficiency of electrical circuits. To understand Ohm's Law, it's important to understand the concepts of current, voltage, and resistance: current is the flow of an electric charge, voltage is the force that drives the current in one direction, and resistance is the opposition of an object to having current pass through it. The formula for Ohm's Law is $E = I \times R$, where E = voltage in volts, I = current in amperes, and R = resistance in ohms; this formula can be used to analyze the voltage, current, and resistance of electricity circuits.

SOURCE VOLTAGE

Another important concept in regards to electrical circuits, source voltage refers to the amount of voltage that is produced by the power source and applied to the circuit. In other words, source voltage depends on how much electricity a circuit will receive. Source voltage is affected by the amount of resistance within the electrical circuit; it can also affect the amount of current, as the current is typically affected by both voltage and resistance. Resistance is not affected by voltage or current, however, but can reduce the amounts of both voltage and current to a electrical circuits.

HYDRAULIC CIRCUIT

A **hydraulic circuit** is a system comprising an interconnected set of discrete components that transport liquid. The purpose of this system may be to control where fluid flows (as in a network of tubes of coolant in a thermodynamic system) or to control fluid pressure (as in hydraulic amplifiers). The approach of describing a fluid system in terms of discrete components is inspired by the success of electrical circuit theory. Just as electric circuit theory works when elements are discrete and linear, hydraulic circuit theory works best when the elements (passive component such as pipes or transmission lines or active components such as power packs or pumps) are discrete and linear. This usually means that hydraulic circuit analysis works best for long, thin tubes with discrete pumps, as found in chemical process flow systems or microscale devices.

Components

The circuit comprises the following components:

- Active components
  - o Hydraulic power pack
- Transmission lines
  - o Hydraulic hoses
- Passive components
  - o Hydraulic cylinders

PNEUMATIC CIRCUIT

A **pneumatic circuit** is an interconnected set of components that convert compressed gas (usually air) into mechanical work. In the normal sense of the term, the circuit must include a compressor or compressor-fed tank.

Components

The circuit comprises the following components:

- Active components
  - o Compressor
- Transmission lines
  - o Air tank
  - o Pneumatic hoses
  - o Open atmosphere (for returning the spent gas to the compressor)
  - o Valves
- Passive components
  - o Pneumatic cylinders
  - o Service Unit
- FRL - Filter Regulator and Lubricator

Pneumatic Cylinder

In general based on application pneumatic cylinder is selected which are single acting cylinder, it will have single port in cylinder were extension of cylinder is by compressed air and retraction by means of open coiled spring. In double acting cylinder two ports are available both extension and retraction by means compressed air

Direction Control Valve (DCV)

The direction control valve is used to control the direction of flow of compressed air. Usually classified into normally open (NO)and normally closed (NC)

valves. The normally open valves will permit flow from inlet port of valve to outlet port normally the flow will be cut by changing the position of the valve. The normally closed valves will not permit flow from inlet port of valve to outlet port normally the flow will be permitted only by changing the position of the valve. In general valves are designated as 2/2 DCV, 3/2DCV, 5/2 DCV,5/3 DCV etc. In which the first numerical indicates number of ports and second numerical indicates number of position To change the position, the valves are generally actuated by:

- Pedal Operated
- Push button operated
- Spring operated
- Solenoid operated
- By using Pneumatic source itself etc.

The other auxiliary valves are:

- Two pressure valve (And Valve): Usually two valve actuators are used when both the push buttons are pressed at a time the air flow takes place if either any one is pressed at a time air flow will not take place in valve outlet. Generally used in mechanical press and machine tools to ensure operator's hands are safe during operation.
- OR Valve : Usually two valve actuators are used when either one push button is pressed the air flow takes place.
- Check valve: Allows air flow in one direction
- Quick exhaust valve: The valve construction is OR valve with exhaust port,ensures quick return of cylinder therefore cycle time reduces
- Flow control valve:
- Time delay valve:
- Pressure relief valve etc.

The following devices operate using compressed gases, but are not normally thought of as being pneumatic circuits:

- Guns
- Rockets
- Refrigerators
- Internal combustion engines
- Scuba sets

CIRCUIT DIAGRAM

A **circuit diagram** (also known as an **electrical diagram, elementary diagram,** or **electronic schematic**) is a simplified conventional graphical representation of an electrical circuit. A pictorial circuit diagram uses simple images of components, while a schematic diagram shows the components of the circuit as simplified standard symbols; both types show the connections between the devices, including power and signal connections. Arrangement of the components interconnections on the diagram does not correspond to their physical locations in the finished device.

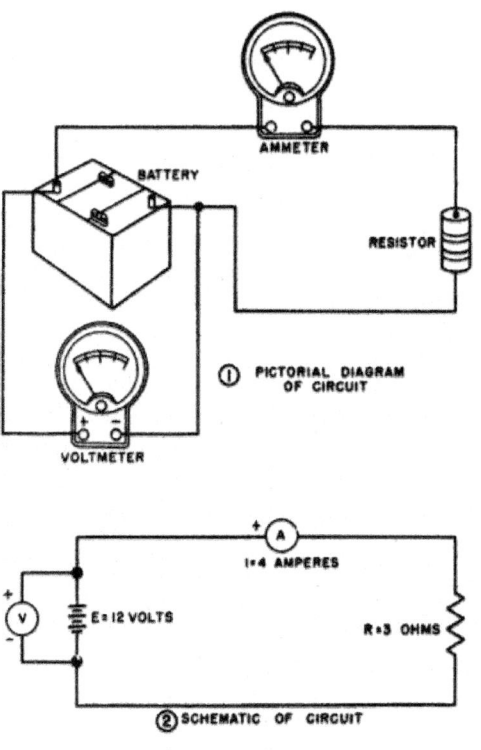

Comparison of pictorial and schematic styles of circuit diagrams

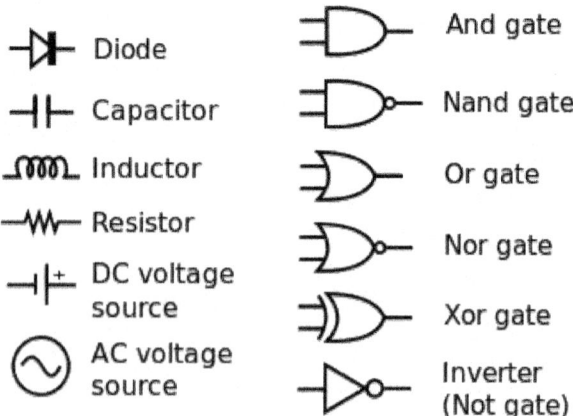

Common schematic diagram symbols

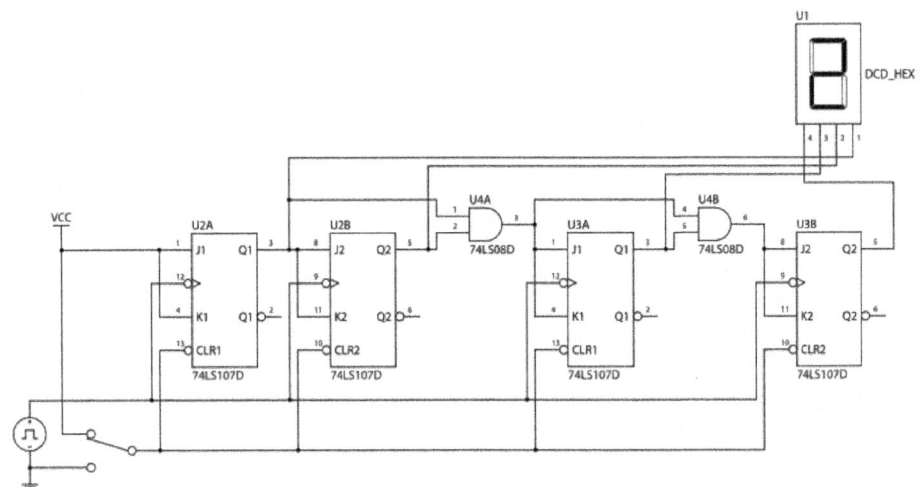

The circuit diagram for a four-bit TTL counter, a type of state machine.

Unlike a block diagram or layout diagram, a circuit diagram shows the actual wire connections being used. The diagram does not show the physical arrangement of components. A drawing meant to depict what the physical arrangement of the wires and the components they connect is called "artwork" or "layout" or the "physical design."

Circuit diagrams are used for the design (circuit design), construction (such as PCB layout), and maintenance of electrical and electronic equipment.

In computer science, circuit diagrams are especially useful when visualizing different expressions using Boolean algebra.

Symbols

Circuit diagrams are pictures with symbols that have differed from country to country and have changed over time, but are now to a large extent internationally standardized. Simple components often had symbols intended to represent some feature of the physical construction of the device. For example, the symbol for a resistor shown here dates back to the days when that component was made from a long piece of wire wrapped in such a manner as to not produce inductance, which would have made it a coil. These wirewound resistors are now used only in high-power applications, smaller resistors being cast from *carbon composition* (a mixture of carbon and filler) or fabricated as an insulating tube or chip coated with a metal film. The internationally standardized symbol for a resistor is therefore now simplified to an oblong, sometimes with the value in ohms written inside, instead of the zig-zag symbol. A less common symbol is simply a series of peaks on one side of the line representing the conductor, rather than back-and-forth as shown here.

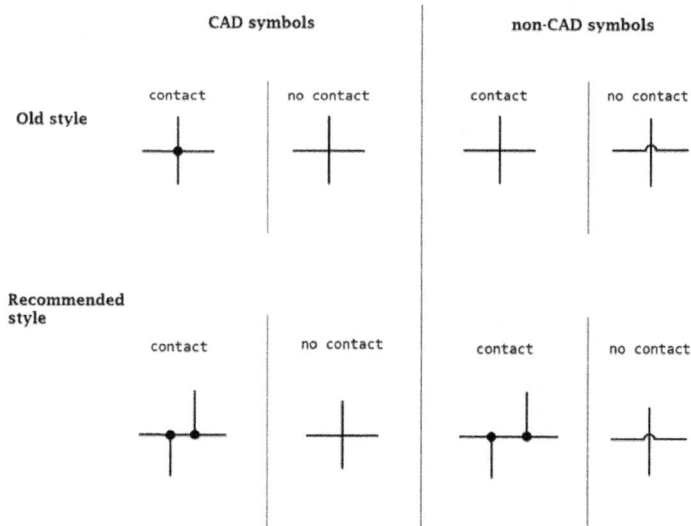

Wire Crossover Symbols for Circuit Diagrams. Note that the CAD symbol for insulated crossing wires is exactly the same as the older, non-CAD symbol for non-insulated crossing wires. To avoid confusion, the wire "jump" (semi-circle) symbol for insulated wires in non-CAD schematics is recommended (as opposed to using the CAD-style symbol for no connection), so as to avoid confusion with the original, older style symbol, which means the exact opposite. The newer, recommended style for 4-way wire connections in both CAD and non-CAD schematics is to stagger the joining wires into T-junctions.

The linkages between leads were once simple crossings of lines. With the arrival of computerized drafting, the connection of two intersecting wires was shown by a crossing of wires with a "dot" or "blob" to indicate a connection. At the same time, the crossover was simplified to be the same crossing, but without a "dot". However, there was a danger of confusing the wires that were connected and not connected in this manner, if the dot was drawn too small or accidentally omitted (*e.g.* the "dot" could disappear after several passes through a copy machine). As such, the modern practice for representing a 4-way wire connection is to draw a straight wire and then to draw the other wires staggered along it with "dots" as connections, so as to form two separate T-junctions that brook no confusion and are clearly not a crossover.

For crossing wires that are insulated from one another, a small semi-circle symbol is commonly used to show one wire "jumping over" the other wire (similar to how jumper wires are used).

A common, hybrid style of drawing combines the T-junction crossovers with "dot" connections and the wire "jump" semi-circle symbols for insulated crossings. In this manner, a "dot" that is too small to see or that has accidentally disappeared can still be clearly differentiated from a "jump".

On a circuit diagram, the symbols for components are labelled with a descriptor or reference designator matching that on the list of parts. For example, C1 is the first capacitor, L1 is the first inductor, Q1 is the first transistor, and R1 is the first resistor (note that this is not written as a subscript, as in R_1, L_1,...). Often the value or type designation of the component is given on the diagram beside the part, but detailed specifications would go on the parts list.

Organization of Drawings

It is a usual although not universal convention that schematic drawings are organized on the page from left to right and top to bottom in the same sequence as the flow of the main signal or power path. For example, a schematic for a radio receiver might start with the antenna input at the left of the page and end with the loudspeaker at the right. Positive power supply connections for each stage would be shown towards the top of the page, with grounds, negative supplies, or other return paths towards the bottom. Schematic drawings intended for maintenance may have the principle signal paths highlighted to assist in understanding the signal flow through the circuit. More complex devices have multi-page schematics and must rely on cross-reference symbols to show the flow of signals between the different sheets of the drawing.

Detailed rules for the preparation of circuit diagrams (and other document kinds used in electrotechnology) are provided in the International standard IEC 61082-1.

Relay logic line diagrams (also called ladder logic diagrams) use another common standardized convention for organizing schematic drawings, with a vertical power supply "rail" on the left and another on the right, and components strung between them like the rungs of a ladder.

Artwork

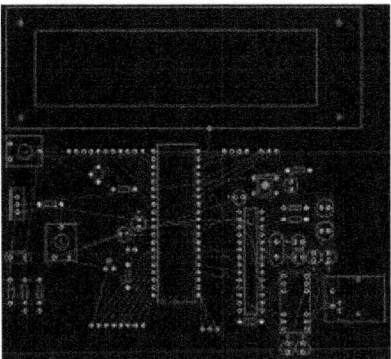

A rat's nest.

Once the schematic has been made, it is converted into a layout that can be fabricated onto a printed circuit board (PCB). The layout is usually started by the process of schematic capture. The result is what is known as a rat's nest. The rat's nest is a jumble of wires (lines) criss-crossing each other to their destination nodes. These wires are routed either manually or by the use of electronics design automation (EDA) tools. The EDA tools arrange and rearrange the placement of components and find paths for tracks to connect various nodes. This results in the final layout artwork for the integrated circuit or printed circuit board.

A generalized design flow would be as follows:

Schematic → Schematic capture → Rat's nest → Routing → Artwork → PCB development & etching → Component mounting → Testing

Chapter 8

PNEUMATIC LOGIC CIRCUITS

Electrical and electronic devices, such as relay logic circuits, programmable controllers, or computers, normally control fluid power circuits. Fluid power systems can also be controlled with "Air Logic." These controls perform any function normally handled by relays, pressure or vacuum switches, time delays, counters, and limit switches. While the circuitry is similar, compressed air is the control medium instead of electrical current.

Environments high in dust or moisture are excellent places for air logic controls because practically no danger from explosion or electrical shock is possible even in these atmospheres. Water can splash on the controls with no effect on the operation. If there is danger of explosion, air controls can not ignite the materials involved.

Air logic can also be used on machines that have cylinders or fluid motors, but no type of electrical device. In such instances, two services are required because the machine is powered by air but controlled electrically. In cases where electrical and mechanical maintenance come under different labor grades, air logic is also ideal because different technicians work on different aspects of the machine -- one works on the circuit and the other handles the machine parts that are electrically driven.

Air logic does have its disadvantages; most common is the lack of understanding of how the components work and how to read the schematic drawing. If an air controlled machine fails, very few persons can work on it. Also, air logic with long control lines will have a noticeably slower cycle. Control lines longer than 10 to 15 ft fill and exhaust slowly when compared to electrical signals. In addition, air quality must be above average for long life.

Air logic controls are basically miniaturized 3- and 4-way air valves. The actions of the valves provide on or off functions like relays or switches. They also exhaust the spent signal. The symbols used for air logic are similar to electronic symbols. Some manufacturers use modified electrical symbols and ladder diagrams to show circuitry.

The following is an explanation of the basic logic components showing the ANSI logic symbol and ISO graphic symbol for a comparable directional control valve.

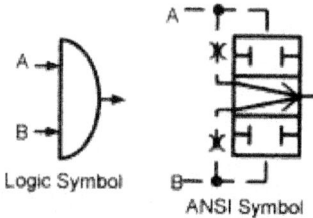

Figure 2-1: Passive "and" element.

AND, OR, AND NOT SYMBOLS

This ensures that two functions have completed before there is a command to continue the cycle. This can also be described by saying "this input, this input, and that input must be present before getting an output. Connect "and" inputs in a series when using more than two inputs. The first "and" receives signals' one and two while the output of this element hooks to one input of the second "and." The other input of the second "and" receives the third signal, making three inputs necessary before giving an output.

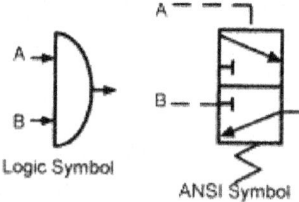

Figure 2-2: Active "and" element.

Some manufacturers supply both types of elements. This gives you **Figure 2-1** "and" with **Figure 2-2** designated as "yes." The difference in elements is that the "and" in **Figure 2-1** uses the lower of the two inputs as an output. This is a passive element. In contrast, a "yes" element has two inputs which obtain an output, but the designer has the choice of which input pairs with the output. Using this feature can amplify a weak signal. The weak signal pilots the valve open while the through signal comes from a full pressure supply. The "yes" in this situation is an active element.

Figure 2-3 shows the symbol for an "or" element. A shuttle valve serves the same purpose as an "or" element. Both inputs to an "or" element provide an output. A pilot signal from two different sources can pass through to start the next function. This can also be described by saying this signal or that signal provides an output. An "or" element differs from an inline "tee" because an "or" passes either input to the output but does not allow the inputs to pass to eachother.

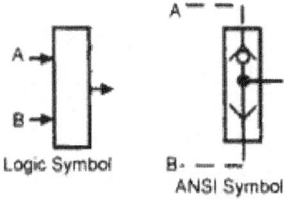

Figure 2-3: "Or" element

Stacking "or" elements allows for more than two inputs. Use an extra "or" element for each input after the first two signals.

Figure 2-4 shows the symbol for a "not" element, which is a normally open 3-way valve. An input signal or pressure supply will go through the valve until there is a pilot signal at port A. Pressurizing port A blocks supply and exhausts the output signal to atmosphere. "Not" elements will block a signal or supply as long as there is pilot pressure on the A port. The "not" always returns to a normally open condition without a pilot signal.

Replace a limit switch with a "not" element to indicate a cylinder is at the end of stroke. Pressure from the cylinder port goes to port A of the "not" element, holding it closed. As the cylinder moves to the work, pressure stays steady because of the meter-out flow control. When the cylinder contacts the work, the signal on port A drops, the "not" element opens and sends a signal to start the next operation.

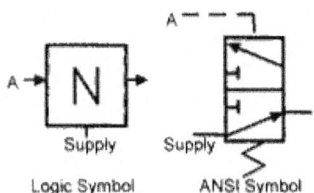

Figure 2-4: "Not" element.

The cylinder can stop at any position and the "not" output signal will indicate its nonmovement. This will always happen whether the cylinder stopped where it should have or if it stalled by some other means. Because this can happen, take care when using a "not" element to replace a limit switch. In contrast, this feature can be advantageous when clamping different sized parts. Use a "not" element for applications where different work locations stop the cylinder.

Most manufacturers supply a different pilot ratio for a "not" element used as a limit switch. The valve function is the same but it shifts at much lower pressure. Some manufacturers make a special "not" element that mounts directly to a cylinder port. A port-mounted meter-out flow control used in conjunction with this special "not" element makes a compact installation.

Caution! Pressure control valves only show pressure buildup. When a positive location must be made, use limit valves.

Flip-flop Circuits

"Flip flop" elements, with their symbol shown in Figure 2-5 are double piloted 5-way valves that send supply air to either outlet port with a signal at pilot ports S or R. Supply can be system pressure or air from another logic element. The main use for a "flip flop" is to eliminate the first pilot signal to a directional control valve. This allows a second signal on the directional valve's opposite pilot port to shift it back. "Flip flops," sometimes called "memory" elements, stay in their last shifted position even with no air supply. Whether the signal maintains or drops out, output from the "flip flop" stays the same.

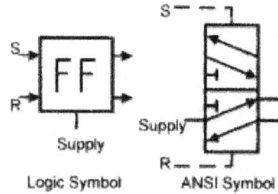

Figure 2-5: "Flip flop" element.

The S and R signals stand for "set" and "reset." The "set" signal shifts the "flip flop" for a function; whether the signal continues or not, the element stays shifted. The "reset" signal returns the "flip flop" back to its original position until the next cycle.

"Flip flop" can also be used to set up a new cycle, allowing the operator to momentarily push the start buttons. Use this same "flip flop" to eliminate unwanted signals and set up the circuit for cycle completion as required.

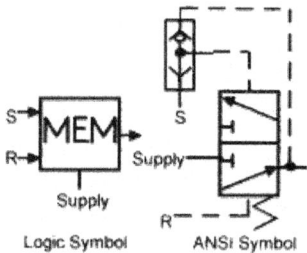

Figure 2-6: "Memory" element.

Figure 2-6 shows another valve actually called a "memory" element, which is a normally closed 3-way valve with a built in shuttle valve. The shuttle valve uses the "memory's" output air to hold it shifted once it receives an S "set" signal. A momentary "set" signal gives continuous pilot output. An R "reset" signal shifts the "memory" element to normally closed and exhausts output air. In addition, turning supply pressure off returns a "memory" element to its start position.

There are three different types of time delays in air logic control. Fixed- or adjustable-time delays are common in both normally closed and normally open configurations. Some time delays use an orifice and accumulator chamber for delays up to one minute. Some manufacturers use air actuated diaphragms and orifices that eliminate system pressure fluctuation inaccuracies.

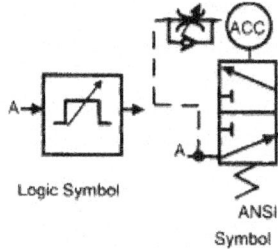

Logic Symbol

ANSI
Symbol

Figure 2-7: "One shot" element.

A "one-shot" timer, shown in **Figure 2-7**, is sometimes called an "impulse timer." A "one shot" timer takes a signal and passes it on to the circuit. At the same time, the input signal goes through an orifice to an accumulator tank. The setting of the orifice and size of the accumulator give a certain time delay before the normally open 3-way valve closes. After a "one shot" times out and closes, it remains closed as long as it has an input signal.

Figure 2-7 shows an adjustable time delay before it loses its output. Leaving off the sloping arrow in the symbols makes it a preset time delay. Times range from to 2 or more sec on valves with preset time delays.

Many circuits uses "one shots" to eliminate opposing signals. When a valve receives a signal to extend to a cylinder, it resists a return pilot signal to itself until loss of the first pilot. Using a "one shot" element drops the extend signal shortly after iniatiation. However, when the short duratoinj signal meets a hard-to-shift valve, the time may not always be long enough to move the valve spool. The cycle will stall if the valve does not have time to shift. For best results, use a "flip flop" to drop unwanted signals after it performs a task. **Figures 2-17 to 2-20** and accompanying text further describe "flip flop" valves.

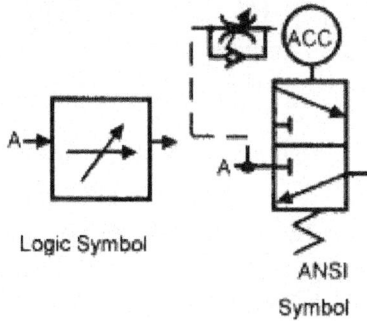

Logic Symbol

ANSI
Symbol

Figure 2-8: "On delay timer" element.

Passing a signal through the element after timing stops is done with an adjustable, normally closed "time-on" time delay. Figure 2-8 shows the symbol for this element. A "time-on" delay is a preset fixed timer without the sloping arrow. Most anti-tie down circuits use a fixed time delay, thus forcing the operator to actuate both palmbuttons concurrently.

The symbol in **Figure 2-8** shows an input *A* moving towards the blocked port of a 3-way directional valve. Signal *A* also moves to a meter-in flow control to fill an accumulator. After the accumulator is filled, pilot pressure shifts the 3-way valve, allowing air to pass on to the next operation. As long as the input signal stays on the time delay stays open.

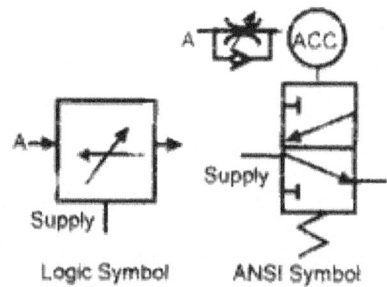

Figure 2-9: "Off delay timer" element.

Some brand of "time-on" delays use shop air to the normally closed port *A* of the 3-way valve while the signal to the timing section comes from another logic element or limit valve. This allows a strong passing signal to travel long distances or to quickly shift several other logic elements.

With a built-in accumulator tank, the time delay length is usually unjer one-to-one and one-half minutes. With added external accumulators, time delays up to 5 min are possible. The repeatability of long time delays using accumulators is poor. Often, diaphragm type timers go to 3 min with good repeatability.

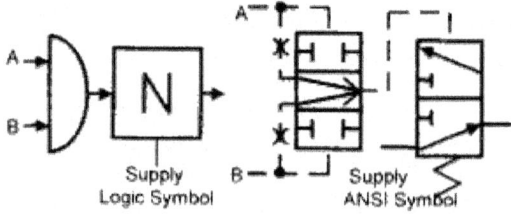

Figure 2-10: "Nand" element.

With a normally open 3-way valve in place of a normally closed 3-way, the delay is "time off." **Figure 2-9** shows the symbol for a "time off" delay timer. A continuous input to the supply gives an output until a set time after receiving a signal at *A*. When *A* receives a signal, the time delay starts and continues timing. When the accumulator fills, it closes the normally open 3-way valve and exhausts the signal. As before, a preset, non-adjustable time delay is available.

"Time-on" and "time-off" delays often are identical in appearance. The part number may be the only way to tell these units apart.

To get different functions, connect air logic elements together like the examples in **Figures 2-10 and 2-11**. These two common pairs might be familiar to anyone using air logic. A "nand" element, shown in **Figure 2-10**, uses an "and" to signal a "not." The term "nand" means "not this and this." As long as there are not signals at *A* and *B*, air passes. If signals *A* and *B* are present, the "not" closes and exhausts the output signal.

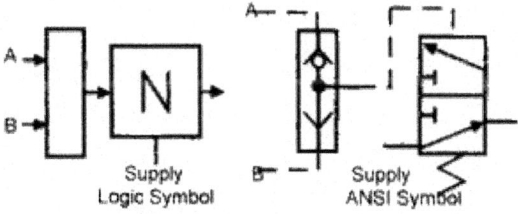

Figure 2-11: "Nor" element.

A "nor" element, shown in Figure 2-11, uses an "or" to signal a "not." The term "nor" means not this or this. As long as there is not a signal at *A* or *B*, air passes through the "not." If a signal is present at either *A* or *B*), the "not closes and exhausts the output signal.

Some other commonly used air logic elements include:

- Amplifiers to detect a low pressure signal (down to 3-in. water column) and send it on as an 80 psi signal.
- Pressure or vacuum sequence elements shift after reaching a set pressure or vacuum.
- Programmable controllers are combination elements that are used to design complex circuits with minimum knowledge of circuit design.
- Air-operated indicators show circuit condition and/or function. Several colors are available but none emit light.

The following text and images depict examples of air logic circuits, showing how some basic circuits perform machine control functions.

Anti-tie Down Air Logic Circuit, Using Logic Symbols

The two-hand, anti-tie down circuit schematic in **Figure 2-12** uses ANSI air logic symbols to simplify schematic drawings. However, most mechanics do not understand the hardware behind the symbols. An electrician may recognize the symbols but often does not understand how air logic functions. So like most problems with hydraulics and pneumatics, changing parts, turning knobs, and swapping lines continues until the machine starts working or an expert is called.

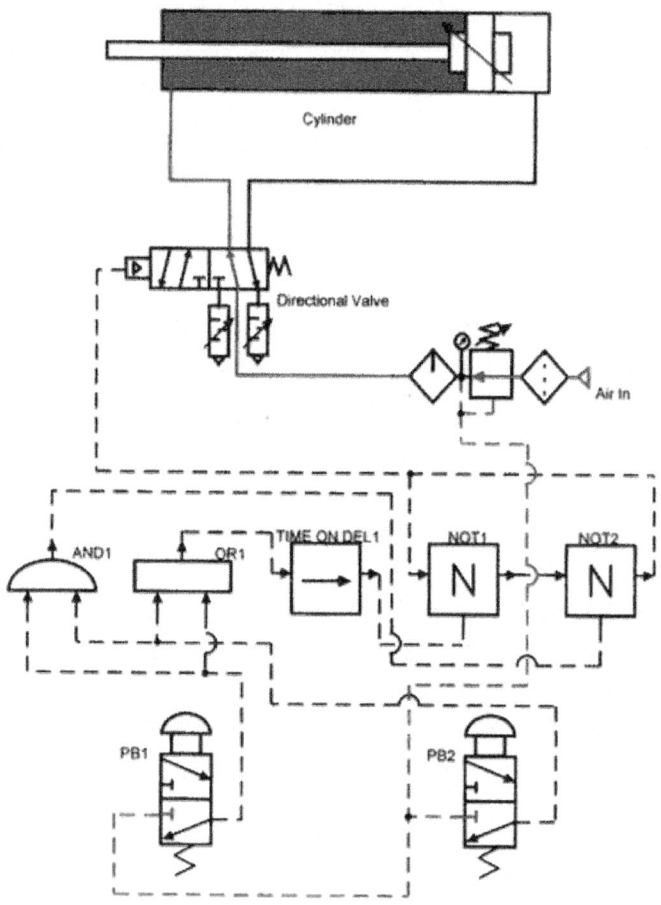

Figure 2-12: Anti-tie down air logic circuit using logic symbols

To make the cylinder in **Figure 2-12** extend, depress both palm buttons at the same time and hold them shifted. Tying either palm button more than one second before actuating the second palm button keeps the cylinder from moving. Depressing the second palm button within one second after the first palm button makes the cylinder extend and stay. Letting up on either or both of the palm buttons causes the cylinder to retract. This means that if the operator tries to use one of his hands to adjust or hold a part, the cylinder retracts. To start another cycle, release both palm buttons to reset the time delay. Both of the operator's hands must stay on the palm buttons when the cylinder is extending.

Notice that the *AND1* and *OR1* elements on the left receive signals from the palm buttons at the same time. *AND1* uses both signals to get an output while *ORr1* gives an output when depressing either palm button.

Actuating palm button *PB2* sends a signal through *OR1* to start *TIME ON DEL1*. After approximately one half to one second, *TIME ON DEL1* opens, sending a signal through normally open *NOT1* to close normally open *NOT2*. Depressing

PB1 after *NOT2* closes gives an *AND1* output, but it cannot go through to shift the directional valve. Actuating either palm button separately blocks the signal to shift the directional valve at *PB2*.

Shifting both palm buttons concurrently sends a signal through *OR1* starting *TIME ON DEL1*. At the same time, an output from *AND1* passes through *NOT2*, shifting the directional valve to extend the cylinder. The output of *NOT2* also closes *NOT1*, blocking the output from *TIME ON DEL1*. Depressing and holding both palm buttons extends the cylinder and keeps it there.

Releasing one palm button while the cylinder is extending drops one output of *AND1*. When *AND1* drops out, the directional valve spring returns, the cylinder retracts, *NOT1* opens, and *TIME ON DEL1* output closes *NOT2*. Depressing the released palm button again leaves the cylinder retracted because *TIME ON DEL1* closes *NOT2*. To start another cycle, release both palm buttons to reset TIME ON DEL1. **Figure 2-13** shows this operation using ISO valve symbols.

Anti-tie Down Air Logic Circuit, Using ISO Symbols

Figures 2-13 to 2-16 show the previous anti-tie down circuit in ISO symbols. Most people understand ISO symbols since they show valve function more clearly.

The circuit is at rest in **Figure 2-13**. *PB1* and *PB2* are not actuated, so there is no signal being sent to the directional valve. This schematic is what the machine supplier sends with his documentation for the machine.

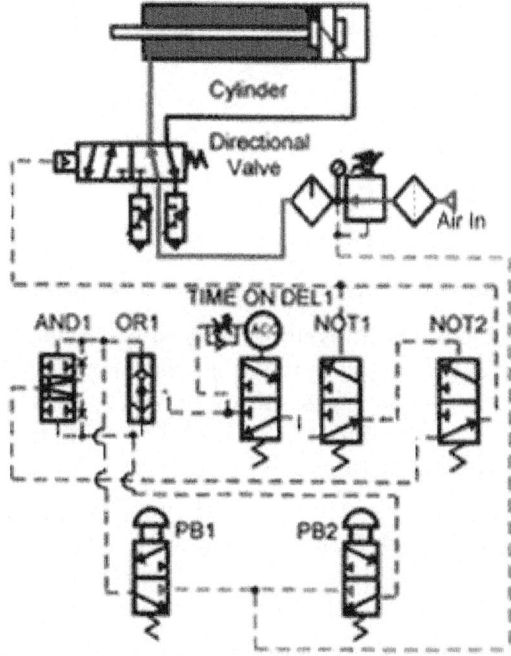

Figure 2-13: Anti-tie down air logic circuit at rest, air on.

Figure 2-14 shows the circuit when depressing only one palm button. Here, *PB2* sends an air signal to one port of the *and* element. The *and* element does not send any output because it needs two signals. The air signal from *PB2* does go through the *or* element and starts *TIME ON DEL1* timing. If *PB1* is not shifted within a short time, *TIME ON DEL 1* times out. When *TIME ON DEL1* times out, it sends a signal through *NOT1* and closes *NOT2*. After *TIME ON DEL1* closes *NOT2*, the signal from *PB1* through the *and* element becomes blocked at *NOT2*. Either palm button gives the same results. This protects the operator because depressing both palm buttons is the only way to cycle the machine. Before a cycle is possible in this condition, the palm buttons must exhaust any air signal from the *or* element, resetting *TIME ON DEL 1*.

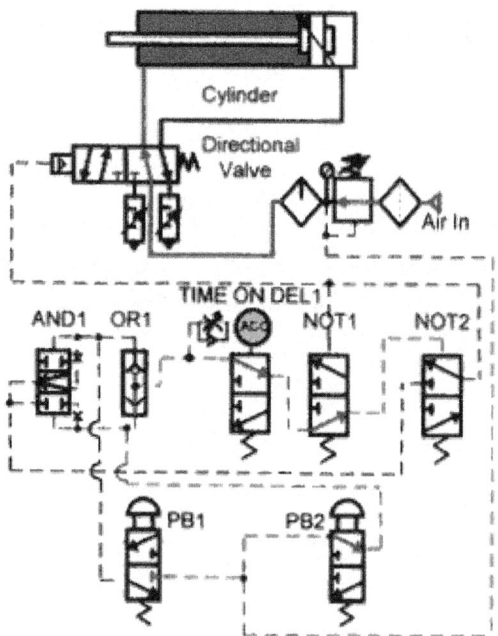

Figure 2-14: Anti-tie down air logic circuit, one palm button depressed.

Depressing both palm buttons simultaneously gives the results shown in **Figure 2-15**. Either one of the signals to the *or* element starts *TIME ON DEL 1* but one of the signals to the *and* element passes through *NOT2*, on to the directional valve. The signal from *NOT2* also pilots *NOT1* closed, blocking the output from *TIME ON DEL 1*. As long as both buttons stay shifted, the cylinder extends and holds.

After *TIME ON DEL 1* times out, the circuit changes to the one shown in **Figure 2-16**. Output from *TIME ON DEL 1* L1 stops at *NOT1* because the start signal from the *and* element is holding it closed. With this circuit the operator has to keep both hands on the palm buttons to make the cylinder extend.

The following circuits show other uses for these elements and how more complex circuits use other logic valves.

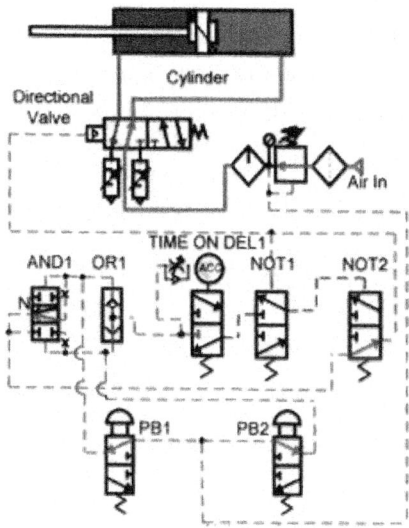

Figure 2-15: Anti-tie down air logic circuit, both palm buttons just actuated

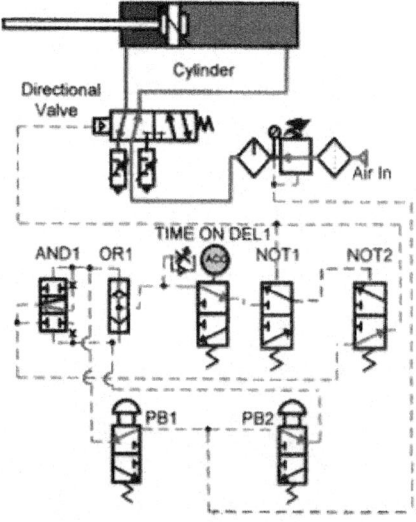

Figure 2-16: Anti-tie down air logic circuit, both palm buttons actuated, time delay timed out

Anti-tie Down, Non-repeat, Flip Flop Air Logic Circuit

Figures 2-17 through 2-20 show a two cylinder circuit with *CYLA* extending *(A+, CYLB* extending *(B+), CYLB* retracting *(B-)*, and *CYLA* retracting (A-).

Notice *CYLB* retracts immediately after extending, which means there would be an extend signal opposing a retract signal if the circuit only has limit valves for control. Using a "one shot" valve to stop the opposing signal works, but is less reliable than the "flip flop," FF circuit shown here.

Figure 2-17 shows both palm buttons depressed, causing the output of the anti-tie down circuit to shift FF. The output of the top port of FF sends a signal to shift a doubled pilot valve and extend CYLA. The FF output will also supply the normally closed port of limit valve LVA1. Shifting the FF also drops the signal from limit valve LVB0 that retracted CYLA. CYLA extends until it contacts limit valve LVA1.

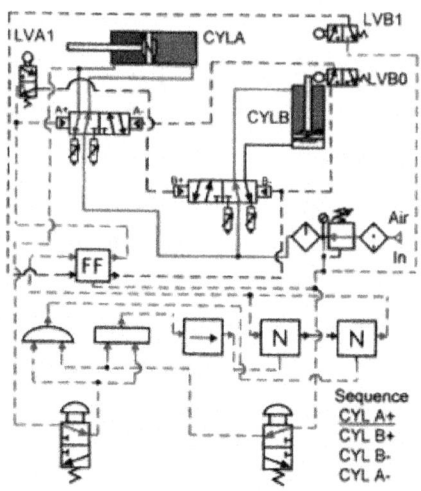

Figure 2-17: Anti-tie down, non-repeat flip flop circuit, cylinder A extending.

When CYLA contacts LVA1, Figure 2-18, air from the top port of FF passes through it and shifts a double piloted valve making CYLB extend. The signal to retract CYLB came from the bottom port of FF that is now exhausting to atmosphere. CYLB continues to extend until it contacts limit valve LVB1.

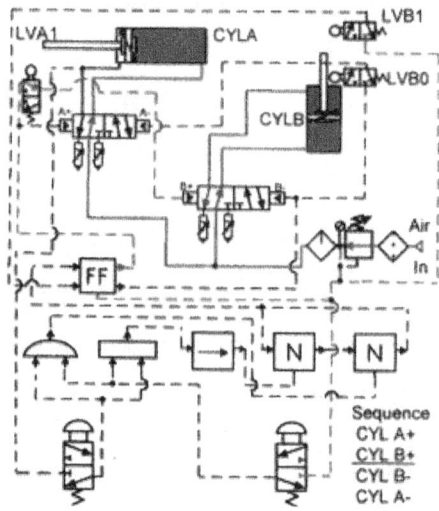

Figure 2-18: Anti-tie down, non-repeat flip flop circuit, cylinder B extending

The normally closed inlet port of *LVB1* has a constant air supply, so when *CYLB* contacts it, **Figure 2-19**, it shifts *FF* back to starting position. A signal from the bottom port of the *FF* shifts a double-piloted valve to retract *CYLB* and supplies air to the normally closed port of limit valve *LVB0*. After *FF* shifts back to the starting condition, it drops the extend signals to both double piloted directional valves. This makes it possible to shift the double-piloted valves to retract the cylinders. *CYLB* continues to retract until it contacts *LVB0*.

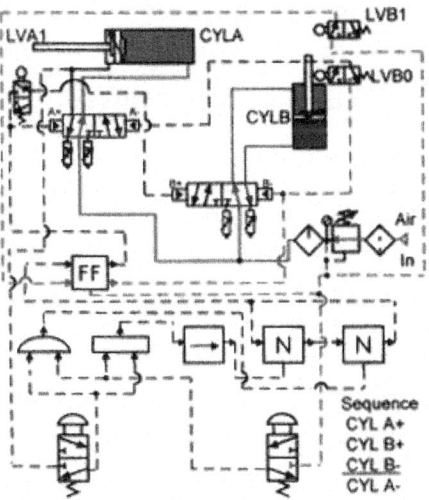

Figure 2-19: Anti-tie down, non-repeat flip flop circuit, cylinder B retracting

A signal from *LVB0* shifts the double-piloted valve to retract *CYLA* as shown in **Figure 2-20**. This cylinder can retract since its extend signal dropped out when *FF* shifted from *LVB0*. *CYLA* retracts to home position and ends the cycle.

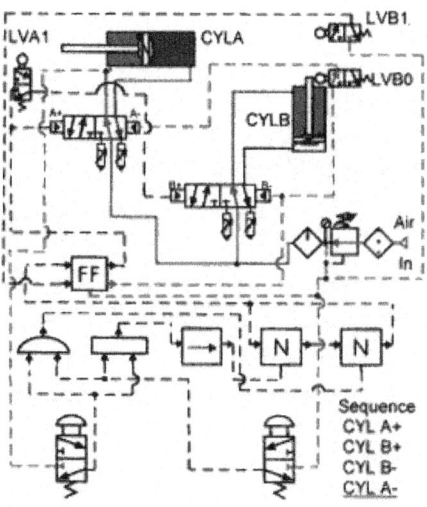

Figure 2-20: Anti-tie down, non-repeat flip flop circuit, cylinder A retracting

The nonrepeat feature is possible because when the circuit is in the at rest position, there is a supply to the left palm button from the rod end port of *CYLA*. After the cycle starts and *CYLA* reaches the end of its stroke, the left palm button loses its supply. Whether the operator lets off the palm buttons or not, loss of air to the left palm button disables the anti-tie down circuit.

With both palm buttons supplied with direct shop air, if the operator kept the palm buttons shifted all during the cycle, the machine would probably stall after *CYLB* extended. The nonrepeat feature adds little cost, but may save lost production.

Using Modified "Not" Elements as Limit Valves

The circuit in **Figure 2-21** operates the same as **Figures 2-17 through 2-20** on the preceding page. The only difference is pressure controlled "not" elements replace limit valves.

"Not" elements can replace limit valves when the movement they are detecting is not critical. "Not" limits operate any time the cylinder has a pressure drop. The pressure drop could be end of stroke or any place the cylinder stops for any reason. **If actuator position is critical, always use limit valves.**

Using a standard "not" to replace limit valves works, but the special low pressure "not" is best. Some manufacturers call this an "inhibitor", others, a "pressure trip release." Whatever the name, the modification causes the valve to shift at a lower differential pressure. This keeps a reduced backpressure at the cylinder port from giving a premature signal.

Using "not" elements in place of limit valves makes installation and plumbing easier, but can make troubleshooting more difficult. Placing the "not" elements in the control box works, but cylinder port mounting is best. No matter the location, they must read the air between the cylinder port and a meter-out flow control. This location ensures they see backpressure when the cylinder is moving.

Because a "not" is normally open, pressure holding the cylinder in position and backpressure from a meter-out flow control when the cylinder is moving give the signal to hold it shut. When the cylinder stops, pressure drops, allowing the "not" to open and send a signal to continue the cycle.

The circuit in **Figure 2-21** uses "not3" to tell *CYLB* to extend, "not5" to tell *CYLB* to retract, and "not4" to tell *CYLA* to retract.

Since the "not" works on loss of pressure, a cylinder with leaking seals can keep it from shifting. After a slowly moving cylinder stops, slow deterioration of pressure may delay the output signal.

Loss pressure valves have many benefits. For example, it does not matter when the cylinder contacts the part. Whether the part is 1 or 20 in. thick, when the cylinder makes contact, there will be an indication. In addition, air pressure changes have little or no effect on them as the "not" only reads minimum pressure. Finally, maximum pressure setting does not affect a "not" like it does a sequence valve.

"Not" elements are a preferred choice over sequence valves because sequence valves only work with meter-in flow controls. Any air cylinder has better control with a meter-out circuit and overrunning loads require meter-out flow control.

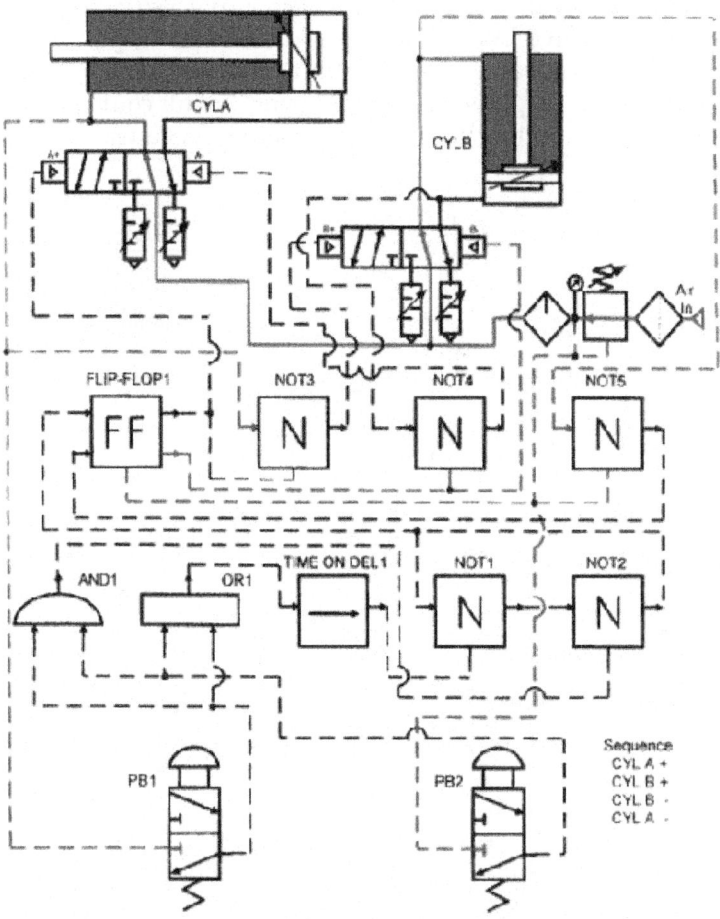

Figure 2-21: Anti-tie down, non-repeat and flip flop circuit, using modified "not" elements as limit valves.

Always use loss of pressure controls with caution since they can operate any time cylinder pressure drops below their minimum shifting pressure.

Anti-tie Down, Non-repeat and Flip Flop Air Logic Circuit with Automatic Cycling Air Drills

Figure 2-22 shows a clamp cylinder, *CYLA*, and three self-contained automatic air powered drills controlled with air logic.

Some circuits clamp a part then start the drills with a "one shot" element. As long as all the drills start there is no problem. However, if any drill fails to cycle, parts may come off the fixture with one or more holes missing. When double drilling is necessary, part costs' and scrap increase. The circuit in **Figure 2-22** eliminates this problem with air logic elements and piping.

When the anti-tie down circuit shifts "flip flop" *FF*, a signal from its top port goes to extend clamp cylinder *CYLA*. *FF* top port output also supplies the normally closed port of limit valve *LVA1*. *CYLA* extends, clamps the part, and shifts limit valve *LVA1*. Use a limit valve here since the drills could sling a loosely clamped part out of the fixture. A pressure operated "not" circuit could allow premature cycling of the drills, resulting in damage and safety concerns.

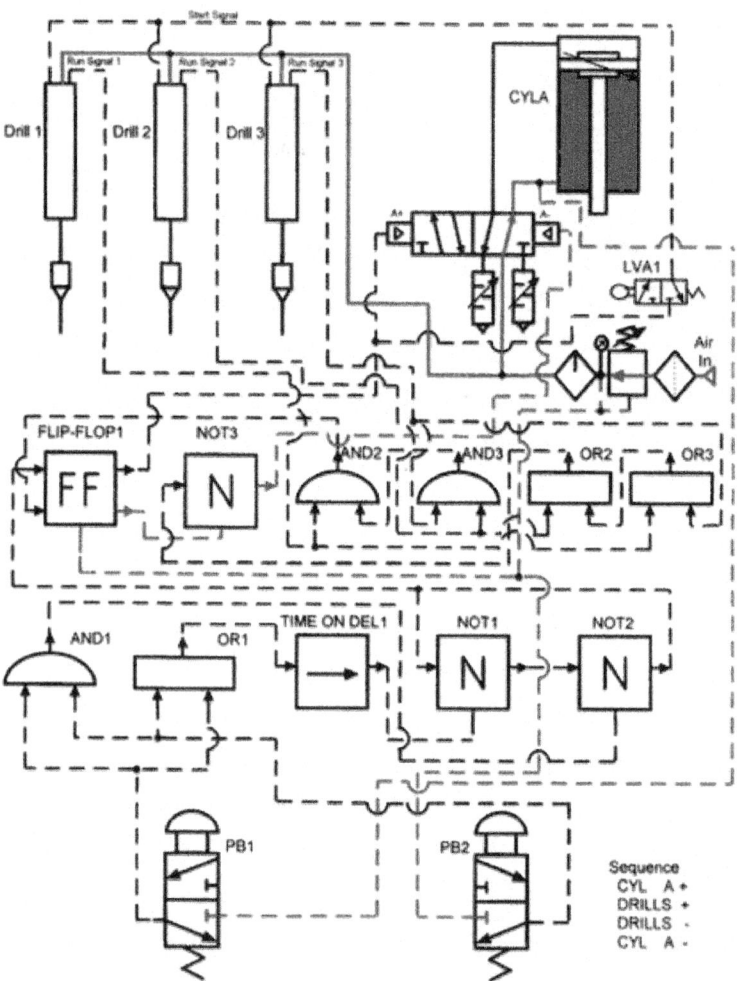

Figure 2-22: Anti-tie down, non-repeat and flip flop circuit, with automatic cycling air drills

After clamping the part, *FF* output goes through limit valve *LVA1* to the drills' start ports. When the drills start they give a run signal when they move from home position. This run signal remains on until the drills fully retract. On most of these types of air-operated drills, the output is the same air that turns the air motor in the drill. One brand of air drill has an output when at rest and exhausts that signal when it starts.

The drill run signals go to the input ports of two "ands" and two "ors". When the two cascaded "ands" have three signals indicating all the drills are moving, their output shifts FF back to its starting position and exhausts the drill start signal. The three inputs to the two cascaded "ors" pass through to close "not3" so the clamp will not open until all drills have retracted. Output from the lower port of FF goes to the inlet of "not3" to set up clamp CYLA open sequence.

The drills will continue forward until they meet their internal limit valves and retract. The run signals drop out as each drill finishes and retracts to home position. When the last drill is home, the run signal from the last "or" element exhausts and "not3" opens. When "not3" opens, its output shifts the clamp valve to retract CYLA.

If starting of one of the drills is sluggish, the run start signal stays on until it moves. If a drill fails to start, the run signal stays on and the running drill stay extended. In either case, the operator knows when a problem exists. If one of the drills hangs in the part, the clamp will not open until the drill is free to retract. For every added drill, use another "and" and "or" element. With air indicators installed in each drill run signal line, picking out a nonrunning drill is easy.

Chapter 9

FLUIDICS

FLUIDICS

Fluidics, or **fluidic logic**, is the use of a fluid to perform analog or digital operations similar to those performed with electronics.

The physical basis of fluidics is pneumatics and hydraulics, based on the theoretical foundation of fluid dynamics. The term *fluidics* is normally used when devices have no moving parts, so ordinary hydraulic components such as hydraulic cylinders and spool valves are not considered or referred to as fluidic devices. The 1960s saw the application of fluidics to sophisticated control systems, with the introduction of the fluidic amplifier.

A jet of fluid can be deflected by a weaker jet striking it at the side. This provides nonlinear amplification, similar to the transistor used in electronic digital logic. It is used mostly in environments where electronic digital logic would be unreliable, as in systems exposed to high levels of electromagnetic interference or ionizing radiation.

Nanotechnology considers fluidics as one of its instruments. In this domain, effects such as fluid-solid and fluid-fluid interface forces are often highly significant. Fluidics have also been used for military applications.

Amplifier

The basic concept of the **fluidic amplifier** is shown here. A fluid supply, which may be air, water, or hydraulic fluid, enters at the bottom. Pressure applied to the control ports C_1 or C_2 deflects the stream, so that it exits *via* either port O_1 or O_2. The stream entering the control ports may be much weaker than the stream being deflected, so the device has gain.

Given this basic device, flip flops and other fluidic logic elements can be constructed. Simple systems of digital logic can thus be built.

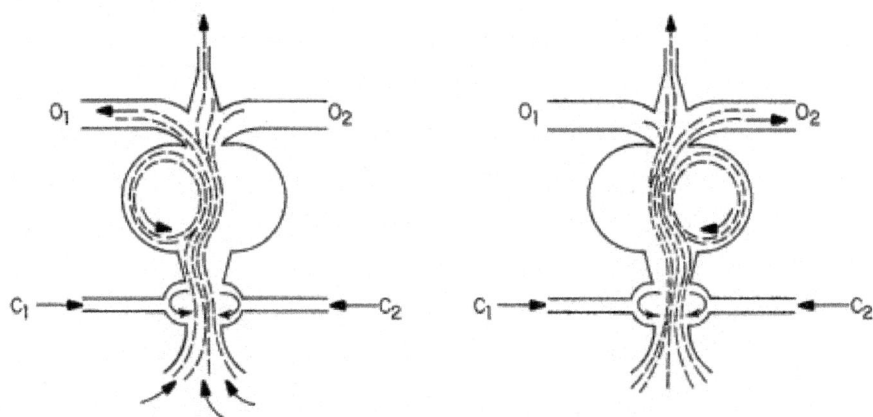

Fluidic amplifier, showing flow in both states. From U.S. Patent #4,000,757.

Fluidic amplifiers typically have bandwidths in the low kilohertz range, so systems built from them are quite slow compared to electronic devices.

Triode

The fluidic triode is an amplification device that uses a fluid to convey the signal.

Although much studied in the laboratory they have few practical applications. Many expect them to be key elements of nanotechnology.

Fluidic triodes were used as the final stage in the main Public Address system at the 1964 New York World's Fair.

The Fluidic Triode was invented in 1962 by Murray O. Meetze, Jr., a high school student in Heath Springs, S.C. He also built a fluid diode, a fluid oscillator and a variety of hydraulic "circuits," including one that has no electronic counterpart. As a result he was invited to the National Science Fair, held this year at the Seattle Century 21 Exposition. There his project won an award.

Elements

Logic gates can be built that use water instead of electricity to power the gating function. These are reliant on being positioned in one orientation to perform correctly. An OR gate is simply two pipes being merged, a NOT gate consists of "A" deflecting a supply stream to produce $\bar{A}$. An inverter could also be implemented with the XOR gate, as $A \text{ XOR } 1 = \bar{A}$.

Bubble logic is another kind of fluidic logic. Bubble logic gates conserve the number of bits entering and exiting the device, since bubbles are neither produced nor destroyed in the logic operation, analogous to billiard ball computer gates.

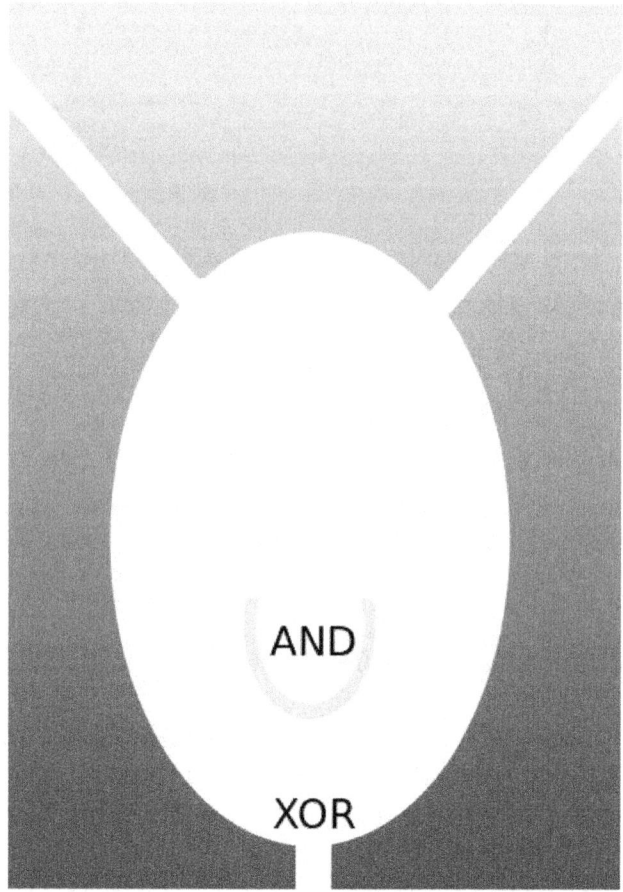

The two gates AND and XOR in one module. The bucket in the center collects the AND output, and the output at the bottom is A XOR B.

Uses

Fluidic components appear in some hydraulic and pneumatic systems, including some automotive automatic transmissions. As digital logic has become more accepted in industrial control, the role of fluidics in industrial control has declined.

In the consumer market, fluidically controlled products are increasing in both popularity and presence, installed in items ranging from toy spray gun through shower heads (the shower head video features a computer generated animation of a fluidic and how it creates oscillating waves), and hot tub jets; all providing oscillating streams or air and/or water.

Research

Fluidic injection is being researched for use in aircraft to control direction, in two ways: circulation control and thrust vectoring. In both, larger more complex mechanical parts are replaced by fluidic systems, in which larger forces in fluids are diverted by smaller jets or flows of fluid intermittently, to change the direction of vehicles.

In circulation control, near the trailing edges of wings, aircraft flight control systems such as ailerons, elevators, elevons, flaps and flaperons are replaced by slots which emit fluid flows.

In thrust vectoring, in jet engine nozzles, swiveling parts are replaced by slots which inject fluid flows into jets. Such systems divert thrust *via* fluid effects. Tests show that air forced into a jet engine exhaust stream can deflect thrust up to 15 degrees.

In such uses, fluidics is desirable for lower: mass, cost (up to 50% less), drag (up to 15% less during use), inertia (for faster, stronger control response), complexity (mechanically simpler, fewer or no moving parts or surfaces, less maintenance), and radar cross section for stealth. This will likely be used in many unmanned aerial vehicles (UAVs), 6th generation fighter aircraft, and ships.

BOOLEAN ALGEBRA

In mathematics and mathematical logic, **Boolean algebra** is the subarea of algebra in which the values of the variables are the truth values *true* and *false*, usually denoted 1 and 0 respectively. Instead of elementary algebra where the values of the variables are numbers, and the main operations are addition and multiplication, the main operations of Boolean algebra are the conjunction *and*, denoted ∧, the disjunction *or*, denoted ∨, and the negation *not*, denoted ¬.

Boolean algebra was introduced in 1854 by George Boole in his book *An Investigation of the Laws of Thought*. According to Huntington the term "Boolean algebra" was first suggested by Sheffer in 1913.

Boole's first book *The Mathematical Analysis of Logic* published in 1847 included the original theory. This was proposed as a Mathematical language dealing with the questions of logic which is now needed in the design of modern digital equipment, and now exists as a core data type in all modern programming languages generally abbreviated to as type bool, representing true or false within assertion logic.

Boolean algebra has been fundamental in the development of digital electronics. It is also used in set theory and statistics.

History

Boole's algebra predated the modern developments in abstract algebra and mathematical logic; it is however seen as connected to the origins of both fields. In an abstract setting, Boolean algebra was perfected in the late 19th century by Jevons, Schröder, Huntington, and others until it reached the modern conception of an (abstract) mathematical structure. For example, the empirical observation that one can manipulate expressions in the algebra of sets by translating them into expressions in Boole's algebra is explained in modern terms by saying that the algebra of sets is *a* Boolean algebra (note the indefinite article). In fact, M. H. Stone proved in 1936 that every Boolean algebra is isomorphic to a field of sets.

In the 1930s, while studying switching circuits, Claude Shannon observed that one could also apply the rules of Boole's algebra in this setting, and he introduced **switching algebra** as a way to analyze and design circuits by algebraic means in terms of logic gates. Shannon already had at his disposal the abstract mathematical apparatus, thus he cast his switching algebra as the two-element Boolean algebra. In circuit engineering settings today, there is little need to consider other Boolean algebras, thus "switching algebra" and "Boolean algebra" are often used interchangeably. Efficient implementation of Boolean functions is a fundamental problem in the design of combinatorial logic circuits. Modern electronic design automation tools for VLSI circuits often rely on an efficient representation of Boolean functions known as (reduced ordered) binary decision diagrams (BDD) for logic synthesis and formal verification.

Logic sentences that can be expressed in classical propositional calculus have an equivalent expression in Boolean algebra. Thus, **Boolean logic** is sometimes used to denote propositional calculus performed in this way. Boolean algebra is not sufficient to capture logic formulas using quantifiers, like those from first order logic. Although the development of mathematical logic did not follow Boole's program, the connection between his algebra and logic was later put on firm ground in the setting of algebraic logic, which also studies the algebraic systems of many other logics. The problem of determining whether the variables of a given Boolean (propositional) formula can be assigned in such a way as to make the formula evaluate to true is called the Boolean satisfiability problem (SAT), and is of importance to theoretical computer science, being the first problem shown to be NP-complete. The closely related model of computation known as a Boolean circuit relates time complexity (of an algorithm) to circuit complexity.

Values

Whereas in elementary algebra expressions denote mainly numbers, in Boolean algebra they denote the truth values *false* and *true*. These values are represented with the bits (or binary digits), namely 0 and 1. They do not behave like the integers 0 and 1, for which $1 + 1 = 2$, but may be identified with the elements of the two-element field GF(2), for which $1 + 1 = 0$ with + serving as the Boolean operation XOR.

Boolean algebra also deals with functions which have their values in the set $\{0, 1\}$. A sequence of bits is a commonly used such function. Another common example is the subsets of a set E: to a subset F of E is associated the indicator function that takes the value 1 on F and 0 outside F.

As with elementary algebra, the purely equational part of the theory may be developed without considering explicit values for the variables.

Operations

Basic Operations

The basic operations of Boolean algebra are as follows.

- And (conjunction), denoted x∧y (sometimes x AND y or Kxy), satisfies x∧y = 1 if x = y = 1 and x∧y = 0 otherwise.
- Or (disjunction), denoted x∨y (sometimes x OR y or Axy), satisfies x∨y = 0 if x = y = 0 and x∨y = 1 otherwise.
- Not (negation), denoted ¬x (sometimes NOT x, Nx or !x), satisfies ¬x = 0 if x = 1 and ¬x = 1 if x = 0.

If the truth values 0 and 1 are interpreted as integers, these operation may be expressed with the ordinary operations of the arithmetic:

$$x \wedge y = x \times y$$
$$x \vee y = x + y - (x \times y)$$
$$\neg x = 1 - x$$

Alternatively the values of $x \wedge y$, $x \vee y$, and $\neg x$ can be expressed by tabulating their values with truth tables as follows.

x	y	$x \wedge y$	$x \vee y$	x	$\neg x$
0	0	0	0	0	1
1	0	0	1	1	0
0	1	0	1		
1	1	1	1		

One may consider that only the negation and one of the two other operations are basic, because of the following identities that allow to define the conjunction in terms of the negation and the disjunction, and vice versa:

$$x \wedge y = \neg(\neg x \vee \neg y)$$
$$x \wedge y = \neg(\neg x \vee \neg y)$$

Derived Operations

The three Boolean operations described above are referred to as basic, meaning that they can be taken as a basis for other Boolean operations that can be built up from them by **composition,** the manner in which operations are combined or compounded. Operations composed from the basic operations include the following examples:

$$x \rightarrow y = \neg x \vee y$$

$$x \oplus y = (x \vee y) \wedge \neg(x \wedge y)$$
$$x \equiv y = \neg(x \oplus y)$$

These definitions give rise to the following truth tables giving the values of these operations for all four possible inputs.

x	y	$x \to y$	$x \oplus y$	$x \equiv y$
0	0	1	0	1
1	0	0	1	0
0	1	1	1	0
1	1	1	0	1

The first operation, $x \to y$, or Cxy, is called **material implication**. If x is true then the value of $x \to y$ is taken to be that of y. But if x is false then the value of y can be ignored; however the operation must return *some* truth value and there are only two choices, so the return value is the one that entails less, namely *true*. (Relevance logic addresses this by viewing an implication with a false premise as something other than either true or false.)

The second operation, $x \oplus y$, or Jxy, is called **exclusive or** to distinguish it from disjunction as the inclusive kind. It excludes the possibility of both x and y. Defined in terms of arithmetic it is addition mod 2 where $1 + 1 = 0$.

The third operation, the complement of exclusive or, is **equivalence** or Boolean equality: $x \equiv y$, or Exy, is true just when x and y have the same value. Hence $x \oplus y$ as its complement can be understood as $x \neq y$, being true just when x and y are different. Its counterpart in arithmetic mod 2 is $x + y + 1$.

Given two operands, each with two possible values, there are $2^2 = 4$ possible combinations of inputs. Because each output can have two possible values, there are a total of $2^4 = 16$ possible binary Boolean operations.

Laws

A **law** of Boolean algebra is an identity such as $x \vee (y \vee z) = (x \vee y) \vee z$ between two Boolean terms, where a **Boolean term** is defined as an expression built up from variables and the constants 0 and 1 using the operations $\wedge$, $\vee$, and $\neg$. The concept can be extended to terms involving other Boolean operations such as $\oplus$, $\to$, and $\equiv$, but such extensions are unnecessary for the purposes to which the laws are put. Such purposes include the definition of a Boolean algebra as any model of the Boolean laws, and as a means for deriving new laws from old as in the derivation of $x\vee(y \wedge z) = x\vee(z \wedge y)$ from $y \wedge z = z \wedge y$ as treated in the section on axiomatization.

Monotone Laws

Boolean algebra satisfies many of the same laws as ordinary algebra when one matches up $\vee$ with addition and $\wedge$ with multiplication. In particular the following laws are common to both kinds of algebra:

Associativity of $\vee$ $x \vee (y \vee z) = (x \vee y) \vee z$

Associativity of ∧	$x \wedge (y \wedge z) = (x \wedge y) \wedge z$
Commutativity of ∨	$x \vee y = y \vee x$
Commutativity of ∧	$x \wedge y = y \wedge x$
Distributivity of ∧ over ∨	$x \wedge (y \vee z) = (x \wedge y) \vee (x \wedge 2)$
Identity for ∨	$x \vee 0 = x$
Identity for ∧	$x \wedge 1 = x$
Annihilator for ∧	$x \wedge 0 = 0$

Boolean algebra however obeys some additional laws, in particular the following:

Idempotence of ∨	$x \vee x = x$
Idempotence of ∧	$x \wedge x = x$
Absorption 1	$x \wedge (x \vee y) = x$
Absorption 2	$x \vee (x \wedge y) = x$
Distributivity of ∨ over ∧	$x \wedge (y \vee z) = (x \wedge y) \, A \, (x \vee z)$
Annihilator for ∨	$x \vee 1 = 1$

A consequence of the first of these laws is $1 \vee 1 = 1$, which is false in ordinary algebra, where $1 + 1 = 2$. Taking $x = 2$ in the second law shows that it is not an ordinary algebra law either, since $2 \times 2 = 4$. The remaining four laws can be falsified in ordinary algebra by taking all variables to be 1, for example in Absorption Law 1 the left hand side is $1(1 + 1) = 2$ while the right hand side is 1, and so on.

All of the laws treated so far have been for conjunction and disjunction. These operations have the property that changing either argument either leaves the output unchanged or the output changes in the same way as the input. Equivalently, changing any variable from 0 to 1 never results in the output changing from 1 to 0. Operations with this property are said to be **monotone**. Thus the axioms so far have all been for monotonic Boolean logic. Nonmonotonicity enters *via* complement ¬ as follows.

Nonmonotone Laws

The complement operation is defined by the following two laws.

Complementation 1 $x \wedge \neg x = 0$

Complementation 2 $x \vee \neg x = 1$

All properties of negation including the laws below follow from the above two laws alone.

In both ordinary and Boolean algebra, negation works by exchanging pairs of elements, whence in both algebras it satisfies the double negation law

Double negation $\neg(\neg x) = x$

But whereas *ordinary algebra* satisfies the two laws

$$(-x)(-y) = xy$$
$$(-x) + (-y) = -(x + y)$$

Boolean algebra satisfies De Morgan's laws:

De Morgan 1 $\neg\, x \wedge \neg y = \neg(x \vee y)$

De Morgan 2 $\neg\, x \vee \neg y = \neg(x \wedge y)$

Completeness

The laws listed above define Boolean algebra, in the sense that they entail the rest of the subject. The laws *Complementation* 1 and 2, together with the monotone laws, suffice for this purpose and can therefore be taken as one possible *complete* set of laws or axiomatization of Boolean algebra. Every law of Boolean algebra follows logically from these axioms. Furthermore Boolean algebras can then be defined as the models of these axioms as treated in the section thereon.

To clarify, writing down further laws of Boolean algebra cannot give rise to any new consequences of these axioms, nor can it rule out any model of them. In contrast, in a list of some but not all of the same laws, there could have been Boolean laws that did not follow from those on the list, and moreover there would have been models of the listed laws that were not Boolean algebras.

This axiomatization is by no means the only one, or even necessarily the most natural given that we did not pay attention to whether some of the axioms followed from others but simply chose to stop when we noticed we had enough laws, treated further in the section on axiomatizations. Or the intermediate notion of axiom can be sidestepped altogether by defining a Boolean law directly as any **tautology**, understood as an equation that holds for all values of its variables over 0 and 1. All these definitions of Boolean algebra can be shown to be equivalent.

Boolean algebra has the interesting property that $x = y$ can be proved from any non-tautology. This is because the substitution instance of any non-tautology obtained by instantiating its variables with constants 0 or 1 so as to witness its non-tautologyhood reduces by equational reasoning to $0 = 1$. For example the non-tautologyhood of $x \wedge y = x$ is witnessed by $x = 1$ and $y = 0$ and so taking this as an axiom would allow us to infer $1 \wedge 0 = 1$ as a substitution instance of the axiom and hence $0 = 1$. We can then show $x = y$ by the reasoning $x = x \wedge 1 = x \wedge 0 = 0 = 1$ $= y \vee 1 = y \vee 0 = y$.

Duality Principle

There is nothing magical about the choice of symbols for the values of Boolean algebra. We could rename 0 and 1 to say α and β, and as long as we did so consistently throughout it would still be Boolean algebra, albeit with some obvious cosmetic differences.

But suppose we rename 0 and 1 to 1 and 0 respectively. Then it would still be Boolean algebra, and moreover operating on the same values. However it would not be identical to our original Boolean algebra because now we find ∨ behaving the way ∧ used to do and vice versa. So there are still some cosmetic differences to show that we've been fiddling with the notation, despite the fact that we're still using 0s and 1s.

But if in addition to interchanging the names of the values we also interchange the names of the two binary operations, *now* there is no trace of what we have done. The end product is completely indistinguishable from what we started with. We might notice that the columns for $x \wedge y$ and $x \vee y$ in the truth tables had changed places, but that switch is immaterial.

When values and operations can be paired up in a way that leaves everything important unchanged when all pairs are switched simultaneously, we call the members of each pair **dual** to each other. Thus 0 and 1 are dual, and ∧ and ∨ are dual. The **Duality Principle**, also called De Morgan duality, asserts that Boolean algebra is unchanged when all dual pairs are interchanged.

One change we did not need to make as part of this interchange was to complement. We say that complement is a **self-dual** operation. The identity or do-nothing operation x is also self-dual. A more complicated example of a self-dual operation is $(x \wedge y) \vee (y \wedge z) \vee (z \wedge x)$. There is no self-dual binary operation. A composition of self-dual operations is a self-dual operation. For example, if $f(x,y,z) = (x \wedge y) \vee (y \wedge z) \vee (z \wedge x)$, then $f(f(x,y,z),x,t)$ is a self-dual operation of four arguments x,y,z,t.

The principle of duality can be explained from a group theory perspective by fact that there are exactly four functions that are one-to-one mappings (automorphisms) of the set of Boolean polynomials back to itself: the identity function, the complement function, the dual function and the contradual function (complemented dual). These four functions form a group under function composition, isomorphic to the Klein four-group, acting on the set of Boolean polynomials.

Diagrammatic Representations

Venn Diagrams

A Venn diagram is a representation of a Boolean operation using shaded overlapping regions. There is one region for each variable, all circular in the examples here. The interior and exterior of region x corresponds respectively to the values 1 (true) and 0 (false) for variable x. The shading indicates the value of the operation for each combination of regions, with dark denoting 1 and light 0.

The three Venn diagrams in the figure below represent respectively conjunction $x \wedge y$, disjunction $x \vee y$, and complement $\neg x$.

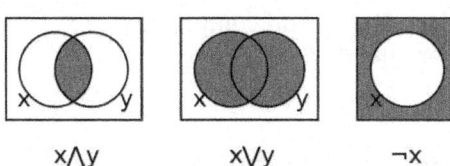

$x \wedge y$ $x \vee y$ $\neg x$

Venn diagrams for conjunction, disjunction, and complement.

For conjunction, the region inside both circles is shaded to indicate that $x \wedge y$ is 1 when both variables are 1. The other regions are left unshaded to indicate that $x \wedge y$ is 0 for the other three combinations.

The second diagram represents disjunction $x \vee y$ by shading those regions that lie inside either or both circles. The third diagram represents complement $\neg x$ by shading the region *not* inside the circle.

While we have not shown the Venn diagrams for the constants 0 and 1, they are trivial, being respectively a white box and a dark box, neither one containing a circle. However we could put a circle for x in those boxes, in which case each would denote a function of one argument, x, which returns the same value independently of x, called a constant function. As far as their outputs are concerned, constants and constant functions are indistinguishable; the difference is that a constant takes no arguments, called a *zeroary* or *nullary* operation, while a constant function takes one argument, which it ignores, and is a *unary* operation.

Venn diagrams are helpful in visualizing laws. The commutativity laws for $\wedge$ and $\vee$ can be seen from the symmetry of the diagrams: a binary operation that was not commutative would not have a symmetric diagram because interchanging x and y would have the effect of reflecting the diagram horizontally and any failure of commutativity would then appear as a failure of symmetry.

Idempotence of $\wedge$ and $\vee$ can be visualized by sliding the two circles together and noting that the shaded area then becomes the whole circle, for both $\wedge$ and $\vee$.

To see the first absorption law, $x \wedge (x \vee y) = x$, start with the diagram in the middle for $x \vee y$ and note that the portion of the shaded area in common with the x circle is the whole of the x circle. For the second absorption law, $x \vee (x \wedge y) = x$, start with the left diagram for $x \wedge y$ and note that shading the whole of the x circle results in just the x circle being shaded, since the previous shading was inside the x circle.

The double negation law can be seen by complementing the shading in the third diagram for $\neg x$, which shades the x circle.

To visualize the first De Morgan's law, $(\neg x) \wedge (\neg y) = \neg(x \vee y)$, start with the middle diagram for $x \vee y$ and complement its shading so that only the region outside both circles is shaded, which is what the right hand side of the law describes. The result is the same as if we shaded that region which is both outside the x circle *and* outside the y circle, *i.e.* the conjunction of their exteriors, which is what the left hand side of the law describes.

The second De Morgan's law, $(\neg x) \vee (\neg y) = \neg(x \wedge y)$, works the same way with the two diagrams interchanged.

The first complement law, $x \wedge \neg x = 0$, says that the interior and exterior of the x circle have no overlap. The second complement law, $x \vee \neg x = 1$, says that everything is either inside or outside the x circle.

Digital Logic Gates

Digital logic is the application of the Boolean algebra of 0 and 1 to electronic hardware consisting of logic gates connected to form a circuit diagram. Each gate implements a Boolean operation, and is depicted schematically by a shape indicating the operation. The shapes associated with the gates for conjunction (AND-gates), disjunction (OR-gates), and complement (inverters) are as follows.

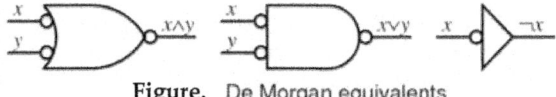

Figure. Logic gates

The lines on the left of each gate represent input wires or *ports*. The value of the input is represented by a voltage on the lead. For so-called "active-high" logic, 0 is represented by a voltage close to zero or "ground", while 1 is represented by a voltage close to the supply voltage; active-low reverses this. The line on the right of each gate represents the output port, which normally follows the same voltage conventions as the input ports.

Complement is implemented with an inverter gate. The triangle denotes the operation that simply copies the input to the output; the small circle on the output denotes the actual inversion complementing the input. The convention of putting such a circle on any port means that the signal passing through this port is complemented on the way through, whether it is an input or output port.

The Duality Principle, or De Morgan's laws, can be understood as asserting that complementing all three ports of an AND gate converts it to an OR gate and vice versa, as shown in Figure below. Complementing both ports of an inverter however leaves the operation unchanged.

Figure. De Morgan equivalents

More generally one may complement any of the eight subsets of the three ports of either an AND or OR gate. The resulting sixteen possibilities give rise to only eight Boolean operations, namely those with an odd number of 1's in their truth table. There are eight such because the "odd-bit-out" can be either 0 or 1 and can go in any of four positions in the truth table. There being sixteen binary Boolean operations, this must leave eight operations with an even number of 1's in their truth tables. Two of these are the constants 0 and 1 (as binary operations that ignore both their inputs); four are the operations that depend nontrivially on exactly one of their two inputs, namely x, y, $\neg x$, and $\neg y$; and the remaining two are $x \oplus y$ (XOR) and its complement $x \equiv y$.

Boolean Algebras

The term "algebra" denotes both a subject, namely the subject of algebra, and an object, namely an algebraic structure. Whereas the foregoing has addressed the subject of Boolean algebra, this section deals with mathematical objects called Boolean algebras, defined in full generality as any model of the Boolean laws. We begin with a special case of the notion definable without reference to the laws, namely concrete Boolean algebras, and then give the formal definition of the general notion.

Concrete Boolean Algebras

A **concrete Boolean algebra** or field of sets is any nonempty set of subsets of a given set X closed under the set operations of union, intersection, and complement relative to X.

(As an aside, historically X itself was required to be nonempty as well to exclude the degenerate or one-element Boolean algebra, which is the one exception to the rule that all Boolean algebras satisfy the same equations since the degenerate algebra satisfies every equation. However this exclusion conflicts with the preferred purely equational definition of "Boolean algebra," there being no way to rule out the one-element algebra using only equations — $0 \neq 1$ does not count, being a negated equation. Hence modern authors allow the degenerate Boolean algebra and let X be empty.)

Example 1. The power set 2^X of X, consisting of all subsets of X. Here X may be any set: empty, finite, infinite, or even uncountable.

Example 2. The empty set and X. This two-element algebra shows that a concrete Boolean algebra can be finite even when it consists of subsets of an infinite set. It can be seen that every field of subsets of X must contain the empty set and X. Hence no smaller example is possible, other than the degenerate algebra obtained by taking X to be empty so as to make the empty set and X coincide.

Example 3. The set of finite and cofinite sets of integers, where a cofinite set is one omitting only finitely many integers. This is clearly closed under complement, and is closed under union because the union of a cofinite set with any set is cofinite, while the union of two finite sets is finite. Intersection behaves like union with "finite" and "cofinite" interchanged.

Example 4. For a less trivial example of the point made by Example 2, consider a Venn diagram formed by n closed curves partitioning the diagram into 2^n regions, and let X be the (infinite) set of all points in the plane not on any curve but somewhere within the diagram. The interior of each region is thus an infinite subset of X, and every point in X is in exactly one region. Then the set of all 2^{2n} possible unions of regions (including the empty set obtained as the union of the empty set of regions and X obtained as the union of all 2^n regions) is closed under union, intersection, and complement relative to X and therefore forms a concrete Boolean algebra. Again we have finitely many subsets of an infinite set forming a concrete Boolean algebra, with Example 2 arising as the case $n = 0$ of no curves.

Subsets as Bit Vectors

A subset Y of X can be identified with an indexed family of bits with index set X, with the bit indexed by $x \in X$ being 1 or 0 according to whether or not $x \in Y$. (This is the so-called characteristic function notion of a subset.) For example a 32-bit computer word consists of 32 bits indexed by the set {0,1,2,...,31}, with 0 and 31 indexing the low and high order bits respectively. For a smaller example, if $X = \{a,b,c\}$ where a, b, c are viewed as bit positions in that order from left to right, the eight subsets {}, {c}, {b}, {b,c}, {a}, {a,c}, {a,b}, and {a,b,c} of X can be identified with the respective bit vectors 000, 001, 010, 011, 100, 101, 110, and 111. Bit vectors indexed by the set of natural numbers are infinite sequences of bits, while those indexed by the reals in the unit interval [0,1] are packed too densely to be able to write conventionally but nonetheless form well-defined indexed families (imagine coloring every point of the interval [0,1] either black or white independently; the black points then form an arbitrary subset of [0,1]).

From this bit vector viewpoint, a concrete Boolean algebra can be defined equivalently as a nonempty set of bit vectors all of the same length (more generally, indexed by the same set) and closed under the bit vector operations of bitwise $\wedge$, $\vee$, and $\neg$, as in $1010 \wedge 0110 = 0010$, $1010 \vee 0110 = 1110$, and $\neg 1010 = 0101$, the bit vector realizations of intersection, union, and complement respectively.

The Prototypical Boolean Algebra

The set {0,1} and its Boolean operations as treated above can be understood as the special case of bit vectors of length one, which by the identification of bit vectors with subsets can also be understood as the two subsets of a one-element set. We call this the **prototypical** Boolean algebra, justified by the following observation.

The laws satisfied by all nondegenerate concrete Boolean algebras coincide with those satisfied by the prototypical Boolean algebra.

This observation is easily proved as follows. Certainly any law satisfied by all concrete Boolean algebras is satisfied by the prototypical one since it is concrete. Conversely any law that fails for some concrete Boolean algebra must have failed at a particular bit position, in which case that position by itself furnishes a one-bit counterexample to that law. Nondegeneracy ensures the existence of at least one bit position because there is only one empty bit vector.

The final goal of the next section can be understood as eliminating "concrete" from the above observation. We shall however reach that goal *via* the surprisingly stronger observation that, up to isomorphism, all Boolean algebras are concrete.

Boolean Algebras: The Definition

The Boolean algebras we have seen so far have all been concrete, consisting of bit vectors or equivalently of subsets of some set. Such a Boolean algebra consists of a set and operations on that set which can be *shown* to satisfy the laws of Boolean algebra.

Instead of showing that the Boolean laws are satisfied, we can instead postulate a set X, two binary operations on X, and one unary operation, and *require* that those operations satisfy the laws of Boolean algebra. The elements of X need not be bit vectors or subsets but can be anything at all. This leads to the more general *abstract* definition.

A **Boolean algebra** is any set with binary operations ∧ and ∨ and a unary operation ¬ thereon satisfying the Boolean laws.

For the purposes of this definition it is irrelevant how the operations came to satisfy the laws, whether by fiat or proof. All concrete Boolean algebras satisfy the laws, whence every concrete Boolean algebra is a Boolean algebra according to our definitions. This axiomatic definition of a Boolean algebra as a set and certain operations satisfying certain laws or axioms *by fiat* is entirely analogous to the abstract definitions of group, ring, field *etc.* characteristic of modern or abstract algebra.

Given any complete axiomatization of Boolean algebra, such as the axioms for a complemented distributive lattice, a sufficient condition for an algebraic structure of this kind to satisfy all the Boolean laws is that it satisfy just those axioms. The following is therefore an equivalent definition.

A **Boolean algebra** is a complemented distributive lattice.

The section on axiomatization lists other axiomatizations, any of which can be made the basis of an equivalent definition.

Representable Boolean Algebras

Although every concrete Boolean algebra is a Boolean algebra, not every Boolean algebra need be concrete. Let n be a square-free positive integer, one not divisible by the square of an integer, for example 30 but not 12. The operations of greatest common divisor, least common multiple, and division into n (that is, $\neg x = n/x$), can be shown to satisfy all the Boolean laws when their arguments range over the positive divisors of n. Hence those divisors form a Boolean algebra. These divisors are not subsets of a set, making the divisors of n a Boolean algebra that is not concrete according to our definitions.

However if we *represent* each divisor of n by the set of its prime factors, we find that this nonconcrete Boolean algebra is isomorphic to the concrete Boolean algebra consisting of all sets of prime factors of n, with union corresponding to least common multiple, intersection to greatest common divisor, and complement to division into n. So this example while not technically concrete is at least "morally" concrete *via* this representation, called an isomorphism. This example is an instance of the following notion.

A Boolean algebra is called **representable** when it is isomorphic to a concrete Boolean algebra.

The obvious next question is answered positively as follows.

Every Boolean algebra is representable.

That is, up to isomorphism, abstract and concrete Boolean algebras are the same thing. This quite nontrivial result depends on the Boolean prime ideal theorem, a choice principle slightly weaker than the axiom of choice, and is treated in more detail in the article Stone's representation theorem for Boolean algebras.

This strong relationship implies a weaker result strengthening the observation in the previous subsection to the following easy consequence of representability.

The laws satisfied by all Boolean algebras coincide with those satisfied by the prototypical Boolean algebra.

It is weaker in the sense that it does not of itself imply representability. Boolean algebras are special here, for example a relation algebra is a Boolean algebra with additional structure but it is not the case that every relation algebra is representable in the sense appropriate to relation algebras.

Axiomatizing Boolean Algebra

The above definition of an abstract Boolean algebra as a set and operations satisfying "the" Boolean laws raises the question, what are those laws? A simple-minded answer is "all Boolean laws," which can be defined as all equations that hold for the Boolean algebra of 0 and 1. Since there are infinitely many such laws this is not a terribly satisfactory answer in practice, leading to the next question: does it suffice to require only finitely many laws to hold?

In the case of Boolean algebras the answer is yes. In particular the finitely many equations we have listed above suffice. We say that Boolean algebra is **finitely axiomatizable** or **finitely based.**

Can this list be made shorter yet? Again the answer is yes. To begin with, some of the above laws are implied by some of the others. A sufficient subset of the above laws consists of the pairs of associativity, commutativity, and absorption laws, distributivity of $\wedge$ over $\vee$ (or the other distributivity law — one suffices), and the two complement laws. In fact this is the traditional axiomatization of Boolean algebra as a complemented distributive lattice.

By introducing additional laws not listed above it becomes possible to shorten the list yet further. In 1933 Edward Huntington showed that if the basic operations are taken to be $x \vee y$ and $\neg x$, with $x \wedge y$ considered a derived operation (*e.g. via* De Morgan's law in the form $x \wedge y = \neg(\neg x \vee \neg y)$), then the equation $\neg(\neg x \vee \neg y) \vee \neg(\neg x \vee y)$ $= x$ along with the two equations expressing associativity and commutativity of $\vee$ completely axiomatized Boolean algebra. When the only basic operation is the binary NAND operation $\neg(x \wedge y)$, Stephen Wolfram has proposed in his book *A New Kind of Science* the single axiom $(((xy)z)(x((xz)x))) = z$ as a one-equation axiomatization of Boolean algebra, where for convenience here xy denotes the NAND rather than the AND of x and y.

Propositional Logic

Propositional logic is a logical system that is intimately connected to Boolean algebra. Many syntactic concepts of Boolean algebra carry over to propositional logic with only minor changes in notation and terminology, while the semantics of propositional logic are defined *via* Boolean algebras in a way that the tautologies (theorems) of propositional logic correspond to equational theorems of Boolean algebra.

Syntactically, every Boolean term corresponds to a **propositional formula** of propositional logic. In this translation between Boolean algebra and propositional logic, Boolean variables x,y... become **propositional variables** (or **atoms**) $P,Q,...$, Boolean terms such as $x \vee y$ become propositional formulas $P \vee Q$, 0 becomes *false* or $\bot$, and 1 becomes *true* or T. It is convenient when referring to generic propositions to use Greek letters Φ, Ψ,... as metavariables (variables outside the language of propositional calculus, used when talking *about* propositional calculus) to denote propositions.

The semantics of propositional logic rely on **truth assignments**. The essential idea of a truth assignment is that the propositional variables are mapped to elements of a fixed Boolean algebra, and then the **truth value** of a propositional formula using these letters is the element of the Boolean algebra that is obtained by computing the value of the Boolean term corresponding to the formula. In classical semantics, only the two-element Boolean algebra is used, while in Boolean-valued semantics arbitrary Boolean algebras are considered. A **tautology** is a propositional formula that is assigned truth value 1 by every truth assignment of its propositional variables to an arbitrary Boolean algebra (or, equivalently, every truth assignment to the two element Boolean algebra).

These semantics permit a translation between tautologies of propositional logic and equational theorems of Boolean algebra. Every tautology Φ of propositional logic can be expressed as the Boolean equation $\Phi = 1$, which will be a theorem of Boolean algebra. Conversely every theorem $\Phi = \Psi$ of Boolean algebra corresponds to the tautologies $(\Phi \vee \neg\Psi) \wedge (\neg\Phi \vee \Psi)$ and $(\Phi \wedge \Psi) \vee (\neg\Phi \wedge \neg\Psi)$. If $\rightarrow$ is in the language these last tautologies can also be written as $(\Phi \rightarrow \Psi) \wedge (\Psi \rightarrow \Phi)$, or as two separate theorems $\Phi \rightarrow \Psi$ and $\Psi \rightarrow \Phi$; if $\equiv$ is available then the single tautology $\Phi \equiv \Psi$ can be used.

Applications

One motivating application of propositional calculus is the analysis of propositions and deductive arguments in natural language. Whereas the proposition "if $x = 3$ then $x+1 = 4$" depends on the meanings of such symbols as + and 1, the proposition "if $x = 3$ then $x = 3$" does not; it is true merely by virtue of its structure, and remains true whether "$x = 3$" is replaced by "$x = 4$" or "the moon is made of green cheese." The generic or abstract form of this tautology is "if P then P", or in the language of Boolean algebra, "$P \rightarrow P$".

Replacing P by $x = 3$ or any other proposition is called **instantiation** of P by that proposition. The result of instantiating P in an abstract proposition is called an **instance** of the proposition. Thus "$x = 3 \rightarrow x = 3$" is a tautology by virtue of being an instance of the abstract tautology "$P \rightarrow P$". All occurrences of the instantiated variable must be instantiated with the same proposition, to avoid such nonsense as $P \rightarrow x = 3$ or $x = 3 \rightarrow x = 4$.

Propositional calculus restricts attention to abstract propositions, those built up from propositional variables using Boolean operations. Instantiation is still possible within propositional calculus, but only by instantiating propositional variables by abstract propositions, such as instantiating Q by $Q{\to}P$ in $P{\to}(Q{\to}P)$ to yield the instance $P{\to}((Q{\to}P){\to}P)$.

(The availability of instantiation as part of the machinery of propositional calculus avoids the need for metavariables within the language of propositional calculus, since ordinary propositional variables can be considered within the language to denote arbitrary propositions. The metavariables themselves are outside the reach of instantiation, not being part of the language of propositional calculus but rather part of the same language for talking about it that this sentence is written in, where we need to be able to distinguish propositional variables and their instantiations as being distinct syntactic entities.)

Deductive Systems for Propositional Logic

An axiomatization of propositional calculus is a set of tautologies called axioms and one or more inference rules for producing new tautologies from old. A *proof* in an axiom system A is a finite nonempty sequence of propositions each of which is either an instance of an axiom of A or follows by some rule of A from propositions appearing earlier in the proof (thereby disallowing circular reasoning). The last proposition is the **theorem** proved by the proof. Every nonempty initial segment of a proof is itself a proof, whence every proposition in a proof is itself a theorem. An axiomatization is **sound** when every theorem is a tautology, and **complete** when every tautology is a theorem.

Sequent Calculus

Propositional calculus is commonly organized as a Hilbert system, whose operations are just those of Boolean algebra and whose theorems are Boolean tautologies, those Boolean terms equal to the Boolean constant 1. Another form is sequent calculus, which has two sorts, propositions as in ordinary propositional calculus, and pairs of lists of propositions called sequents, such as $A{\vee}B$, $A{\wedge}C$,... $\vdash A$, $B{\to}C$,.... The two halves of a sequent are called the antecedent and the succedent respectively. The customary metavariable denoting an antecedent or part thereof is Γ, and for a succedent Δ; thus $\Gamma,A \vdash \Delta$ would denote a sequent whose succedent is a list Δ and whose antecedent is a list Γ with an additional proposition A appended after it. The antecedent is interpreted as the conjunction of its propositions, the succedent as the disjunction of its propositions, and the sequent itself as the entailment of the succedent by the antecedent.

Entailment differs from implication in that whereas the latter is a binary *operation* that returns a value in a Boolean algebra, the former is a binary *relation* which either holds or does not hold. In this sense entailment is an *external* form of implication, meaning external to the Boolean algebra, thinking of the reader of the sequent as also being external and interpreting and comparing antecedents and succedents in some Boolean algebra. The natural interpretation of $\vdash$ is as $\leq$ in the partial order of the Boolean algebra defined by $x \leq y$ just when $x \vee y = y$. This ability to mix external implication $\vdash$ and internal implication $\rightarrow$ in the one logic is among the essential differences between sequent calculus and propositional calculus.

Applications

Two-valued Logic

Boolean algebra as the calculus of two values is fundamental to digital logic, computer programming, and mathematical logic, and is also used in other areas of mathematics such as set theory and statistics.

Digital logic codes its symbols in various ways: as voltages on wires in high-speed circuits and capacitive storage devices, as orientations of a magnetic domain in ferromagnetic storage devices, as holes in punched cards or paper tape, and so on. Now it is possible to code more than two symbols in any given medium. For example one might use respectively 0, 1, 2, and 3 volts to code a four-symbol alphabet on a wire, or holes of different sizes in a punched card. In practice however the tight constraints of high speed, small size, and low power combine to make noise a major factor. This makes it hard to distinguish between symbols when there are many of them at a single site. Rather than attempting to distinguish between four voltages on one wire, digital designers have settled on two voltages per wire, high and low. To obtain four symbols one uses two wires, and so on.

Programmers programming in machine code, assembly language, and other programming languages that expose the low-level digital structure of the data registers operate on whatever symbols were chosen for the hardware, invariably bit vectors in modern computers for the above reasons. Such languages support both the numeric operations of addition, multiplication, *etc.* performed on words interpreted as integers, as well as the logical operations of disjunction, conjunction, *etc.* performed bit-wise on words interpreted as bit vectors. Programmers therefore have the option of working in and applying the laws of either numeric algebra or Boolean algebra as needed. A core differentiating feature is carry propagation with the former but not the latter.

Other areas where two values is a good choice are the law and mathematics. In everyday relaxed conversation, nuanced or complex answers such as "maybe" or "only on the weekend" are acceptable. In more focused situations such as a court of law or theorem-based mathematics however it is deemed advantageous to frame questions so as to admit a simple yes-or-no answer — is the defendant guilty or not guilty, is the proposition true or false — and to disallow any other answer. However much of a straitjacket this might prove in practice for the respondent, the principle of the simple yes-no question has become a central feature of both judicial and mathematical logic, making two-valued logic deserving of organization and study in its own right.

A central concept of set theory is membership. Now an organization may permit multiple degrees of membership, such as novice, associate, and full. With sets however an element is either in or out. The candidates for membership in a set work just like the wires in a digital computer: each candidate is either a member or a nonmember, just as each wire is either high or low.

Algebra being a fundamental tool in any area amenable to mathematical treatment, these considerations combine to make the algebra of two values of fundamental importance to computer hardware, mathematical logic, and set theory.

Two-valued logic can be extended to multi-valued logic, notably by replacing the Boolean domain $\{0, 1\}$ with the unit interval $[0,1]$, in which case rather than only taking values 0 or 1, any value between and including 0 and 1 can be assumed. Algebraically, negation (NOT) is replaced with $1 - x$, conjunction (AND) is replaced with multiplication (xy), and disjunction (OR) is defined *via* De Morgan's law. Interpreting these values as logical truth values yields a multi-valued logic, which forms the basis for fuzzy logic and probabilistic logic. In these interpretations, a value is interpreted as the "degree" of truth – to what extent a proposition is true, or the probability that the proposition is true.

Boolean Operations

The original application for Boolean operations was mathematical logic, where it combines the truth values, true or false, of individual formulas.

Natural languages such as English have words for several Boolean operations, in particular conjunction (*and*), disjunction (*or*), negation (*not*), and implication (*implies*). *But not* is synonymous with *and not*. When used to combine situational assertions such as "the block is on the table" and "cats drink milk," which naively are either true or false, the meanings of these logical connectives often have the meaning of their logical counterparts. However with descriptions of behavior such as "Jim walked through the door", one starts to notice differences such as failure of commutativity, for example the conjunction of "Jim opened the door" with "Jim walked through the door" in that order is not equivalent to their conjunction in the other order, since *and* usually means *and then* in such cases. Questions can be similar: the order "Is the sky blue, and why is the sky blue?" makes more sense than the reverse order. Conjunctive commands about behavior are like behavioral assertions, as in *get dressed and go to school*. Disjunctive commands such *love me or leave me* or *fish or cut bait* tend to be asymmetric *via* the implication that one alternative is less preferable. Conjoined nouns such as *tea and milk* generally describe aggregation as with set union while *tea or milk* is a choice. However context can reverse these senses, as in *your choices are coffee and tea* which usually means the same as *your choices are coffee or tea* (alternatives). Double negation as in "I don't not like milk" rarely means literally "I do like milk" but rather conveys some sort of hedging, as though to imply that there is a third possibility. "Not not P" can be loosely interpreted as "surely P", and although *P* necessarily implies "not not *P*" the converse is suspect in English, much as with intuitionistic logic. In view of the highly idiosyncratic usage of conjunctions in natural languages, Boolean algebra cannot be considered a reliable framework for interpreting them.

Boolean operations are used in digital logic to combine the bits carried on individual wires, thereby interpreting them over $\{0,1\}$. When a vector of n identical binary gates are used to combine two bit vectors each of n bits, the individual bit operations can be understood collectively as a single operation on values from a Boolean algebra with 2^n elements.

Naive set theory interprets Boolean operations as acting on subsets of a given set X. As we saw earlier this behavior exactly parallels the coordinate-wise combinations of bit vectors, with the union of two sets corresponding to the disjunction of two bit vectors and so on.

The 256-element free Boolean algebra on three generators is deployed in computer displays based on raster graphics, which use bit blit to manipulate whole regions consisting of pixels, relying on Boolean operations to specify how the source region should be combined with the destination, typically with the help of a third region called the mask. Modern video cards offer all $2^{23} = 256$ ternary operations for this purpose, with the choice of operation being a one-byte (8-bit) parameter. The constants SRC = 0xaa or 10101010, DST = 0xcc or 11001100, and MSK = 0xf0 or 11110000 allow Boolean operations such as (SRC^DST)&MSK (meaning XOR the source and destination and then AND the result with the mask) to be written directly as a constant denoting a byte calculated at compile time, 0x60 in the (SRC^DST)&MSK example, 0x66 if just SRC^DST, *etc.* At run time the video card interprets the byte as the raster operation indicated by the original expression in a uniform way that requires remarkably little hardware and which takes time completely independent of the complexity of the expression.

Solid modeling systems for computer aided design offer a variety of methods for building objects from other objects, combination by Boolean operations being one of them. In this method the space in which objects exist is understood as a set S of voxels (the three-dimensional analogue of pixels in two-dimensional graphics) and shapes are defined as subsets of S, allowing objects to be combined as sets *via* union, intersection, *etc.* One obvious use is in building a complex shape from simple shapes simply as the union of the latter. Another use is in sculpting understood as removal of material: any grinding, milling, routing, or drilling operation that can be performed with physical machinery on physical materials can be simulated on the computer with the Boolean operation $x \wedge \neg y$ or $x - y$, which in set theory is set difference, remove the elements of y from those of x. Thus given two shapes one to be machined and the other the material to be removed, the result of machining the former to remove the latter is described simply as their set difference.

Boolean Searches

Search engine queries also employ Boolean logic. For this application, each web page on the Internet may be considered to be an "element" of a "set". The following examples use a syntax supported by Google.

- Doublequotes are used to combine whitespace-separated words into a single search term.
- Whitespace is used to specify logical AND, as it is the default operator for joining search terms:

"Search term 1" "Search term 2"

- The OR keyword is used for logical OR:

"Search term 1" OR "Search term 2"

- The minus sign is used for logical NOT (AND NOT):

"Search term 1" − "Search term 2"

TRUTH TABLE

A **truth table** is a mathematical table used in logic — specifically in connection with Boolean algebra, boolean functions, and propositional calculus — to compute the functional values of logical expressions on each of their functional arguments, that is, on each combination of values taken by their logical variables (Enderton, 2001). In particular, truth tables can be used to tell whether a propositional expression is true for all legitimate input values, that is, logically valid.

Practically, a truth table is composed of one column for each input variable, and one final column for all of the possible results of the logical operation that the table is meant to represent. Each row of the truth table therefore contains one possible configuration of the input variables (for instance, A=true B=false), and the result of the operation for those values.

Binary Operations

Logical Identity

Logical identity is an operation on one logical value, typically the value of a proposition, that produces a value of *true* if its operand is true and a value of *false* if its operand is false.

The truth table for the logical identity operator is as follows:

Logical Identity	
p	*p*
Operand	*Value*
T	T
F	F

Logical Negation

Logical negation is an operation on one logical value, typically the value of a proposition, that produces a value of *true* if its operand is false and a value of *false* if its operand is true.

The truth table for **NOT p** (also written as ¬p, **Np**, **Fpq**, or ~p) is as follows:

Logical Negation	
P	¬p
T	F
F	T

Binary Operations

Truth Table For All Binary Logical Operators

Here is a truth table giving definitions of all 16 of the possible truth functions of two binary variables:

P	Q	F⁰	NOR¹	Xq²	¬p³	↛⁴	¬q⁵	XOR⁶	NAND⁷	AND⁸	XNOR⁹	q¹⁰	if/then¹¹	p¹²	then/if¹³	OR¹⁴	T¹⁵
T	T	F	F	F	F	F	F	F	F	T	T	T	T	T	T	T	T
T	F	F	F	F	F	T	T	T	T	F	F	F	F	T	T	T	T
F	T	F	F	T	T	F	F	T	T	F	F	T	T	F	F	T	T
F	F	F	T	F	T	F	T	F	T	F	T	F	T	F	T	F	T
Com		✓	✓					✓	✓	✓	✓					✓	✓
L id			F				F		T	T	T,F	T				F	
R id				F			F		T	T				T,F	T	F	

where T = true and F = false. The **Com** row indicates whether an operator, **op**, is commutative - **P op Q = Q op P**. The **L id** row shows the operator's left identity if it has one - a value **I** such that **I op Q = Q**. The **R id** row shows the operator's right identity if it has one - a value **I** such that **P op I = P**.

Key:

					Operation name
0	(F F F F)(p, q)	F	False		Contradiction
1	(F F F T)(p, q)	NOR	↓		Logical NOR
2	(F F T F)(p, q)	Xq	p ↚ q		Converse nonimplication
3	(F F T T)(p, q)	Np	¬p		Negation
4	(F T F F)(p, q)	Xp	p ↛ q		Material nonimplication
5	(F T F T)(p, q)	Nq	¬q		Negation
6	(F T T F)(p, q)	XOR	⊕		Exclusive disjunction
7	(F T T T)(p, q)	NAND	↑		Logical NAND
8	(T F F F)(p, q)	AND	∧		Logical conjunction
9	(T F F T)(p, q)	XNOR	If and only if		Logical biconditional
10	(T F T F)(p, q)	q	q		Projection function
11	(T F T T)(p, q)	XNp	if/then		Material implication
12	(T T F F)(p, q)	p	p		Projection function
13	(T T F T)(p, q)	XNq	then/if		Converse implication
14	(T T T F)(p, q)	OR	∨		Logical disjunction
15	(T T T T)(p, q)	T	true		Tautology

Logical operators can also be visualized using Venn diagrams.

Logical Conjunction

Logical conjunction is an operation on two logical values, typically the values of two propositions, that produces a value of *true* if both of its operands are true.

The truth table for **p AND q** (also written as **p ∧ q, Kpq, p & q,** or **p .q**) is as follows:

Logical Conjunction		
p	*q*	*p ∧ q*
T	T	T
T	F	F
F	T	F
F	F	F

In ordinary language terms, if both *p* and *q* are true, then the conjunction *p ∧ q* is true. For all other assignments of logical values to *p* and to *q* the conjunction *p ∧ q* is false.

It can also be said that if *p*, then *p ∧ q* is *q*, otherwise *p ∧ q* is *p*.

Logical Disjunction

Logical disjunction is an operation on two logical values, typically the values of two propositions, that produces a value of *true* if at least one of its operands is true.

The truth table for **p OR q** (also written as **p ∨ q, Apq, p || q,** or **p + q**) is as follows:

Logical Disjunction		
p	*q*	*p ∨ q*
T	T	T
T	F	T
F	T	T
F	F	F

Stated in English, if *p*, then *p ∨ q* is *p*, otherwise *p ∨ q* is *q*.

Logical Implication

Logical implication or the material conditional are both associated with an operation on two logical values, typically the values of two propositions, that produces a value of *false* just in the singular case the first operand is true and the second operand is false.

The truth table associated with the material conditional **if p then q** (symbolized as **p → q**) and the logical implication **p implies q** (symbolized as **p ⇒ q,** or **Cpq**) is as follows:

Logical Implication		
p	q	$p \to q$
T	T	T
T	F	F
F	T	T
F	F	T

It may also be useful to note that $p \to q$ is equivalent to $\neg p \lor q$.

Logical Equality

Logical equality (also known as biconditional) is an operation on two logical values, typically the values of two propositions, that produces a value of *true* if both operands are false or both operands are true.

The truth table for **p XNOR q** (also written as $p \leftrightarrow q$, **Epq**, $p = q$, or $p \equiv q$) is as follows:

Logical Equality		
p	q	$p \equiv q$
T	T	T
T	F	F
F	T	F
F	F	T

So p EQ q is true if p and q have the same truth value (both true or both false), and false if they have different truth values.

Exclusive Disjunction

Exclusive disjunction is an operation on two logical values, typically the values of two propositions, that produces a value of *true* if one but not both of its operands is true.

The truth table for **p XOR q** (also written as $p \oplus q$, **Jpq**, or $p \neq q$) is as follows:

Exclusive Disjunction		
p	q	$p \oplus q$
T	T	F
T	F	T
F	T	T
F	F	F

For two propositions, **XOR** can also be written as $(p = 1 \land q = 0) \lor (p = 0 \land q = 1)$.

Logical NAND

The logical NAND is an operation on two logical values, typically the values of two propositions, that produces a value of *false* if both of its operands are true. In other words, it produces a value of *true* if at least one of its operands is false.

The truth table for **p NAND q** (also written as **p ↑ q, Dpq,** or **p | q**) is as follows:

Logical NAND		
p	*q*	*p ↑ q*
T	T	F
T	F	T
F	T	T
F	F	T

It is frequently useful to express a logical operation as a compound operation, that is, as an operation that is built up or composed from other operations. Many such compositions are possible, depending on the operations that are taken as basic or "primitive" and the operations that are taken as composite or "derivative".

In the case of logical NAND, it is clearly expressible as a compound of NOT and AND.

The negation of a conjunction: ¬(*p* ∧ *q*), and the disjunction of negations: (¬*p*) ∨ (¬*q*) can be tabulated as follows:

p	*q*	*p ∧ q*	¬(*p ∧ q*)	¬*p*	¬*q*	(¬*p*) ∨ (¬*q*)
T	T	T	F	F	F	F
T	F	F	T	F	T	T
F	T	F	T	T	F	T
F	F	F	T	T	T	T

Logical NOR

The logical NOR is an operation on two logical values, typically the values of two propositions, that produces a value of *true* if both of its operands are false. In other words, it produces a value of *false* if at least one of its operands is true. ↓ is also known as the Peirce arrow after its inventor, Charles Sanders Peirce, and is a Sole sufficient operator.

The truth table for **p NOR q** (also written as **p ↓ q, Xpq,** or **p ⊥ q**) is as follows:

Logical NOR		
p	*q*	*p ↓ q*
T	T	F
T	F	F
F	T	F
F	F	T

The negation of a disjunction $\neg(p \lor q)$, and the conjunction of negations $(\neg p) \land (\neg q)$ can be tabulated as follows:

p	q	$p \lor q$	$\neg(p \lor q)$	$\neg p$	$\neg q$	$(\neg p) \land (\neg q)$
T	T	T	F	F	F	F
T	F	T	F	F	T	F
F	T	T	F	T	F	F
F	F	F	T	T	T	T

Inspection of the tabular derivations for NAND and NOR, under each assignment of logical values to the functional arguments p and q, produces the identical patterns of functional values for $\neg(p \land q)$ as for $(\neg p) \lor (\neg q)$, and for $\neg(p \lor q)$ as for $(\neg p) \land (\neg q)$. Thus the first and second expressions in each pair are logically equivalent, and may be substituted for each other in all contexts that pertain solely to their logical values.

This equivalence is one of De Morgan's laws.

Applications

Truth tables can be used to prove many other logical equivalences. For example, consider the following truth table:

Logical Equivalence : $(p \to q) = (\neg p \lor q)$				
P	q	$\neg p$	$\neg p \lor q$	$p \to q$
T	T	F	T	T
T	F	F	F	F
F	T	T	T	T
F	F	T	T	T

This demonstrates the fact that $p \to q$ is logically equivalent to $\neg p \lor q$.

Truth Table For Most Commonly Used Logical Operators

Here is a truth table giving definitions of the most commonly used 6 of the 16 possible truth functions of 2 binary variables (P,Q are thus boolean variables):

P	Q	$P \land Q$	$P \lor Q$	$P \underline{\lor} Q$	$P \Delta Q$	$P \Rightarrow Q$	$P \Leftarrow Q$	$P \Longleftrightarrow Q$
T	T	T	T	F	T	T	T	T
T	F	F	T	T	F	F	T	F
F	T	F	T	T	F	T	F	F
F	F	F	F	F	T	T	T	T

Key:

T = true, F = false

$\land$ = AND (logical conjunction)

$\vee$ = OR (logical disjunction)

$\underline{\vee}$ = XOR (exclusive or)

$\underline{\triangle}$ = XNOR (exclusive nor)

$\longrightarrow$ = conditional "if-then"

$\longleftarrow$ = conditional "(then)-if"

$\longleftrightarrow$ biconditional or "if-and-only-if" is logically equivalent to $\triangle$: XNOR (exclusive nor).

Logical operators can also be visualized using Venn diagrams.

Condensed Truth Tables For Binary Operators

For binary operators, a condensed form of truth table is also used, where the row headings and the column headings specify the operands and the table cells specify the result. For example Boolean logic uses this condensed truth table notation:

$\wedge$	F	T	$\vee$	F	T
F	F	F	**F**	F	T
T	F	T	**T**	T	T

This notation is useful especially if the operations are commutative, although one can additionally specify that the rows are the first operand and the columns are the second operand. This condensed notation is particularly useful in discussing multi-valued extensions of logic, as it significantly cuts down on combinatoric explosion of the number of rows otherwise needed. It also provides for quickly recognizable characteristic "shape" of the distribution of the values in the table which can assist the reader in grasping the rules more quickly.

Truth Tables in Digital Logic

Truth tables are also used to specify the functionality of hardware look-up tables (LUTs) in digital logic circuitry. For an n-input LUT, the truth table will have 2^n values (or rows in the above tabular format), completely specifying a boolean function for the LUT. By representing each boolean value as a bit in a binary number, truth table values can be efficiently encoded as integer values in electronic design automation (EDA) software. For example, a 32-bit integer can encode the truth table for a LUT with up to 5 inputs.

When using an integer representation of a truth table, the output value of the LUT can be obtained by calculating a bit index k based on the input values of the LUT, in which case the LUT's output value is the kth bit of the integer. For example, to evaluate the output value of a LUT given an array of n boolean input values, the bit index of the truth table's output value can be computed as follows: if the ith input is true, let $Vi = 1$, else let $Vi = 0$. Then the kth bit of the binary representation of the truth table is the LUT's output value, where $k = V0*2^0 + V1*2^1 + V2*2^2 +... + Vn*2^n$.

Truth tables are a simple and straightforward way to encode boolean functions, however given the exponential growth in size as the number of inputs increase, they are not suitable for functions with a large number of inputs. Other representations which are more memory efficient are text equations and binary decision diagrams.

Applications of Truth Tables in Digital Electronics

In digital electronics and computer science (fields of applied logic engineering and mathematics), truth tables can be used to reduce basic boolean operations to simple correlations of inputs to outputs, without the use of logic gates or code. For example, a binary addition can be represented with the truth table:

A	B		C	R
1	1		1	0
1	0		0	1
0	1		0	1
0	0		0	0

where

A = First Operand

B = Second Operand

C = Carry

R = Result

This truth table is read left to right:

- Value pair (A,B) equals value pair (C,R).
- Or for this example, A plus B equal result R, with the Carry C.

Note that this table does not describe the logic operations necessary to implement this operation, rather it simply specifies the function of inputs to output values.

With respect to the result, this example may be arithmetically viewed as modulo 2 binary addition, and as logically equivalent to the exclusive-or (exclusive disjunction) binary logic operation.

In this case it can be used for only very simple inputs and outputs, such as 1s and 0s. However, if the number of types of values one can have on the inputs increases, the size of the truth table will increase.

For instance, in an addition operation, one needs two operands, A and B. Each can have one of two values, zero or one. The number of combinations of these two values is 2×2, or four. So the result is four possible outputs of C and R. If one were to use base 3, the size would increase to 3×3, or nine possible outputs.

The first "addition" example above is called a half-adder. A full-adder is when the carry from the previous operation is provided as input to the next adder. Thus, a truth table of eight rows would be needed to describe a full adder's logic:

A	B	C*	C	R
0	0	0	0	0
0	1	0	0	1
1	0	0	0	1
1	1	0	1	0
0	0	1	0	1
0	1	1	1	0
1	0	1	1	0
1	1	1	1	1

Same as previous, but..

C* = Carry from previous adder

History

Irving Anellis has done the research to show that C.S. Peirce appears to be the earliest logician to devise a truth table matrix. From the summary of his paper:

In 1997, John Shosky discovered, on the verso of a page of the typed transcript of Bertrand Russell's 1912 lecture on "The Philosophy of Logical Atomism" truth table matrices. The matrix for negation is Russell's, alongside of which is the matrix for material implication in the hand of Ludwig Wittgenstein. It is shown that an unpublished manuscript identified as composed by Peirce in 1893 includes a truth table matrix that is equivalent to the matrix for material implication discovered by John Shosky. An unpublished manuscript by Peirce identified as having been composed in 1883-84 in connection with the composition of Peirce's "On the Algebra of Logic: A Contribution to the Philosophy of Notation" that appeared in the American Journal of Mathematics in 1885 includes an example of an indirect truth table for the conditional.

LOGIC GATE

In electronics, a **logic gate** is an idealized or physical device implementing a Boolean function; that is, it performs a logical operation on one or more logical inputs, and produces a single logical output. Depending on the context, the term may refer to an **ideal logic gate**, one that has for instance zero rise time and unlimited fan-out, or it may refer to a non-ideal physical device.

Logic gates are primarily implemented using diodes or transistors acting as electronic switches, but can also be constructed using electromagnetic relays (relay logic), fluidic logic, pneumatic logic, optics, molecules, or even mechanical elements. With amplification, logic gates can be cascaded in the same way that

Boolean functions can be composed, allowing the construction of a physical model of all of Boolean logic, and therefore, all of the algorithms and mathematics that can be described with Boolean logic.

Logic circuits include such devices as multiplexers, registers, arithmetic logic units (ALUs), and computer memory, all the way up through complete microprocessors, which may contain more than 100 million gates. In practice, the gates are made from field-effect transistors (FETs), particularly MOSFETs (metal–oxide–semiconductor field-effect transistors).

Compound logic gates AND-OR-Invert (AOI) and OR-AND-Invert (OAI) are often employed in circuit design because their construction using MOSFETs is simpler and more efficient than the sum of the individual gates.

In reversible logic, Toffoli gates are used.

Electronic Gates

To build a functionally complete logic system, relays, valves (vacuum tubes), or transistors can be used. The simplest family of logic gates using bipolar transistors is called resistor-transistor logic (RTL). Unlike simple diode logic gates (which do not have a gain element), RTL gates can be cascaded indefinitely to produce more complex logic functions. RTL gates were used in early integrated circuits. For higher speed and better density, the resistors used in RTL were replaced by diodes resulting in diode-transistor logic (DTL). Transistor-transistor logic (TTL) then supplanted DTL. As integrated circuits became more complex, bipolar transistors were replaced with smaller field-effect transistors (MOSFETs). To reduce power consumption still further, most contemporary chip implementations of digital systems now use CMOS logic. CMOS uses complementary (both n-channel and p-channel) MOSFET devices to achieve a high speed with low power dissipation.

For small-scale logic, designers now use prefabricated logic gates from families of devices such as the TTL 7400 series by Texas Instruments, the CMOS 4000 series by RCA, and their more recent descendants. Increasingly, these fixed-function logic gates are being replaced by programmable logic devices, which allow designers to pack a large number of mixed logic gates into a single integrated circuit. The field-programmable nature of programmable logic devices such as FPGAs has removed the 'hard' property of hardware; it is now possible to change the logic design of a hardware system by reprogramming some of its components, thus allowing the features or function of a hardware implementation of a logic system to be changed.

Electronic logic gates differ significantly from their relay-and-switch equivalents. They are much faster, consume much less power, and are much smaller (all by a factor of a million or more in most cases). Also, there is a fundamental structural difference. The switch circuit creates a continuous metallic path for current to flow (in either direction) between its input and its output. The semiconductor logic gate, on the other hand, acts as a high-gain voltage amplifier, which sinks a tiny current at its input and produces a low-impedance voltage at its output.

It is not possible for current to flow between the output and the input of a semi-conductor logic gate.

Another important advantage of standardized integrated circuit logic families, such as the 7400 and 4000 families, is that they can be cascaded. This means that the output of one gate can be wired to the inputs of one or several other gates, and so on. Systems with varying degrees of complexity can be built without great concern of the designer for the internal workings of the gates, provided the limitations of each integrated circuit are considered.

The output of one gate can only drive a finite number of inputs to other gates, a number called the 'fanout limit'. Also, there is always a delay, called the 'propagation delay', from a change in input of a gate to the corresponding change in its output. When gates are cascaded, the total propagation delay is approximately the sum of the individual delays, an effect which can become a problem in high-speed circuits. Additional delay can be caused when a large number of inputs are connected to an output, due to the distributed capacitance of all the inputs and wiring and the finite amount of current that each output can provide.

Symbols

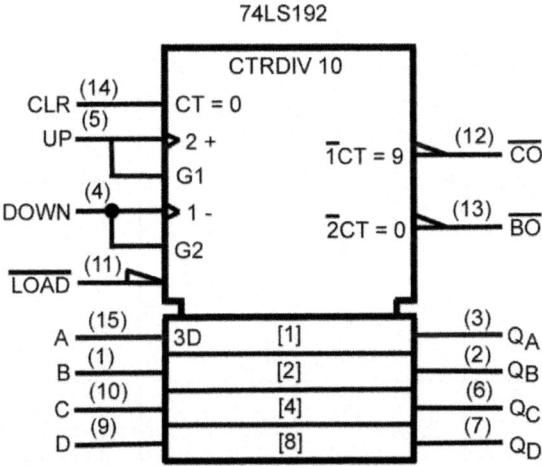

A synchronous 4-bit up/down decade counter symbol in accordance with ANSI/IEEE Std. 91-1984 and IEC Publication 60617-12.

There are two sets of symbols for elementary logic gates in common use, both defined in ANSI/IEEE Std 91-1984 and its supplement ANSI/IEEE Std 91a-1991. The "distinctive shape" set, based on traditional schematics, is used for simple drawings, and derives from MIL-STD-806 of the 1950s and 1960s. It is sometimes unofficially described as "military", reflecting its origin. The "rectangular shape" set, based on IEC 60617-12 and other early industry standards, has rectangular outlines for all types of gate and allows representation of a much wider range of devices than is possible with the traditional symbols. The IEC's system has been adopted by other standards, such as EN 60617-12:1999 in Europe and BS EN 60617-12:1999 in the United Kingdom.

The goal of IEEE Std 91-1984 was to provide a uniform method of describing the complex logic functions of digital circuits with schematic symbols. These functions were more complex than simple AND and OR gates. They could be medium scale circuits such as a 4-bit counter to a large scale circuit such as a microprocessor. IEC 617-12 and its successor IEC 60617-12 do not explicitly show the "distinctive shape" symbols, but do not prohibit them. These are, however, shown in ANSI/IEEE 91 with this note: "The distinctive-shape symbol is, according to IEC Publication 617, Part 12, not preferred, but is not considered to be in contradiction to that standard." This compromise was reached between the respective IEEE and IEC working groups to permit the IEEE and IEC standards to be in mutual compliance with one another.

In the 1980s, schematics were the predominant method to design both circuit boards and custom ICs known as gate arrays. Today custom ICs and the field-programmable gate array are typically designed with Hardware Description Languages (HDL) such as Verilog or VHDL.

Type	Distinctive shape	Rectangular shape	Boolean algebra between A & B	Truth table		
AND		&	$A.B$ or A & B	INPUT		OUTPUT
				A	B	A AND B
				0	0	0
				0	1	0
				1	0	0
				1	1	1
OR		≥1	$A + B$	INPUT		OUTPUT
				A	B	A OR B
				0	0	0
				0	1	1
				1	0	1
				1	1	1
NOT		1	$\overline{A}$ or $\sim A$	INPUT	OUTPUT	
				A	NOT A	
				0	1	
				1	0	

In electronics a NOT gate is more commonly called an inverter. The circle on the symbol is called a *bubble*, and is used in logic diagrams to indicate a logic negation between the external logic state and the internal logic state (1 to 0 or vice versa). On a circuit diagram it must be accompanied by a statement asserting that the *positive logic convention* or *negative logic convention* is being used (high voltage level = 1 or high voltage level = 0, respectively). The *wedge* is used in circuit diagrams to directly indicate an active-low (high voltage level = 0) input or output without requiring a uniform convention throughout the circuit diagram. This is called *Direct Polarity Indication*. Both the *bubble* and the *wedge* can be used on distinctive-shape and rectangular-shape symbols on circuit diagrams, depending on the logic convention used. On pure logic diagrams, only the *bubble* is meaningful.

				INPUT		OUTPUT
NAND		&	$\overline{A \cdot B}$ or $\overline{A \vert B}$	A	B	A NAND B
				0	0	1
				0	1	1
				1	0	1
				1	1	0

				INPUT		OUTPUT
NOR		≥1	$\overline{A + B}$ or $\overline{A - B}$	A	B	A NOR B
				0	0	1
				0	1	0
				1	0	0
				1	1	0

				INPUT		OUTPUT
XOR		=1	$A \oplus B$	A	B	A XOR B
				0	0	0
				0	1	1
				1	0	1
				1	1	0

				INPUT		OUTPUT
XNOR		=1	$\overline{A \oplus B}$ or $A \odot B$	A	B	A XNOR B
				0	0	1
				0	1	0
				1	0	0
				1	1	1

Two more gates are the exclusive-OR or XOR function and its inverse, exclusive-NOR or XNOR. The two input Exclusive-OR is true only when the two input values are *different*, false if they are equal, regardless of the value. If there are more than two inputs, the gate generates a true at its output if the number of trues at its input is *odd*. In practice, these gates are built from combinations of simpler logic gates.

Universal Logic Gates

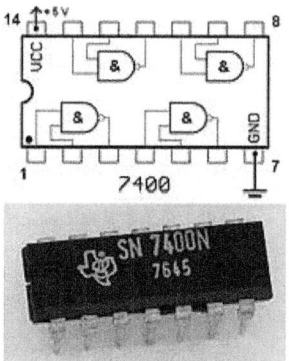

The 7400 chip, containing four NANDs. The two additional pins supply power (+5 V) and connect the ground.

Charles Sanders Peirce showed that NOR gates alone (or alternatively NAND gates alone) can be used to reproduce the functions of all the other logic gates, but his work on it was unpublished until 1933. The first published proof was by Henry M. Sheffer in 1913, so the NAND logical operation is sometimes called Sheffer stroke; the logical NOR is sometimes called *Peirce's arrow*. Consequently, these gates are sometimes called *universal logic gates*.

De Morgan Equivalent Symbols

By use of De Morgan's theorem, an *AND* function is identical to an *OR* function with negated inputs and outputs. Likewise, an *OR* function is identical to an *AND* function with negated inputs and outputs. A NAND gate is equivalent to an OR gate with negated inputs, and a NOR gate is equivalent to an AND gate with negated inputs.

This leads to an alternative set of symbols for basic gates that use the opposite core symbol (*AND* or *OR*) but with the inputs and outputs negated. Use of these alternative symbols can make logic circuit diagrams much clearer and help to show accidental connection of an active high output to an active low input or vice-versa. Any connection that has logic negations at both ends can be replaced by a negationless connection and a suitable change of gate or vice-versa. Any connection that has a negation at one end and no negation at the other can be made easier to interpret by instead using the De Morgan equivalent symbol at either of the two ends. When negation or polarity indicators on both ends of a connection match, there is no logic negation in that path (effectively, bubbles "cancel"), making it easier to follow logic states from one symbol to the next. This is commonly seen in real logic diagrams - thus the reader must not get into the habit of associating the shapes exclusively as OR or AND shapes, but also take into account the bubbles at both inputs and outputs in order to determine the "true" logic function indicated.

A De Morgan symbol can show more clearly a gate's primary logical purpose and the polarity of its nodes that are considered in the "signaled" (active, on) state. Consider the simplified case where a two-input NAND gate is used to drive a motor when either of its inputs are brought low by a switch. The "signaled" state (motor on) occurs when either one OR the other switch is on. Unlike a regular NAND symbol, which suggests AND logic, the De Morgan version, a two negative-input OR gate, correctly shows that OR is of interest. The regular NAND symbol has a bubble at the output and none at the inputs (the opposite of the states that will turn the motor on), but the De Morgan symbol shows both inputs and output in the polarity that will drive the motor.

De Morgan's theorem is most commonly used to implement logic gates as combinations of only NAND gates, or as combinations of only NOR gates, for economic reasons.

Data Storage

Logic gates can also be used to store data. A storage element can be constructed by connecting several gates in a "latch" circuit. More complicated designs that use clock signals and that change only on a rising or falling edge of the clock are called edge-triggered "flip-flops". The combination of multiple flip-flops in parallel, to store a multiple-bit value, is known as a register. When using any of these gate setups the overall system has memory; it is then called a sequential logic system since its output can be influenced by its previous state(s).

These logic circuits are known as computer memory. They vary in performance, based on factors of speed, complexity, and reliability of storage, and many different types of designs are used based on the application.

Three-state Logic Gates

A tristate buffer can be thought of as a switch. If B is on, the switch is closed.
If B is off, the switch is open.

A three-state logic gate is a type of logic gate that can have three different outputs: high , low (L) and high-impedance (Z). The high-impedance state plays no role in the logic, which is strictly binary. These devices are used on buses of the CPU to allow multiple chips to send data. A group of three-states driving a line with a suitable control circuit is basically equivalent to a multiplexer, which may be physically distributed over separate devices or plug-in cards.

In electronics, a high output would mean the output is sourcing current from the positive power terminal (positive voltage). A low output would mean the output is sinking current to the negative power terminal (zero voltage). High impedance would mean that the output is effectively disconnected from the circuit.

History and Development

The binary number system was refined by Gottfried Wilhelm Leibniz and he also established that by using the binary system, the principles of arithmetic and logic could be combined. In an 1886 letter, Charles Sanders Peirce described how logical operations could be carried out by electrical switching circuits. Eventually, vacuum tubes replaced relays for logic operations. Lee De Forest's modification, in 1907, of the Fleming valve can be used as AND logic gate. Ludwig Wittgenstein introduced a version of the 16-row truth table as proposition 5.101 of *Tractatus Logico-Philosophicus*. Walther Bothe, inventor of the coincidence circuit, got part of the 1954 Nobel Prize in physics, for the first modern electronic AND gate in 1924. Konrad Zuse designed and built electromechanical logic gates for his computer Z1. Claude E. Shannon introduced the use of Boolean algebra in the analysis and design of switching circuits in 1937. Active research is taking place in molecular logic gates.

Implementations

Since the 1990s, most logic gates are made in CMOS technology (*i.e.* NMOS and PMOS transistors are used). Often millions of logic gates are packaged in a single integrated circuit.

There are several logic families with different characteristics (power consumption, speed, cost, size) such as: RDL (resistor-diode logic), RTL (resistor-transistor logic), DTL (diode-transistor logic), TTL (transistor-transistor logic) and CMOS (complementary metal oxide semiconductor). There are also sub-variants, *e.g.* standard CMOS logic vs. advanced types using still CMOS technology, but with some optimizations for avoiding loss of speed due to slower PMOS transistors.

Non-electronic implementations are varied, though few of them are used in practical applications. Many early electromechanical digital computers, such as the Harvard Mark I, were built from relay logic gates, using electro-mechanical relays. Logic gates can be made using pneumatic devices, such as the Sorteberg relay or mechanical logic gates, including on a molecular scale. Logic gates have been made out of DNA and used to create a computer called MAYA. Logic gates can be made from quantum mechanical effects (though quantum computing usually diverges from boolean design). Photonic logic gates use non-linear optical effects.

In principle any method that leads to a gate that is functionally complete (for example, either a NOR or a NAND gate) can be used to make any kind of digital logic circuit. Note that the use of 3-state logic for bus systems is not needed, and can be replaced by digital multiplexers.

COANDĂ EFFECT

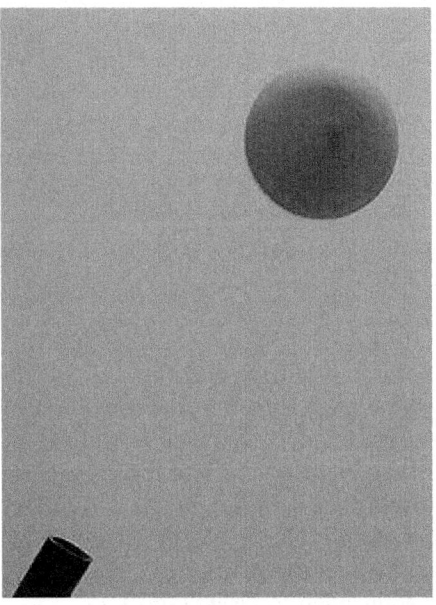

A spinning ping pong ball is held in a diagonal stream of air by the Coandă Effect. The ball "sticks" to the lower side of the air stream, which (in combination with the Magnus effect) stops the ball from falling down. The jet as a whole keeps the ball some distance from the jet exhaust, and gravity prevents it from being blown away.

The **Coandă effect** /kwɑndə/ is the tendency of a fluid jet to be attracted to a nearby surface. The principle was named after Romanian aerodynamics pioneer Henri Coandă, who was the first to recognize the practical application of the phenomenon in aircraft development.

Discovery

An early description of this phenomenon was provided by Thomas Young in a lecture given to The Royal Society in 1800:

The lateral pressure which urges the flame of a candle towards the stream of air from a blowpipe is probably exactly similar to that pressure which eases the inflection of a current of air near an obstacle. Mark the dimple which a slender stream of air makes on the surface of water. Bring a convex body into contact with the side of the stream and the place of the dimple will immediately show the current is deflected towards the body; and if the body be at liberty to move in every direction it will be urged towards the current...

A hundred years later, Henri Coandă identified an application of the effect during experiments with his Coandă-1910 aircraft which mounted an unusual engine designed by Coandă. The motor-driven turbine pushed hot air rearward, and Coandă noticed that the airflow was attracted to nearby surfaces. He discussed this matter with leading aerodynamicist Theodore von Kármán who named it the Coandă effect. In 1934 Coandă obtained a patent in France for a "Method and apparatus for deviation of a fluid into another fluid". The effect was described as the "Deviation of a plain jet of a fluid that penetrates another fluid in the vicinity of a convex wall." This effect is most noticeable near a curved surface, where the air stream has to speed up near the convex side, and hence lose pressure. Natural molecular movement in the air tends to equalise the overall pressure, and pushes in free air, which joins the original jet along the same trajectory as the convex side of the surface. In turn, this generates lift against the convex surface.

Causes

The Coandă effect is a result of entrainment of ambient fluid around the fluid jet. When a nearby wall does not allow the surrounding fluid to be pulled inwards towards the jet (*i.e.*, to be entrained), the jet moves towards the wall instead. The fluid of the jet and the surrounding fluid should be essentially the same substance (a gas jet into a body of gas or a liquid jet into a body of liquid). In one application, a jet of air is blown over the upper surface of an airfoil, which can have a strong influence on the overall lift, especially at high angles of attack when the flow would otherwise separate (stall).

Applications

The Coandă effect has important applications in various high-lift devices on aircraft, where air moving over the wing can be "bent down" towards the ground using flaps and a jet sheet blowing over the curved surface of the top of the wing. The bending of the flow results in aerodynamic lift. The flow from a high speed jet engine mounted in a pod over the wing produces increased lift by dramatically increasing the velocity gradient in the shear flow in the boundary layer. In this velocity gradient particles are blown away from the surface, thus lowering the pressure there. Closely following the work of Coandă on applications of his research, and in particular the work on his "Aerodina Lenticulară," John Frost of Avro Canada also spent considerable time researching the effect, leading to a series of "inside out" hovercraft-like aircraft from which the air exited in a ring around the outside of the aircraft and was directed by being "attached" to a flap-like ring.

This is as opposed to a traditional hovercraft design, in which the air is blown into a central area, the *plenum*, and directed down with the use of a fabric "skirt". Only one of Frost's designs was ever built, the Avrocar.

The VZ-9 AV Avrocar was a Canadian vertical takeoff and landing (VTOL) aircraft developed by Avro Aircraft Ltd. as part of a secret U.S. military project carried out in the early years of the Cold War. The Avrocar intended to exploit the Coandă effect to provide lift and thrust from a single "turborotor" blowing exhaust out the rim of the disk-shaped aircraft to provide anticipated VTOL-like performance. In the air, it would have resembled a flying saucer. Two prototypes were built as "proof-of-concept" test vehicles for a more advanced USAF fighter and also for a U.S. Army tactical combat aircraft requirement.

Avro's 1956 Project 1794 for the US military designed a larger-scale flying saucer based on the Coandă effect and intended to reach speeds between Mach 3 and Mach 4. Project documents remained classified until 2012.

The effect was also implemented during the U.S. Air Force's AMST project. Several aircraft, notably the Boeing YC-14, NASA's Quiet Short-Haul Research Aircraft, and NAL's Asuka research aircraft have been built to take advantage of this effect, by mounting turbofans on the top of the wings to provide high-speed air even at low flying speeds, but to date only one aircraft has gone into production using this system to a major degree, the Antonov An-72 'Coaler'. The Shin Meiwa US-1A flying boat utilizes a similar system, only it directs the propwash from its four turboprop engines over the top of the wing to generate low-speed lift. More uniquely, it incorporates a fifth turboshaft engine inside of the wing center-section solely to provide air for powerful blown flaps. The addition of these two systems gives the aircraft an impressive STOL capability.

The C-17 Globemaster III uses the Coandă effect for a comfortable ride at low flying speeds

The McDonnell Douglas YC-15 and its successor, the Boeing C-17 Globemaster III, also employ the effect. The NOTAR helicopter replaces the conventional propeller tail rotor with a Coandă effect tail.

An important practical use of the Coandă effect is for inclined hydropower screens, which separate debris, fish, *etc.*, otherwise in the input flow to the turbines. Due to the slope, the debris falls from the screens without mechanical clearing, and due to the wires of the screen optimizing the Coandă effect, the water flows though the screen to the penstocks leading the water to the turbines.

The Coandă effect is used in dual-pattern fluid dispensers in automobile windshield washers.

The operation principle of oscillatory flowmeters also relies on the Coandă phenomenon. The incoming liquid enters a chamber that contains 2 "islands". Due to the Coandă effect the main stream splits up and goes under one of the islands. This flow then feeds itself back into the main stream making it split up again, but in the direction of the second isle. This process repeats itself as long as the liquid circulates the chamber, resulting in a self-induced oscillation that is directly proportional to the velocity of the liquid and consequently the volume of substance flowing through the meter. A sensor picks up the frequency of this oscillation and transforms it into an analog signal yielding volume passing through.

In air conditioning the Coandă effect is exploited to increase the throw of a ceiling mounted diffuser. Because the Coandă effect causes air discharged from the diffuser to "stick" to the ceiling, it travels farther before dropping for the same discharge velocity than it would if the diffuser was mounted in free air, without the neighbouring ceiling. Lower discharge velocity means lower noise levels and, in the case of variable air volume (VAV) air conditioning systems, permits greater turn-down ratios. Linear diffusers and slot diffusers that present a greater length of contact with the ceiling exhibit greater Coandă effect.

In cardiovascular medicine, the Coandă effect accounts for the separate streams of blood in the fetal right atrium. It also explains why eccentric mitral regurgitation jets are attracted and dispersed along adjacent left atrial wall surfaces (so called "wall-hugging jets" as seen on echocardiographic color-doppler interrogation). This is clinically relevant because the visual area of these eccentric wall-hugging jets is often underestimated compared to the more readily apparent central jets. In these cases, volumetric methods such as the proximal isovelocity surface area (PISA) method are preferred to quantify the severity of mitral regurgitation.

The Coandă effect is used in medicine as a ventilator.

In meteorology, the Coandă effect theory has also been applied to some air streams flowing out of mountain ranges such as the Carpathian Mountains and Transylvanian Alps, where effects on agriculture and vegetation have been noted. It also appears to be an effect in the Rhone Valley in France and near Big Delta in Alaska.

In Formula One the Coanda effect has been exploited by the McLaren, Sauber, Ferrari and Lotus teams, after the first introduction by Adrian Newey in 2011, to help redirect exhaust gases to run through the rear diffuser with the intention of increasing downforce at the rear of the car. Due to changes in regulations set in place by the FIA from the beginning of the 2014 Formula One season, the intention of redirecting exhaust gases to use the Coandă effect have been negated, due to the mandatory requirement that the car exhaust must not have bodywork directly behind the exit for use of aerodynamic effect.

Demonstration

The Coandă effect can be demonstrated by directing a small jet of air upwards at an angle over a ping pong ball. The jet is drawn to and follows the upper surface of the ball curving around it, due to the (radial) acceleration (slowing and turning) of the air around the ball. With enough airflow, this change in momentum is balanced by the equal and opposite force on the ball supporting its weight. This demonstration can be performed using a vacuum cleaner if the outlet can be attached to the pipe and aimed upwards at an angle.

A common misconception is that Coandă effect is demonstrated when a stream of tap water flows over the back of a spoon held lightly in the stream and the spoon is pulled into the stream. While the flow looks very similar to the air flow over the ping pong ball above, the cause is not really the Coandă effect. Here, because it is a flow of water into air, there is little entrainment of the surrounding fluid (the air) into the jet (the stream of water). This particular demonstration is dominated by surface tension.

Another demonstration is to direct the air flow from, *e.g.*, a vacuum cleaner operating in reverse, tangentially past a round cylinder. A waste basket works well. The air flow seems to "wrap around" the cylinder and can be detected at more than 180° from the incoming flow. Under the right conditions, flow rate, weight of the cylinder, smoothness of the surface it sits on, the cylinder will actually move. Note that the cylinder will not move directly into the flow as a misapplication of the Bernoulli effect would predict, but at a diagonal.

The effect can also be seen by placing a can in front of a lit candle. If one blows directly at the can, the air will bend around it and extinguish the candle.

If two lit candles are placed side-by-side, the heated air from each candle rises and entrains surrounding air. Since both "jets" are trying to entrain common air from the space between the two streams, they are drawn towards each other. This is more apparent if the candles are making a little smoke. This is a demonstration of the Coandă effect without the presence of any surface. In some sense, the plane of symmetry between the two flows can be thought of as the surface. In actual fact this is not the Coanda Effect in action but is in fact the Atmospheric Press in action as putting two candles close together causes an area of warmth between them which warms the air which then rises - leaving the cooler atmosphere to try fill this partial void and so the flames are forced together.

Problems caused

The engineering use of Coandă effect has disadvantages as well as advantages.

In marine propulsion, the efficiency of a propeller or thruster can be severely curtailed by the Coandă effect. The force on the vessel generated by a propeller is a function of the speed, volume and direction of the water jet leaving the propeller. Under certain conditions (*e.g.*, when a ship moves through water) the Coandă effect changes the direction of a propeller jet, causing it to follow the shape of the ship's hull. The side force from a tunnel thruster at the bow of a ship decreases rapidly with forward speed. The side thrust may completely disappear at speeds above about 3 knots.

Chapter 10

TRANSFER DEVICES AND FEEDERS

PRODUCTION LINE

A **production line** is a set of sequential operations established in a factory whereby materials are put through a refining process to produce an end-product that is suitable for onward consumption; or components are assembled to make a finished article.

Typically, raw materials such as metal ores or agricultural products such as foodstuffs or textile source plants (cotton, flax) require a sequence of treatments to render them useful. For metal, the processes include crushing, smelting and further refining. For plants, the useful material has to be separated from husks or contaminants and then treated for onward sale.

History

Early production processes were constrained by the availability of a source of energy, with wind mills and water mills providing power for the crude heavy processes and manpower being used for activities requiring more precision. In earlier centuries, with raw materials, power and people often being in different locations, production was distributed across a number of sites. The concentration of numbers of people in manufactories, and later the factory as exemplified by the cotton mills of Richard Arkwright, started the move towards co-locating individual processes.

Introduction of the Steam Engine

With the development of the steam engine in the latter half of the 18th century, the production elements became less reliant on the location of the power source, and so the processing of goods moved to either the source of the materials or the location of people to perform the tasks. Separate processes for different treatment stages were brought into the same building, and the various stages of refining or manufacture were combined.

Industrial Revolution

With increasing use of steam power, and increasing use of machinery to supplant the use of people, the integrated use of techniques in production lines spurred the industrial revolutions of Europe and the United States.

ASSEMBLY LINE

An Airbus A321 on final assembly line 3 in the Airbus plant at Hamburg Finkenwerder Airport.

An **assembly line** is a manufacturing process (most of the time called a *progressive assembly*) in which parts (usually interchangeable parts) are added as the semi-finished assembly moves from work station to work station where the parts are added in sequence until the final assembly is produced. By mechanically moving the parts to the assembly work and moving the semi-finished assembly from work station to work station, a finished product can be assembled much faster and with much less labor than by having workers carry parts to a stationary piece for assembly.

Assembly lines are the common method of assembling complex items such as automobiles and other transportation equipment, household appliances and electronic goods.

Concept

Assembly lines are designed for the sequential organization of workers, tools or machines, and parts. The motion of workers is minimized to the extent possible. All parts or assemblies are handled either by conveyors or motorized vehicles such as fork lifts, or gravity, with no manual trucking. Heavy lifting is done by machines such as overhead cranes or fork lifts. Each worker typically performs one simple operation.

According to Henry Ford:

The principles of assembly are these:

(1) Place the tools and the men in the sequence of the operation so that each component part shall travel the least possible distance while in the process of finishing.

(2) Use work slides or some other form of carrier so that when a workman completes his operation, he drops the part always in the same place — which place must always be the most convenient place to his hand — and if possible have gravity carry the part to the next workman for his own.

(3) Use sliding assembling lines by which the parts to be assembled are delivered at convenient distances.

Simple Example

Consider the assembly of a car: assume that certain steps in the assembly line are to install the engine, install the hood, and install the wheels (in that order, with arbitrary interstitial steps); only one of these steps can be done at a time. In traditional production, only one car would be assembled at a time. If engine installation takes 20 minutes, hood installation takes five minutes, and wheels installation takes 10 minutes, then a car can be produced every 35 minutes.

In an assembly line, car assembly is split between several stations, all working simultaneously. When one station is finished with a car, it passes it on to the next. By having three stations, a total of three different cars can be operated on at the same time, each one at a different stage of its assembly.

After finishing its work on the first car, the engine installation crew can begin working on the second car. While the engine installation crew works on the second car, the first car can be moved to the hood station and fitted with a hood, then to the wheels station and be fitted with wheels. After the engine has been installed on the second car, the second car moves to the hood assembly. At the same time, the third car moves to the engine assembly. When the third car's engine has been mounted, it then can be moved to the hood station; meanwhile, subsequent cars (if any) can be moved to the engine installation station.

Assuming no loss of time when moving a car from one station to another, the longest stage on the assembly line determines the throughput (20 minutes for the engine installation) so a car can be produced every 20 minutes, once the first car taking 35 minutes has been produced.

History

Before the Industrial Revolution, most manufactured products were made individually by hand. A single craftsman or team of craftsmen would create each part of a product. They would use their skills and tools such as files and knives to create the individual parts. They would then assemble them into the final product, making cut-and-try changes in the parts until they fit and could work together (craft production).

Division of labor was practiced in China where state run monopolies mass-produced metal agricultural implements, china and armor and weapons centuries before it appeared in Europe on the eve of the Industrial Revolution. Adam Smith discussed the division of labour in the manufacture of pins at length in his book *The Wealth of Nations*.

The Venetian Arsenal, dating to about 1104, operated similar to a production line. Ships moved down a canal and were fitted by the various shops they passed. At the peak of its efficiency in the early 16th century, the Venetian Arsenal employed some 16,000 people who could apparently produce nearly one ship each day, and could fit out, arm, and provision a newly built galley with standardized parts on an assembly-line basis. Although the Venice Arsenal lasted until the early Industrial Revolution, production line methods did not become common even then.

Interchangeable Parts

During the early 19th century, the development of machine tools such as the screw-cutting lathe, metal planer, and milling machine, and of toolpath control *via* jigs and fixtures, provided the prerequisites for the modern assembly line by making interchangeable parts a practical reality.

Industrial Revolution

The Industrial Revolution led to a proliferation of manufacturing and invention. Many industries, notably textiles, firearms, clocks and watches, horse-drawn vehicles, railway locomotives, sewing machines, and bicycles, saw expeditious improvement in materials handling, machining, and assembly during the 19th century, although modern concepts such as industrial engineering and logistics had not yet been named.

The automatic flour mill built by Oliver Evans in 1785 was called the beginning of modern bulk material handling by Roe. Evans's mill used a leather belt bucket elevator, screw conveyors, canvas belt conveyors, and other mechanical devices to completely automate the process of making flour. The innovation spread to other mills and breweries.

Probably the earliest industrial example of a linear and continuous assembly process is the Portsmouth Block Mills, built between 1801 and 1803. Marc Isambard Brunel (father of Isambard Kingdom Brunel), with the help of Henry Maudslay and others, designed 22 types of machine tools to make the parts for the rigging blocks used by the Royal Navy. This factory was so successful that it remained in use until the 1960s, with the workshop still visible at HM Dockyard in Portsmouth, and still containing some of the original machinery.

One of the earliest examples of an almost modern factory layout, designed for easy material handling, was the Bridgewater Foundry. The factory grounds were bordered by the Bridgewater Canal and the Liverpool and Manchester Railway. The buildings were arranged in a line with a railway for carrying the work going through the buildings. Cranes were used for lifting the heavy work, which sometimes weighed in the tens of tons. The work passed sequentially through to erection of framework and final assembly.

The first flow assembly line was initiated at the factory of Richard Garrett & Sons, Leiston Works in Leiston in the English county of Suffolk for the manufacture of portable steam engines. The assembly line area was called 'The Long Shop' on account of its length and was fully operational by early 1853. The boiler was brought up from the foundry and put at the start of the line, and as it progressed through the building it would stop at various stages where new parts would be added. From the upper level, where other parts were made, the lighter parts would be lowered over a balcony and then fixed onto the machine on the ground level. When the machine reached the end of the shop, it would be completed.

Late 19th Century Steam and Electric Conveyors

Steam powered conveyor lifts began being used for loading and unloading ships some time in the last quarter of the 19th century. Hounshell shows a ca. 1885 sketch of an electric powered conveyor moving cans through a filling line in a canning factory.

The meatpacking industry of Chicago is believed to be one of the first industrial assembly lines (or dis-assembly lines) to be utilized in the United States starting in 1867. Workers would stand at fixed stations and a pulley system would bring the meat to each worker and they would complete one task. Henry Ford and others have written about the influence of this slaughterhouse practice on the later developments at Ford Motor Company.

20th Century

According to a book entitled *Michigan Yesterday & Today* authored by Robert W. Domm, the modern assembly line and its basic concept is credited to Ransom Olds, who used it to build the first mass-produced automobile, the Oldsmobile Curved Dash. Olds patented the assembly line concept, which he put to work in his Olds Motor Vehicle Company factory in 1901. This development is often overshadowed by Henry Ford, who perfected the assembly line by installing driven conveyor belts that could produce a Model T in ninety-three minutes.

The assembly line developed for the Ford Model T began operation on December 1, 1913. It had immense influence on the world. Despite oversimplistic attempts to attribute it to one man or another, it was in fact a composite development based on logic that took 7 years and plenty of intelligent men. The principal leaders are discussed below.

The basic kernel of an assembly line concept was introduced to Ford Motor Company by William "Pa" Klann upon his return from visiting Swift & Company's slaughterhouse in Chicago and viewing what was referred to as the "disassembly line", where carcasses were butchered as they moved along a conveyor. The efficiency of one person removing the same piece over and over caught his attention. He reported the idea to Peter E. Martin, soon to be head of Ford production, who was doubtful at the time but encouraged him to proceed. Others at Ford have claimed to have put the idea forth to Henry Ford, but Pa Klann's slaughterhouse revelation is well documented in the archives at the Henry Ford Museum and elsewhere, making him an important contributor to the modern automated assembly line concept. The process was an evolution by trial and error of a team consisting primarily of Peter E. Martin, the factory superintendent; Charles E. Sorensen, Martin's assistant; C. Harold Wills, draftsman and toolmaker; Clarence W. Avery; Charles Ebender; and József Galamb. Some of the groundwork for such development had recently been laid by the intelligent layout of machine tool placement that Walter Flanders had been doing at Ford up to 1908.

In 1922 Ford said of his 1913 assembly line:

"I believe that this was the first moving line ever installed. The idea came in a general way from the overhead trolley that the Chicago packers use in dressing beef."

Charles E. Sorensen, in his 1956 memoir *My Forty Years with Ford*, presented a different version of development that was not so much about individual "inventors" as a gradual, logical development of industrial engineering:

"What was worked out at Ford was the practice of moving the work from one worker to another until it became a complete unit, then arranging the flow of these units at the right time and the right place to a moving final assembly line from which came a finished product. Regardless of earlier uses of some of these principles, the direct line of succession of mass production and its intensification into automation stems directly from what we worked out at Ford Motor Company between 1908 and 1913. Henry Ford is generally regarded as the father of mass production. He was not. He was the sponsor of it."

As a result of these developments in method, Ford's cars came off the line in three minute intervals. This was much faster than previous methods, increasing production by eight to one (requiring 12.5 man-hours before, 1 hour 33 minutes after), while using less manpower. It was so successful, paint became a bottleneck. Only japan black would dry fast enough, forcing the company to drop the variety of colors available before 1914, until fast-drying Duco lacquer was developed in 1926. In 1914, an assembly line worker could buy a Model T with four months' pay.

The assembly line technique was an integral part of the diffusion of the automobile into American society. Decreased costs of production allowed the cost of the Model T to fall within the budget of the American middle class. In 1908, the price of a Model T was around $825, and by 1912 it had decreased to around $575. This price reduction is comparable to a reduction from $15,000 to $10,000 in dollar terms from the year 2000.

Ford's complex safety procedures—especially assigning each worker to a specific location instead of allowing them to roam about—dramatically reduced the rate of injury. The combination of high wages and high efficiency is called "Fordism", and was copied by most major industries. The efficiency gains from the assembly line also coincided with the take-off of the United States. The assembly line forced workers to work at a certain pace with very repetitive motions which led to more output per worker while other countries were using less productive methods.

Ford at one point considered suing other car companies because they used the assembly line in their production, but decided against it, realizing it was essential to creation and expansion of the industry as a whole.

In the automotive industry, its success was dominating, and quickly spread worldwide. Ford France and Ford Britain in 1911, Ford Denmark 1923, Ford Germany 1925; in 1919, Vulcan (Southport, Lancashire) was the first native European manufacturer to adopt it. Soon, companies had to have assembly lines, or risk going broke by not being able to compete; by 1930, 250 companies which did not had disappeared.

The massive demand for military hardware in World War II prompted assembly-line techniques in shipbuilding and aircraft production. Thousands of Liberty Ships were built making extensive use of prefabrication, enabling ship assembly to be completed in weeks or even days. After having produced fewer than 3,000 planes for the United States Military in 1939, American aircraft manufacturers built over 300,000 planes in World War II. Vultee pioneered the use of the powered assembly line for aircraft manufacturing. Other companies quickly followed. As William S. Knudsen of the National Defense Advisory Commission observed, "We won because we smothered the enemy in an avalanche of production, the like of which he had never seen, nor dreamed possible."

Sociological Problems

Sociological work has explored the social alienation and boredom that many workers feel because of the repetition of doing the same specialized task all day long. Because workers have to stand in the same place for hours and repeat the same motion hundreds of times per day, repetitive stress injuries are a possible pathology of occupational safety. Industrial noise also proved dangerous. When it was not too high, workers were often prohibited from talking. Charles Piaget, a skilled worker at the LIP factory, recalled that beside being prohibited from speaking, the semi-skilled workers had only 25 centimeters in which to move. Industrial ergonomics later tried to minimize physical trauma.

Improved Working Conditions

In his autobiography Henry Ford mentions several benefits of the assembly line including:

- Workers do no heavy lifting.
- No stooping or bending over.

- No special training required.
- There are jobs that almost anyone can do.
- Provided employment to immigrants.

The gains in productivity allowed Ford to increase worker pay from $1.50 per day to $5.00 per day once employees reached three years of service on the assembly line. Ford continued on to reduce the hourly work week while continuously lowering the Model T price. These goals appear altruistic; however, it has been argued that they were implemented by Ford in order to reduce high employee turnover: when the assembly line was introduced in 1913, it was discovered that "every time the company wanted to add 100 men to its factory personnel, it was necessary to hire 963" in order to counteract the natural distaste the assembly line seems to have inspired.

TRANSFER LINE

A **transfer line** is a manufacturing system which consists of a predetermined sequence of machines connected by an automated material handling system and designed for working on a very small family of parts. Parts can be moved singularly because there's no need for batching when carrying parts between process stations (as opposed to a job shop for example). The line can synchronous, meaning that all parts advance with the same speed, or asynchronous, meaning buffers exist between stations where parts wait to be processed. Not all transfer lines must geometrically be straight lines, for example circular solutions have been developed which make use rotary tables, however using buffers becomes almost impossible.

A crucial problem for this production system is that of **line balancing**: a trade-off between increasing productivity and minimizing cost conserving total processing time.

Advantages

- Easy management: low work in progress and scheduling without simultaneous processing of different products
- Low need for manpower
- Less space needed (compare with job shop)
- Less output variability: no alternative technological cycles and quality control is more effective (less WIP and easier to automate)
- High system saturation: less production mix variability
- Fast lead times

Disadvantages

- Very low flexibility
- Risk of obsolescence: due to new product introduction

- High vulnerability to failures: a failure in a single machine blocks the whole system in very short time

ROTARY FEEDER

Rotary feeders, also known as **rotary airlocks** or **rotary valves**, are commonly used in industrial and agricultural applications as a component in a bulk or specialty material handling system. Rotary feeders are primarily used for discharge of bulk solid material from hoppers/bins, receivers, and cyclones into a pressure or vacuum-driven pneumatic conveying system. Components of a rotary feeder include a rotor shaft, housing, head plates, and packing seals and bearings. Rotors have large vanes cast or welded on and are typically driven by small internal combustion engines or electric motors.

Use

Rotary airlock feeders have wide application in industry wherever dry free-flowing powders, granules, crystals, or pellets are used. Typical materials include: cement, ore, sugar, minerals, grains, plastics, dust, fly ash, flour, gypsum, lime, coffee, cereals, pharmaceuticals, *etc.*

Industries requiring this type include cement, asphalt, chemical, mining, plastics, food, *etc.*

Rotary feeders are ideal for pollution control applications in wood, grain, food, textile, paper, tobacco, rubber, and paint industries, the Standard Series works beneath dust collectors and cyclone separators even with high temperatures and different pressure differentials.

Rotary valves are available with square or round inlet and outlet flanges. Housing can be fabricated out of sheet material or cast. Common materials are cast iron, carbon steel, 304 SS, 316 SS, and other materials. Rotary airlock feeders are often available in standard and heavy duty models, the difference being the head plate and bearing configuration. Heavy duty models use an outboard bearing in which the bearings are moved out away from the head plate. Housing inlet and discharge configurations are termed drop-thru or side entry. Different wear protections are available such as hard chrome or ceramic plating on the inner housing surfaces. Grease and air purge fittings are often provided to prevent contaminants from entering the packing seals.

Uses

- Rotary airlock

The basic use of the rotary airlock feeder is as an airlock transition point, sealing pressurized systems against loss of air or gas while maintaining a flow of material between components with different pressure and suitable for air lock applications ranging from gravity discharge of filters, rotary valves, cyclone dust collectors, and rotary airlock storage devices to precision feeders for dilute phase and continuous dense phase pneumatic convey systems.

- Rotary valve

Rotary airlock feeders/ rotary airlock valves are used in pneumatic conveying systems, dust control equipment, and as volumetric feed-controls.

- Volumetric feeder

Rotary airlock valves are also widely used as volumetric feeders for metering materials at precise flow rates from bins, hoppers, or silos onto conveying or processing systems.

Types

Rotary Airlock Feeders

Drop through rotary airlock feeders are designed for rugged applications that require an outboard bearing style unit where contamination and /or an abrasive product cannot be handled with an inboard bearing style. The outboard bearing feeders is engineered for use in high pressure pneumatic conveying systems, with high temperatures where more of an effective seal is required due to high or excessive wear that is experienced with a simple dust collector.

Blow-thru Rotary Airlock Feeder

The blow-thru rotary airlock feeder is ideal for pneumatic conveying applications in food, grain, chemical, milling, baking, plastics and pharmaceutical industries. The blow-thru airlocks feature a low profile with large capacity. High pressure differentials integral mounting feet, and retrofit competitive units. The blow-thru valves are available with 10-vane open-end rotor; outboard bearings and replaceable shaft seals.

Easy Clean Rotary Feeder

The easy-clean series rotary feeders can be fast and simply disassembled, thoroughly quick cleaned, sanitized and inspected or maintenance in a minimum amount of time without the use of tools or removal from service, thereby reducing downtime and increasing system production. Reassembly without tools is accomplished in minutes. Internal clearances are automatically re-established every time.

The Clean-in-place rotary feeder is a special purpose valve designed for where cross-contamination is a major concern and lengthy shut-downs for clean-out are cost-prohibitive, suited for Dairy, Pharmaceutical industries, Food, Baking, Chemical, Plastics, Paint, and Powder Coating plants.

It is ideal for batch mixing systems such as those handling different colored resins which demand regular cleaning between cycles.

Filter Valve

The filter valve is a low-cost solution designed for light duty dust collector applications.

Knife Rotary Feeder

This type of feeder is used for discharge of secondary fuel as for example: plastics or wood. The knife is cutting the oversize material and is preventing the rotor from blockage.

VIBRATING FEEDER

A **vibratory feeder** is an instrument that uses vibration to "feed" material to a process or machine. Vibratory feeders use both vibration and gravity to move material. Gravity is used to determine the direction, either down, or down and to a side, and then vibration is used to move the material. They are mainly used to transport a large number of smaller objects.

A belt weigher are used only to measure the material flow rate but weigh feeder can measure the flow material and also control or regulate the flow rate by varying the belt conveyor speed like in rsp coke oven

Industries Served

Versatile and rugged vibratory bowl feeders have been extremely used for automatic feeding of small to large and differently shaped industrial parts. They are the oldest but still commonly used automation machine available for aligning and feeding machine parts, electronic parts, plastic parts, chemicals, metallic parts, glass vials, pharmaceuticals, foods, miscellaneous goods *etc.* Available in standard and custom designs, vibratory bowl feeders have been largely purchased by varied industrial sectors for automating high-speed production lines and assembly systems. Some of the industries that use the service of this automation machine include:

- Pharmaceutical
- Automotive
- Electronic
- Food Processing
- Fast Moving Consumable Goods (FMCG)
- Packaging
- Metal working
- Glass
- Foundry
- Steel
- Construction
- Recycling
- Pulp and paper
- Plastics

With these easy-to-use and high performing part feeding machines, customers from varied industrial sectors have achieved lower error rates, lesser power consumption, better profits, better rates of efficiency and less dependency on man power.

Chapter 11

LOCALIZATION OF MOBILE ROBOTS USING ODOMETRY AND AN EXTERNAL VISION SENSOR

Daniel Pizarro*, Manuel Mazo, Enrique Santiso, Marta Marron, David Jimenez, Santiago Cobreces and Cristina Lo sada

Department of Electronics, University of Alcala, NII km 33,600, Alcala de Henares, Spain; E-Mails: mazo@depeca.uah.es (M.M.); santiso@depeca.uah.es (E.S.); marta@depeca.uah.es (M.M.); david.jimenez@depeca.uah.es (D.J.); cobreces@depeca.uah.es (S.C.); losada@depeca.uah.es (C.L.)

*   Author to to whom correspondence should be addressed: E-Mail: pizarro@depeca.uah.es; Tel.: +34-918-856-582. Fax: +34-918-856-591.

ABSTRACT

This paper presents a sensor system for robot localization based on the information obtained from a single camera attached in a fixed place external to the robot. Our approach firstly obtains the 3D geometrical model of the robot based on the projection of its natural appearance in the camera while the robot performs an initialization trajectory. This paper proposes a structure-from-motion solution that uses the odometry sensors inside the robot as a metric reference. Secondly, an online localization method based on a sequential Bayesian inference is proposed, which uses the geometrical model of the robot as a link between image measurements and pose estimation. The online approach is resistant to hard occlusions and the experimental setup proposed in this paper shows its effectiveness in real situations. The proposed approach has many applications in both the industrial and service robot fields.

Keywords

Robotics; vision sensor; intelligent spaces.

1. INTRODUCTION

This paper presents a sensor system composed of a single camera attached to a fixed position and orientation in a bounded environment (indoor workplace) which is observing a mobile robot. The aim of the sensor system is to obtain the orientation and position (*i.e.*, pose) of a mobile robot using both visual information retrieved by the camera and relative odometry readings obtained from the internal sensors of the robot. The camera acquisition and image processing tasks are executed in a specialized hardware, which also controls the behavior and internal sensors of the mobile robot through a wireless channel. The proposed schema allows the robot to perform complex tasks without requiring dedicated processing hardware on it. This approach is sustained in the Intelligent Space [1] concept and it can be equally applied to multiple scenarios, specially both in the industrial field (*e.g.*, automatic crane positioning, autonomous car parking) and in the service fields (*e.g.*,, wheelchair positioning in medical environments or autonomous driving of mobile platforms). The single camera solution presented in this paper allows to cover large areas with less cameras compared to multiple camera configurations where overlapped areas are mandatory. This feature reduces the cost and improves the reliability of the intelligent space philosophy.

In this paper, we suppose that the camera is correctly calibrated and thus the parameters governing the projection model of the camera are previously known. To connect the pose of the robot with information found in the image plane of the camera, we propose to obtain a 3D geometric model of the mobile robot. Such model is composed of several sparse points whose coordinates represent some well-localized points belonging to the physical structure of the robot. These points are determined by image measurements, called image features, which usually correspond to corner-like points due to texture changes or geometry changes such as 3D vertexes. Usually in industrial fields the image features are obtained by including some kind of artificial marker on the structure of the robot (infrared markers or color bands). These methods are very robust and can be used to recognize human activity and models with high degrees of freedom (AICON 3D online [2] or ViconPeak online systems [3]). However, in this paper we want to minimize the required "a priori" knowledge of the robot, so that it is not necessary to place artificial markers or beacons on it to detect its structure in the images.

Independent of the nature of the image features, the information obtained from a camera is naturally ambiguous and thus some sort of extra metric information are required in order to solve such ambiguity. In this work, we propose to use the odometry sensors inside the robot (*i.e.*, wheel velocities) to act as the metric reference.

The general diagram of the algorithm proposed in this paper is shown in Figure 1. It shows a clear division of the processes involved in obtaining the pose of the robot: first we denote as "Initialization of Pose and Geometry" to those processes necessary to start up the system, such as the 3D model of the robot and the initial pose it occupies. The initialization consists of a batch processing algorithm where the robot is commanded to follow a certain trajectory so that the camera is able to track some points of the robot's structure under different viewpoints jointly with the recording of the odometry information. All this information is combined to give the 3D model of the robot and the initial pose it occupies.

Given the initialization information, the second group of processes, named "Sequential Localization", provides the pose of the robot in a sequential manner. It is composed of a pose estimator, given odometry readings and a pose correction block which combines the estimation of the pose with image measurements to accurately give a coherent pose with the measurements. This algorithm operates entirely on-line and thus the pose is available at each time sample.

Both group of processes are supplied with two main sources of information:

1. Image measurements: they consist of the projection in the camera's image plane of certain points of the robot's 3D structure. The measurement process is in charge of searching coherent correspondences through images with different perspective changes due to the movement of the robot.

2. Motion estimation of the robot: The odometry sensors built on-board the robot supply the localization system with an accurate motion estimation in short trajectories but that is prone to accumulative errors in large ones.

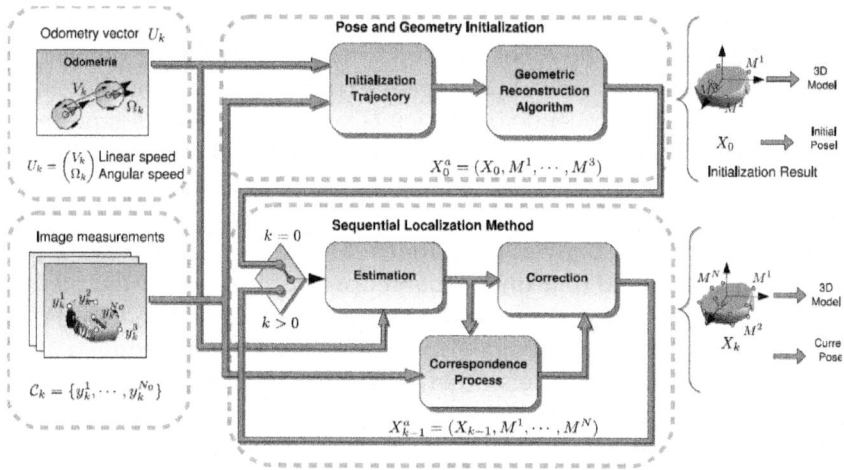

Figure 1. General diagram of the proposed localization system using a vision sensor and odometry readings.

1.1. Previous Works

Despite the inherent potential of using external cameras to localize robots, there are relative few attempts to solve it compared to the approach that consider the camera on-board the robot [4, 5]. However, some examples of robot localization with cameras can be found in the literature, where the robot is equipped with artificial landmarks, either active [6, 7] or passive ones [8, 9]. In other works a model of the robot, either geometrical or of appearance [10, 11], is learnt previously to the tracking task. In [12, 13], the position of static and dynamic objects is obtained by multiple camera fusion inside an occupancy grid. An appearance model is used afterwards to ascertain which object is each robot. Despite the technique used for tracking, the common point of many of the proposals found in the topic comes from the fact that rich knowledge is obtained previously to the tracking, in a supervised task.

In this paper the localization of the robot is obtained in terms of its natural appearance. We propose a full metric and accurate system based on identifying natural features belonging to the geometric structure of the robot. On most natural objects we can find points whose image projection can be tracked in the image plane independently of the position the object occupies and based on local properties found in the image (*i.e.*, lines, corners or color blobs). Those points are considered natural markers, as they serve as reference points in the image plane that can be easily relate with their three-dimensional counterparts. The set of methods focused on tracking natural markers have become a very successful and deeply studied topic in the literature [14, 15], as they represent the basic measurements of most of existing reconstruction methods.

Scene reconstruction from image measurements is a classic and mature discipline in the computer vision field. Among the wide amount of proposals it can be highlighted those grouped under the name "Bundle Adjustment" [16, 17]. Their aim is essentially to estimate, jointly and optimally, the 3D structure of a scene and the camera parameters from a set of images taken under some kind of motion. (*i.e.*, it can be the camera that moves or equally some part of the scene w.r.t. the camera).

In general terms, Bundle Adjustment reconstruction methods are based in iterative optimization methods which try to minimize the image discrepancy between the measured positions of the 3D model and the expected ones using the last iteration solution. The discrepancy balances the contribution of the measurements into the final solution and plays an important role in this paper. Our main contribution is a redefinition of the discrepancy function using a Maximum Likelihood approach which takes into account the statistical distribution of the error. This distribution is especially affected by the odometry errors which are accumulative in long trajectories.

On the other hand, once a geometric model is obtained using a structure-from-motion approach, its pose with respect to a global coordinate origin can be easily retrieved by measuring the projection of the model in the image plane. This problem, commonly known as the Perspective n Point Problem (PnP), has received considerable attention in the literature, where some accurate solutions are found such as [18] or the recent global solution proposed in [19]. In this paper we instead follow a filtering approach, where not only image measurements but also last time pose information and odometry information are used to obtain the pose. This approach, which is based on the use of a Kalman Filter, is much more regular than solving the PnP problem at each time instant and will be described in this paper.

The paper is organized as follows. In Section 2. the notation and mathematical elements used in the paper are described. In Section 3. the description of the initialization algorithms is given. The online Kalman loop is explained in Section 4. and some results in a real environment are shown in Section 5. Conclusions are in Section 6.

2. DEFINITIONS AND NOTATION

This section presents a description of the symbols and mathematical models used in the rest of the paper.

2.1. Robot and Image Measurements

The robot's pose at time k is described by a vector X_k. We suppose that the robot's motion lie on the plane $z = 0$ referred from the world coordinate origin O_W (See Figure 2). The pose vector X_k is described by 3 components (x_k, y_k, a_k), corresponding to two position coordinates x_k, y_k and the rotation angle a_k in the z axis. The motion model $X_k = g(X_{k-1}, U_k)$ defines the relationship between the current pose X_k with respect to the previous time one X_{k-1} and the input U_k given by odometry sensors inside the robot. In our proposal the robot used is a differential wheeled platform and thus $U_k = (Vl_k, \Omega_k)^T$, where Vl_k and Ω_k describe the linear and angular speed of the robot's center of rotation $(O_R$ in Figure 2) respectively The motion model g is then very simple in function of the discretized linear speed Vl_k and the angular speed Ω_k.

The robot's geometry is composed of a sparse set of N 3D points $M = \{M^1, \cdots, M^N\}$ referred from the local coordinate origin O_R described by robot's pose X_k. The points M^i are static in time due to robot's rigidness, and thus, no temporal subindex is required for them. Function $M_{Xk} = t(X_k, M^i)$ uses actual pose X_k to express M^i in the global coordinate origin O_W that X_k is referred to (see Figure 2):

$$M^i_{X_k} = t(X_k, M^i) = R_k M^i + T_k \tag{1}$$

where:

$$R_k = \begin{pmatrix} \cos(\alpha_k) & \sin(\alpha_k) & 0 \\ -\sin(\alpha_k) & \cos(\alpha_k) & 0 \\ 0 & 0 & 1 \end{pmatrix} \qquad T_k = \begin{pmatrix} x_k \\ y_k \\ 0 \end{pmatrix} \tag{2}$$

The augmented vector Xa, which is the state vector of the system, is defined as the concatenation in one column vector of both the pose X_k and the set of static structure points M:

$$X^a_k = \left((X_k)^T \ (M^1)^T \ \cdots \ (M^N)^T \right)^T \tag{3}$$

An augmented motion model g^a is defined as the transition of the state vector:

$$X^a_k = g^a(X^a_{k-1}, U_k) = \left((g(X_{k-1}, U_k))^T \ (M^1)^T \ \cdots \ (M^N)^T \right)^T \tag{4}$$

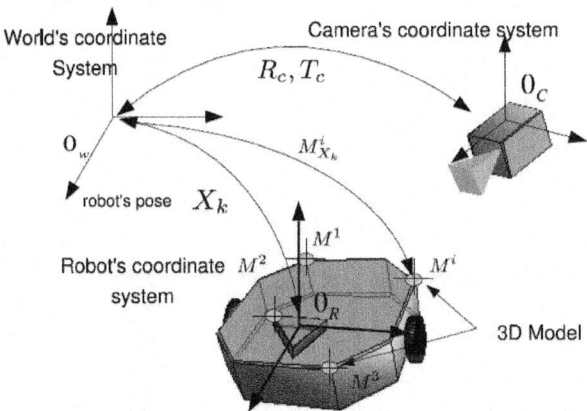

Figure 2. Spatial relationship between world's coordinate origin O_W, robot's coordinate origin O_R and camera's coordinate origin O_C.

It must be noticed that the motion model ga leaves the structure points contained in the state vector untouched, as we suppose that the object is rigid.

The whole set of measurements obtained in the image plane of the camera at time k is denoted by Y_k.

Each measurement inside Y_k is defined by a two-dimensional vector $y_k^i = (u_k^i v_k^i)^T$, describing the projection of each point inside M in the image plane, where i stands for the correspondence with the point M^i. It must be remarked that, in general, Y_k does not contain the image projection of every point in the set M as some of them can be occluded depending on the situation.

The camera is modeled with a zero-skew "pin-hole" model (see [17] for more details), with the following matrix K_c containing the intrinsic parameters:

$$K_c = \begin{pmatrix} f_u & 0 & u_0 \\ 0 & f_v & v_0 \\ 0 & 0 & 1 \end{pmatrix} \tag{5}$$

The extrinsic parameters of the camera (*i.e.,* position and orientation of the camera w.r.t. O_W) are described using the rigid transformation R_c, T_c (*i.e.,* rotation matrix and translation vector). The matrix R_c is defined by three Euler angles $\alpha_c, \beta_c, \gamma_c$. The vector T_c can be decomposed into its three spatial coordinates T_x, T_y, T_z. All calibration parameters can be grouped inside vector P:

$$P = (f_u, f_v, u_0, v_0, \alpha_c, \beta_c, \gamma_c, T_x, T_y, T_z)^T \tag{6}$$

Given a single measurement y_k^i from the set Y_k, its 2D coordinates can be expressed using the "pin-hole" model and the aforementioned calibration parameters:

$$\begin{pmatrix} y_k^i \\ 1 \end{pmatrix} = \lambda_k^i K_c \left(R_c M_{X_k}^i + T_c \right) = \lambda_k^i K_c \left(R_c R_k M^i + R_c T_k + T_c \right) \tag{7}$$

where λ^i_k represents a projective scale factor that can be obtained so that the third element of the right part of Equation (7) is equal to one. It is important to remark that the projection model, although simple, depends in a non-linear fashion w.r.t. Mi due to the factor λ^i_k. For compactness the projection model of Equation (7) can be described as the function h:

$$y_k^i = h(X_k, M^i, P) \tag{8}$$

In the same way, the whole vector Y_k can be expressed with the following function:

$$Y_k = h^a(X^a_k, P) = (h(X_k, M^1, P)^T \cdots h(X_k, M^N, P)^T) \tag{9}$$

2.2. Random Processes

This paper explicitly deals with uncertainties by describing every process as a random variable with a Gaussian distribution whose covariance matrix stands for its uncertainty. The random processes are expressed in bold typography (i.e., X_k is the random process describing the pose and X_k is a realization of it) and each of them are defined by a mean vector and a covariance matrix. Therefore X^a_k is defined by its mean $\hat{X}^a_k$ and covariance Σ^a_k (or simplifying $\mathcal{N}(\hat{X}^a_k, \Sigma^a_k)$). The following processes are considered in the paper:

- Pose $X_k = \mathcal{N}(\hat{X}_k, \Sigma_k)$ and 3D model $M = \mathcal{N}(\hat{M}, \Sigma_M)$ processes. Its joint distribution is encoded
- Measurement process $Y_k = \mathcal{N}(\hat{X}^a_k, \Sigma^a_k)$, whose uncertainty comes from errors in image detection.
- Odometry input values $U_k = \mathcal{N}(\hat{U}_k, \Sigma_{U_k})$, which are polluted by random deviations.

3. INITIALIZATION PROCESS

The initialization step consists of obtaining the initial pose the robot occupies and a 3D model of its structure using visual features (i.e., to obtain an initial guess of X_0^a). The importance of this step is crucial, as the robot's geometric model serves as the necessary link between robot's pose and image measurements. The initialization allows to use afterwards the online approach presented in Section 4.

A delayed initialization approach is proposed, based on collecting image measurements and the corresponding odometry position estimation along a sequence of time (i.e., $k = 1, ..., K$). After taking a sufficient amount of measurements of the object in motion, an iterative optimization method is used to obtain the best solution for X_0^a according to a cost criterion. The odometry readings are used in this step as metric information, which allows to remove the natural ambiguity produced by measurements from a single camera. The main problem of using odometry as a metric reference for reconstruction comes from the accumulative error it presents, which is a very well-known problem in dead-reckoning tasks.

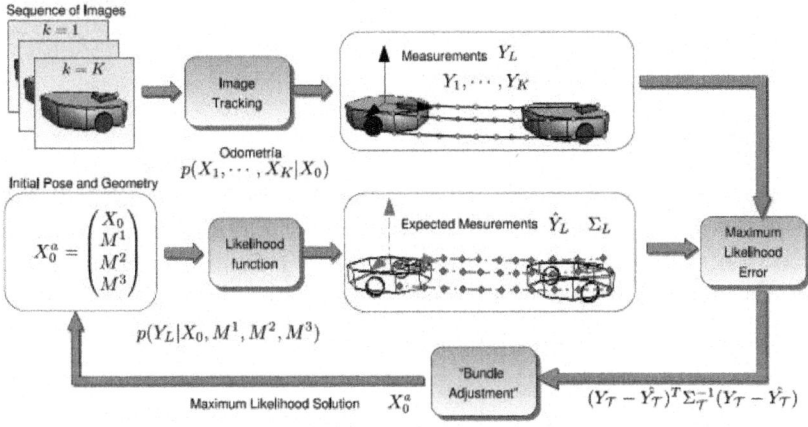

Figure 3. Maximum likelihood initialization by means of reducing the residual of the expected measurements (red diamonds) and the measured image trajectories (blue circles).

The initialization algorithm consists of a Maximum Likelihood (M.L.) estimation of the vector X_0^a, which does not estimate the initialization value itself but its equivalent Gaussian distribution $X_0^a = \mathcal{N}(\hat{X}_0^1, \Sigma_0^a)$ by using a "Bundle Adjustment" method (See Figure 3). The M.L. approach (see [16, 17] for more details) allows to properly tackle the growing drift present in the odometry estimation by giving fair less importance or weight to those instants where the odometry is expected to be more uncertain.

In the following sections, the initialization algorithm is explained in detail.

3.1. Odometry Estimation

A set of odometry readings, $U_1, \cdots, U_K$ are collected over a set of K consecutive time samples $k = 1, \cdots, K$, which corresponds to the initialization trajectory performed by the robot. Using a motion model g given initial position X_0, an estimation of the robot's pose in the whole initialization sequence is obtained as follows:

$$X_1 = g(X_0, U_1)$$
$$X_2 = g(X_1, U_2)$$
$$X_K = g(X_{K-1}, U_K) \tag{10}$$

where we recall that X_k and U_k denote Gaussian processes. Using expression (10), and by propagating the statistical processes through function g, the joint distribution $p(X_1, \cdots, X_K | X_0)$ can be obtained. The propagation of statistical processes through non-linear functions is in general a very complex task, as it requires to solve integral equations. However in this paper we will follow a first order approximation of the non-linear functions centered in the mean of the process (see [20] for more details). Using this technique the Gaussian processes X_k and U_k can be iteratively propagated through function g at the cost of being an approximation. This paper shows that despite being an approximation this approach end in a good estimation of the initialization vector.

By denoting as X to the joint vector containing all poses $X = (X_1^T, \cdots, X_K^T)^T$, the joint distribution of all poses, given the initial state X_0, $p(X \mid X_0)$ is approximated as Gaussian p.d.f. with mean X and covariance matrix Σ_X.

3.2. Image Measurements

In this step we describe how to collect the position in the image plane of different points from the robot's structure during the initialization trajectory. We base our method in the work [15] where the SIFT descriptor is introduced. The process used in the initialization is composed of two steps:

- Feature-based background subtraction.

  We describe the static background of the scene using a single image, from which a set of features, $F_b = \{y_b^1, \cdots, y_b^{Nb}\}$ and its correspondent SIFT descriptors $D_1 = \{d_1^1, \cdots, d_1^{Nd}\}$ are obtained. The sets D_b and F_b are our feature-based background model.

  Given an input image, namely I_k, we find the sets D_k and F_k. We consider that a feature $y_k^i \in F_k, d_k^i \in D_k$ belongs to the background if we can find a feature $yj_b \in F_b, d_b^j \in D_b$ such that $|y_k^i - y_b^j| < R_{max}$ and $|d_k^i - d_b^i| < d_{max}$. This method although simple shows to be very effective and robust in real environments (See Figure 4).

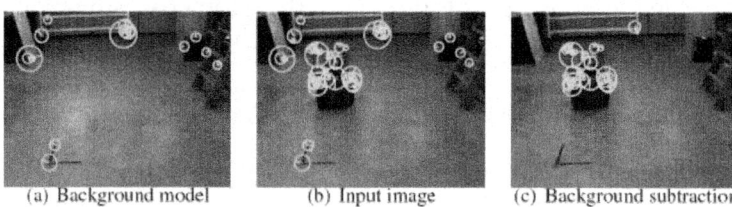

(a) Background model   (b) Input image   (c) Background subtraction

Figure 4. Feature-based background subtraction method.

- Supervised tracking algorithm.

Given the first image of the initialization trajectory, namely I_1, we obtain the set of N features F_1, once they are cleaned by the aforementioned background subtraction method. We propose a method to track them in the whole initialization trajectory.

By only using SIFT descriptors and its tracking in consecutive frames does not produce stable tracks, specially when dealing with highly irregular objects where many of the features are due to corners. We thus propose to use a classical feature tracking approach based on the Kanade–Lucas–Tomasi (KLT) algorithm [14]. To avoid degeneracy in the tracks, which is a very common problem in those kind of trackers, we use the SIFT descriptors to remove those segments of the tracks that clearly do not correspond to the feature tracked. This can be done easily by checking distance between the descriptors in the track. The threshold used must be chosen experimentally so that it does not eliminate useful parts of the tracks. In Figure 5a we can see the tracks obtained by the KLT tracker without degeneracy supervision. In Figure 5b the automatically removed segments are displayed.

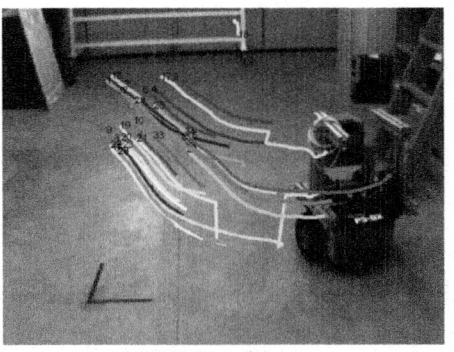
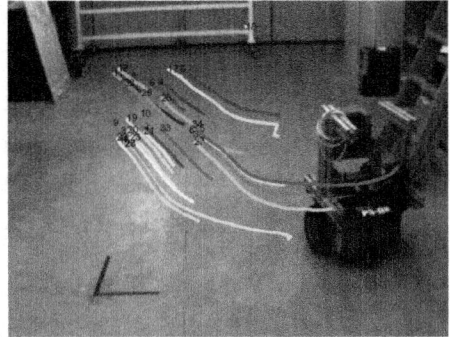

(a) KLT tracking   (b) SIFT-supervised algorithm tracking

Figure 5. Comparison between the KLT tracker and the SIFT-supervised KLT version.

3.3. Likelihood Function

Using the feature-based algorithm proposed before a set of measurements in the whole trajectory, $Y_1, \cdots, Y_K$, is obtained, where each vector Y_k contains the projection of N points from robot's structure at time sample k. The set of N points, $M^1, \cdots, M^N$, jointly with the initial pose X_0, represent the initialization unknowns. As the robot's motion does not depend of its structure, the following statistical independence is true:

$$p(X_1, \cdots, X_K | X_0) = p(X_1, \cdots, X_K | X_0, M^1, \cdots, M^N) = p(X_1, \cdots, X_K | X_0^a) \quad (11)$$

If all trajectories are packed into vector Y_L, the following function put in relation Y_L with the distribution of expression (11):

$$\mathbf{Y_L} = \begin{pmatrix} \mathbf{Y}_1 \\ \vdots \\ \mathbf{Y_K} \end{pmatrix} = \begin{pmatrix} h^a(\mathbf{X_1^a}, P) + \mathbf{V}_1 \\ \vdots \\ h^a(\mathbf{X_K^a}, P) + \mathbf{V_K} \end{pmatrix}$$

where V_k represents the uncertainty in image detection. Using distribution showed in (11), and propagating statistics through function (12), the following likelihood distribution is obtained:

$$p(Y_L | X_0, M^1, \cdots, M^N) = p(Y_L | X_0^a) \quad (13)$$

The likelihood function (13) is represented by a Gaussian distribution using a first order approximation of (12). It is thus defined by a mean $\hat{Y}_L$ and a covariance matrix Σ_L:

$$p(Y_L | X_0^a) = \frac{1}{\sqrt{|\Sigma_L| 2^N}} e^{\frac{1}{2}(Y_L - \hat{Y}_L)^T \Sigma_L^{-1}(Y_L - \hat{Y}_L)} \quad (14)$$

3.4. Maximum Likelihood Approach

The likelihood function (14), parametrized by its covariance matrix Σ_L and its mean $\hat{Y}_{L'}$ is dependent of the conditional parameters and unknown values $X^a{}_0$.

The "Maximum Likelihood" estimation consists of finding the values for $X^a{}_0$ that maximize the likelihood function:

$$\max_{X_0, M^1, \cdots, M^N} p(Y_L | X_0, M^1, \cdots, M^N)$$

$$(15)$$

In its Gaussian approximation and by taking logarithms, it is converted into the following minimization problem:

$$\min_{X_0, M^1, \cdots, M^N} (Y_L - \hat{Y}_L)^T \Sigma_L^{-1} (Y_L - \hat{Y}_L)$$

$$(16)$$

where Y_L is the realization of the process (*i.e.*, the set of measurements from the image) and $\hat{Y}_L$ are the expected measurements given a value of the parameters $X_0, M^1, ..., M^N$.

The configuration of Σ_L is ruled by the expected uncertainty in the measurements and the statistical relation between them. Usually all cross-correlated terms of Σ_L are non-zero, which has an important effect in the sparsity of the Hessian used inside the optimization algorithm and consequently in the computational complexity of the problem (see [21] for more details).

The covariance matrix Σ_L can be approximated assuming either independence between time samples (discarding cross-correlation terms between time indexes) or total independence between time and each measurement (discarding all cross-correlation terms except the 2 × 2 boxes belonging to a single measurement). In Table 1, the different cost functions are shown depending on the discarded terms of Σ_L. The different approximations of Σ_L have a direct impact in the accuracy and the reconstruction error, and the results will be shown in Section 5.

Table 1. Different cost functions depending on the approximations of Σ_L.

Type of Approximation	cost function
Complete Correlated matrix Σ_L (M.C.)	$(Y_L - \hat{Y}_L)^T \Sigma_L^{-1} (Y_L - \hat{Y}_L)$
$2N \times 2N$ block approximation of Σ_L (M.B.)	$\sum_{k=1}^{K} (Y_k - \hat{Y}_k)^T \Sigma_{Y_k}^{-1} (Y_k - \hat{Y}_k)$
2×2 block approximation of Σ_L (M.D.)	$\sum_{k=1}^{K} \sum_{i=1}^{N} (y_k^i - \hat{y}_k^i)^T \Sigma_{y_k^i}^{-1} (y_k^i - \hat{y}_k^i)$
Identity approximation of Σ_L (M.I.)	$\sum_{k=1}^{K} \sum_{i=1}^{N} (y_k^i - \hat{y}_k^i)^T (y_k^i - \hat{y}_k^i)$

Intuitively, if Σ_L results to be a identity matrix, the cost function is reduced to a simple image residual minimization extensively used in Bundle Adjustment techniques, where in principle all cost differences y_k^i - $\hat{y}_k^i$ have equal importance:

$$\min_{X_0^a} \sum_{i=1}^{N} \sum_{k=1}^{K_i} |(y_k^i - \hat{y}_k^i)|^2$$

$$(17)$$

The result of minimizing Equation (16) instead of the usual (17) show significant improvements in the reconstruction error. In Section 5. a comparison is shown between the initialization accuracy under the different assumptions of the matrix $\Sigma_{L'}$ from its diagonal version to the complete correlated form. The minimum of (16) is obtained using the Levenberg-Mardquardt [17] iterative optimization method.

We suppose that during the measurement step there is a low probability of encountering outliers in the features. This argument can be very optimistic in some real configurations where more objects appear in the scene and produce occlusions or false matchings. For those cases, all cost functions presented in this section can be modified to be robust against outliers by using M-Estimators. We refer the reader to [17] for more details.

3.5. Initialization before Optimization

The use of an iterative optimization method for obtaining the minimum of (16) requires a setup value from which to start iterating. An initial estimation of X_0^a close to its correct value reduces the probability to fall in a local extrema of the cost function.

The method proposed in this paper to give an initial value for X_0^a consists of a non-iterative and exact method, which gives directly an accurate solution for X_0^a in absence of noise in odometry values. This method is based on the assumption that the robot moves in a plane and thus only the angle over its Z axis is needed. As explained below, this assumption allows to solve the problem with a rank deficient Linear Least Squares approximation, which is solved using the Singular Value Decomposition of the system's matrix and imposing a constraint that ensures a valid rotation. Its development is briefly introduced next.

For a point Mi of robot's model and at time k of the initialization trajectory, the image measurement y_k^i results from the following projection:

$$\begin{pmatrix} y_k^i \\ 1 \end{pmatrix} = \lambda_k^i K_c (R_c (R_k^\Delta R_0 M^i + T_0 + R_0 T_k^\Delta) + T_c) \tag{18}$$

where the matrix R_k^Δ and vector T_k^Δ represent the rotation and position of the robot in the floor plane given by the odometry supposing that $X_0 = (0\ 0\ 0)^T$. The rotation matrix R_0 and the offset T_0 correspond with the initial pose $X_0 = (x_0^1\ x_0^2\ \alpha_0)$ in form of rigid transformation:

$$R_0 = \begin{pmatrix} cos(\alpha_0) & -sin(\alpha_0) & 0 \\ sin(\alpha_0) & cos(\alpha_0) & 0 \\ 0 & 0 & 1 \end{pmatrix} \qquad T_k = \begin{pmatrix} x_0^1 \\ x_0^2 \\ 0 \end{pmatrix} \tag{18}$$

where R_0 is a non-linear function of the orientation angle. The expression (18) depends non-linearly of vector X_0^a and so a new parametrization is proposed jointly with a transformation which removes the projective parameter λ_k^i.

The point M^i is replaced by the rotated $M_{x_0}^i = R_0 M^i$, removing thus the product between unknowns. The orientation angle in the pose is substituted by parameters

$a = \cos(\alpha_0)$ and $b = \sin(\alpha_0)$, with the constraint $a^2 + b^2 = 1$. Using the new parameterization the expression (18) is transformed in the following:

$$\begin{pmatrix} y_k^i \\ 1 \end{pmatrix} = \lambda K_c (R_c (R_k^\Delta M_{X_0}^i + T_0 + L_{T_k^\Delta} \begin{pmatrix} a \\ b \\ 0 \end{pmatrix}) + T_c) \tag{20}$$

with $L_{T_k^\Delta}$:

$$L_{T_k^\Delta} = \begin{pmatrix} x_k^{1,\Delta} & -x_k^{2,\Delta} & 0 \\ -x_k^{2,\Delta} & x_k^{1,\Delta} & 0 \\ 0 & 0 & 1 \end{pmatrix} \tag{21}$$

where $x_k^{1,\Delta}$ y $x_k^{2,\Delta}$ the two first coordinates of T_k^Δ.

The new unknown vector, which correspond to the new parametrization of X_0^a, is:

$$\Phi = \begin{pmatrix} x_0^1 & x_0^2 & a & b & (M_{X_0}^1)^T, \cdots, (M_{X_0}^N)^T \end{pmatrix}^T \tag{22}$$

The expression (20) is decomposed in the following elements:

$$\begin{pmatrix} u_k^i \\ v_k^i \end{pmatrix} = y_k^i \qquad \begin{pmatrix} U_k^i \\ V_k^i \\ S_k^i \end{pmatrix} = K(R_c(R_k^\Delta M_{X_0}^i + T_0 + L_{T_k^\Delta}) + T_c) \tag{23}$$

where U_k^i, V_k^i y S_k^i are linear in terms of Φ but not in terms of y_k^i.

$$U_k^i = L_{U_k}^i \Phi + b_{U_k}^i \qquad V_k^i = L_{V_k}^i \Phi + b_{V_k}^i \qquad S_k^i = L_{S_k}^i \Phi + b_{S_k}^i \tag{24}$$

The projective scale is removed from the transformation by using vector product properties:

$$\begin{pmatrix} u_k^i \\ v_k^i \\ 1 \end{pmatrix} = \lambda_k^i \begin{pmatrix} U_k^i \\ V_k^i \\ S_k^i \end{pmatrix} \rightarrow \begin{pmatrix} u_k^i \\ v_k^i \\ 1 \end{pmatrix} \times \begin{pmatrix} U_k^i \\ V_k^i \\ S_k^i \end{pmatrix} = \begin{pmatrix} 0 \\ 0 \\ 0 \end{pmatrix} \tag{25}$$

where two lineally independent equations are obtained for Φ.

$$\begin{cases} v_k^i S_k^i - V_k^i = 0 \\ -u_k^i S_k^i + U_k^i = 0 \end{cases} \rightarrow \begin{cases} (v_k^i L_{S_k}^i - L_{V_k}^i)\Phi = v_k^i b_{S_k}^i - b_{V_k}^i \\ (-u_k^i L_{S_k}^i + L_{U_k}^i)\Phi = -u_k^i b_{S_k}^i + b_{U_k}^i \end{cases} \tag{26}$$

Using all measurements inside Y_L, a lineal system of equations is obtained in terms of Φ:

$$A\Phi = B \qquad A = \begin{pmatrix} (v_1^1 L_{S_1}^1 - L_{V_1}^1) \\ (-u_k^1 L_{S_1}^1 + L_{U_1}^1) \\ \vdots \\ (v_K^N L_{S_K}^N - L_{V_K}^N) \\ (-u_K^N L_{S_K}^N + L_{U_K}^N) \end{pmatrix} \qquad B = \begin{pmatrix} v_1^1 b_{S_1}^1 - b_{V_1}^1 \\ -u_k^1 b_{S_1}^1 + b_{U_1}^1 \\ \vdots \\ v_K^N b_{S_K}^N - b_{V_K}^N \\ -u_K^N b_{S_K}^N + b_{U_K}^N \end{pmatrix} \tag{27}$$

It is straightforward to show that system of (27) has a single-parameter family of solutions. If Φ_0 is a possible solution, then $\Phi_0 + \psi\Delta$ is a solution for any $\psi \in R$, with Δ:

$$\Delta = \left(T_{c,x^1} \quad T_{c,x^2} \quad a \quad b \quad \begin{pmatrix} 0 \\ 0 \\ T_{c,x^3} \end{pmatrix}^T \quad \cdots \quad \begin{pmatrix} 0 \\ 0 \\ T_{c,x^3} \end{pmatrix}^T \right) \tag{28}$$

In fact, if Δ is normalized, it matches up with the eigen-vector associated to the zero eigenvalue of matrix $A^T A$.

Using the constraint that $a^2 + b^2 = 1$, and the singular value decomposition of matrix A, an exact solution of system (20) is obtained.

Once Φ is available, the solution X_0^a is obtained by inverting the parametrization:

$$\alpha_0 = \tan^{-1}(a, b) \qquad M^i = R_0^{-1} M_0^i \qquad X_0^a = \left(x_0^1 \quad x_0^2 \quad \alpha_0 \quad (M^1)^T \quad \cdots \quad (M^N)^T \right)^T \tag{29}$$

This method, although exact, is prone to error due to odometry noise and does not benefit from the Maximum Likelihood metric. However it is valid as a method to give an initial value for X_0^a before using the iterative approach.

3.6. Degenerate Configurations

The kind of trajectory performed by the robot during initialization has direct influence in the solution of X_0^a. There are three kinds of movements that yields degenerate solutions:

- Straight motion: there is no information about the center of rotation of the robot and thus the pose has multiple solutions.
- Rotational motion around robot axis: the following one-parameter (*i.e.*, n) family of solutions gives identical measurements in the image plane:

$$M_n^i = nM^i + (n-1)\begin{pmatrix} 0 \\ 0 \\ \bar{T}_c(3) \end{pmatrix} \qquad T_{0n} = nT_0 + (n-1)\begin{pmatrix} \bar{T}_c(1) \\ \bar{T}_c(2) \\ 0 \end{pmatrix} \tag{31}$$

with $\bar{T}_c = R_c^T T_{c'}$.

$$X_k^a(n) = \left(T_{0n} \quad \alpha_0 \quad M_n^1 \quad \cdots \quad M_n^N \right) \tag{31}$$

- Circular trajectory: under a purely circular trajectory the following one-parameter family of initialization vectors gives identical measurements in the image plane:

$$M_n^i = nM^i + R_0^T \qquad T_{0n} = nT_0 + (n-1)R_c^T T_c \tag{32}$$

$$X_k^a(n) = \left(T_{0n} \quad \alpha_0 \quad M_n^1 \quad \cdots \quad M_n^N \right) \tag{33}$$

In practical cases it has been proved to be effective enough to combine straight trajectories with circular motion to avoid degeneracies in the solution of X_0^a.

3.7. Obtaining the Gaussian Equivalent of the Solution

Once the minimum of (16) is reached we suppose that the resulting value of X_0^a is the mean of the distribution X_0^a. This section describes how to also obtain the covariance matrix.

The covariance matrix Σ_0^a of the optimized parameters is easily obtained by using a local approximation of the term $Y_L - \hat{Y}_L$ in the vicinity of the minimum $\hat{X}_0^a$ using the following expression:

$$\Sigma_0^a = \left(J^T \Sigma_L^{-1} J \right)^{-1} \tag{34}$$

where J is the Jacobian matrix of $\hat{Y}$ with respect to parameters X_0^a. The Jacobian is available from the optimization method, in which is used to compute the iteration steps.

4. ONLINE ROBOT LOCALIZATION

In this section the solution to X_k^a given the last pose information is derived. The fact that last frame information is available and the assumption of soft motion between frames allows to greatly simplify the problem.

A special emphasis is given to the fact that any process handled by the system is considered a random entity, in fact a Gaussian distribution defined at each case by its mean vector and covariance matrix. The problem of obtaining pose and structure, encoded in X_k^a given image observations Y_k and the previous time estimation X_{k-1}^a is viewed from the point of view of statistical inference, which means searching for the posterior probability distribution $p(X_k^a | Y_1 \ldots, Y_k)$. That distribution gives the best estimation of X_k^a given all the past knowledge available. In Figure 6, a brief overview of the online method is presented.

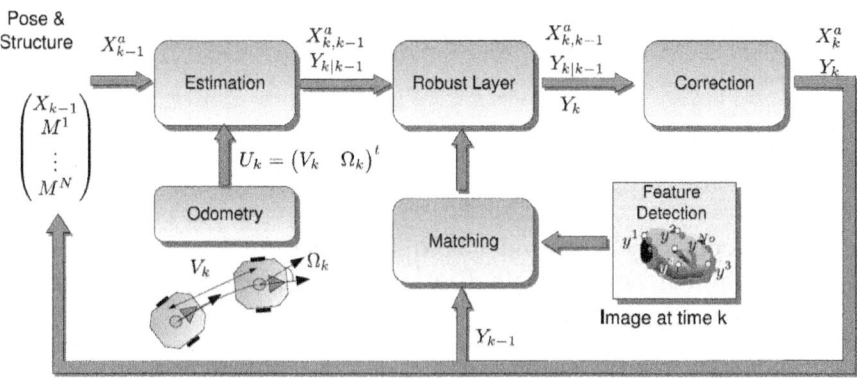

Figure 6. Overview of the online algorithm.

The online approach is divided into three steps:

- **Estimation Step:** using the previous pose distribution $p(X^a_{k-1}|Y_1, \cdots, Y_{k-1})$, defined by its mean $\hat{X}^a_{k-1}$ and covariance matrix Σ^a_{k-1} and the motion model g^a, a Gaussian distribution which infers the next state is given $p(X^a_k|Y_1, \cdots, Y_{k-1})$.

- **Robust Layer:** the correspondence between image measurements and the 3D model of the robot easily fails, so a number of wrongly correspondences or outliers pollute the measurement vector Y_k. Using a robust algorithm and the information contained in the state vector, the outliers are discarded before the next step.

- Correction Step: using an outlier-free measurement vector, we are confident to use all the information available to obtain the target posterior distribution $p(X^a_k|Y_1, \cdots, Y_k)$.

In all three steps it is required to propagate statistic processes over non-linear functions (f and h). As was stated in the previous section we show how to face the problem using first order expansions as it offers more compactness and is more readable. As a consequence the "Estimation" and "Correction" steps are solved using the so called Extended Kalman Filter (EKF) equations, which have been already implemented in problems of similar complexity such as visual Simultaneous Localization and Mapping (SLAM) [5].

4.1. Estimation Step

The estimation step uses the motion models available to infer the next pose of the robot which implies an increment in uncertainty. Starting from the last pose distribution $p(X^a_{k-1}|Y_1, \cdots, Y_{k-1}) = N(\hat{X}^a_{k-1}, \Sigma^a_{k-1})$, the motion model g^a and the noise included in odometry values, the following update is obtained:

$$
\begin{aligned}
\hat{X}^a_{k|k-1} &= g^a(\hat{X}^a_k, U_k) \\
\Sigma^a_{k|k-1} &= J^T_x \Sigma^a_{k-1} J_x + J^T_U \Sigma_W J_U,
\end{aligned}
\tag{35}
$$

where $\hat{X}^a_{k|k-1}$ and $\Sigma^a_{k|k-1}$ are the mean and covariance matrix of distribution:

$$
p(X^a_k|Y_1, \cdots, Y_{k-1})
$$

The matrices J_X and J_U are the first derivatives of the function g^a with respect to X^a_{k-1} and U_k respectively. Usually J_X in odometry systems is the identity therefore, at this step, the covariance matrix $\Sigma^a_{k|k-1}$ results to be bigger in terms of eigenvalues, which means uncertainty.

4.2. Correction Step

The correction step removes the added uncertainty in the estimation by using image measurements. It passes from the distribution $p(X^a_k|Y_1, \cdots, Y_{k-1})$ to the target distribution $p(X^a_k|Y_1, \cdots, Y_k)$, which includes the last measurement.

Using the estimation shown in (35), and knowing the correspondence between measurements with the camera and structure point of the state vector, the estimated measurement is given:

$$Y_{k|k-1} = h^a(X_k^a) \tag{36}$$

$$\Sigma_{Y_{k|k-1}} = J_h^T \Sigma_{k|k-1}^a J_h + \Sigma_V \tag{37}$$

$$\Sigma_{X^a Y} = \Sigma_{k|k-1}^a J_h \tag{38}$$

where J_h is the Jacobian matrix of the function h^a with respect to X_k^a and Σ_V is block diagonal matrix with Σ_v on each block. Function h^a performs the projection in the image plane of the camera of all visible points that form up the measurement vector Y_k. The correction step itself is a linear correction of $X_k^a{}_{|k-1}$ and $\Sigma_{k|k-1}^a$ by means of the Kalman gain K_G:

$$K_G = \Sigma_{X^a Y} \Sigma_{Y_{k|k-1}}^{-1} \tag{39}$$

$$X_k^a = X_{k|k-1}^a + K_G(Y_k - Y_{k|k-1}) \tag{40}$$

$$\Sigma_k^a = \Sigma_{k|k-1}^a - K_G \Sigma_{X^a Y}^T \tag{41}$$

As it is stated in (41) the resulting Σ^a is reduced compared to Σ_{k-1}^a which means that after the correction step, the uncertainty is "smaller".

4.3. Robust Layer

The robust layer has the objective of removing bad measurements from Y_k to avoid inconsistent updates of X_k^a in the correction step. In this paper we propose to include the Random Sample Consensus (RANSAC) algorithm [22] between the estimation and correction step of the filter. The general idea is to found among the measured data Y_k a set which agrees in the pose X_k, using the 3D model obtained in X_{k-1}^a.

The interest of applying RANSAC in a sequential update approach resides on several reasons: firstly it allows to efficiently discard outliers from Y_k preventing algorithm's degeneracy, which happens even if the motion model is accurate. Secondly, compared to online robust approaches, where a robust cost function is optimized, RANSAC allows not to break the Kalman filter approach, as it only cleans the measurement vector of outliers. Furthermore we have observed experimentally that the RANSAC algorithm can be very fast between iterations, as only a few outliers are inside the data. (We use the RANSAC implementation described in [17], which implements a dynamical computation of the outlier probability).

The RANSAC method proposed in the commented framework obtains the consensus pose X_k from the set of measurements Y_k and the 3D model available in X_{k-1}^a using the algorithm presented in [18]. For a robot moving in a plane, as it is the case with the mobile robot considered in this paper, it is enough to use a minimum of 2 correspondences between the model and the measurement which makes the RANSAC very fast for removing outliers.

4.4. Image Measurements

Contrary to the initialization case, in this step we have an accurate prediction of the tracked points in the image plane at each time instant, namely vector $Y_{k|k-1}$. Using such prediction we can easily match the 3D points in the state vector with measurements taken in a measurement set T_k, using the SIFT detector applied to current image I_k. The matching is done in terms of distance in the image plane. Let $y_{k|k-1}$ a feature inside vector $Y_{k|k-1}$ and y_k^j a candidate obtained using SIFT method. We conclude that they are matched if $|y_k^j - y_{ik|k-1}|^2_M < \tau_{max}$, where $||_M$ states for the Mahalanobis distance using the covariance $\Sigma_{yk|k-1}$ computed from the matrix $\Sigma_{Yk|k-1}$. The Mahalanobis distance allows to take into account the uncertainty predicted for $y_{k|k-1}^i$ and also helps to select a threshold τ_{max} with a probabilistic criterion.

5. RESULTS

This section describes the experimental setup developed to support the theoretical algorithms proposed in this paper. The experiments are divided in those performed over synthetic data and those run in a real implementation of the Intelligent Space in the University of Alcala (ISPACE-UAH) [23]. In both kind of experiments the same camera parameters are used, derived from the real device used in its real placement. The single camera consists of a CCD based sensor with resolution of 640 × 480 pixels and a physical size of 1/2 (around 8 mm diagonal). The optical system is chosen with a focal length of 6.5 mm which gives around 45° of Field of View (FOV). The camera is connected to a processing and acquisition node through a IEEE1394 port, which support 15 fps in RGB image format acquisition. The intrinsic parameters are the following:

$$f_u = 636.7888 \; f_v = 637.5610 \; u_0 = 313.3236 \; v_0 = 210.6894 \qquad (42)$$

The camera is placed with respect a global coordinate origin, as it is displayed in Figure 7. Camera calibration is performed previously to the localization task, using checkerboards as calibration patterns and the "Matlab Calibration Toolbox" implemented by [24]. The distortion parameters of the camera are not considered in this case. As it can be shown in Figure 7, the camera is placed in oblique angle, which is specially useful for covering large areas with less FOV requirements (less distortion) compared to overhead configurations.

The robotic platform is connected to the same processing node controlling the camera by means of a wireless network. The camera acquisition and the odometry values obtained from the robot are synchronized. The control loop of the robot is internal, and it is prepared to follow position landmarks based on odometry feedback at faster sampling frequency than the image localization system (15 fps).

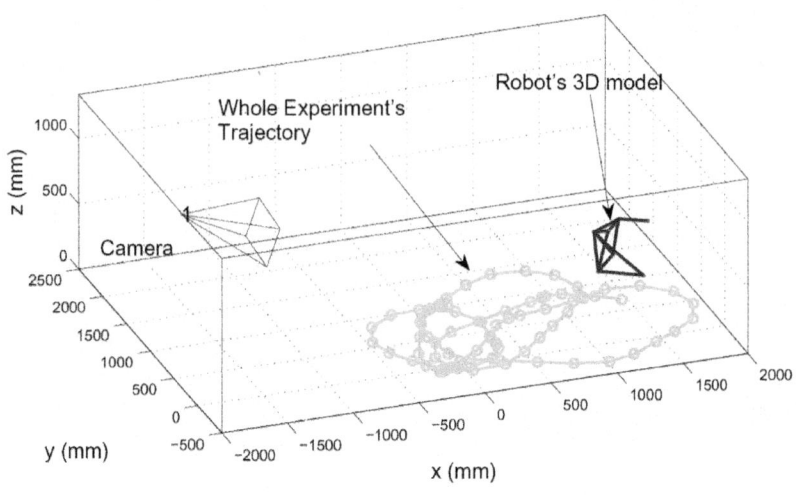

Figure 7. Geometric distribution of the camera and robot's trajectory.

Therefore, for each image acquisition cycle, the odometry system obtain several readings that are used by the internal loop control. In this paper the localization information provided by the cameras is not included in the control loop of the robot.

5.1. Simulated Data

In this experiment the robot structure is simulated with a set of $N = 10$ points distributed randomly inside a cylinder with radius $R = 0.5m$ and height $h = 1m$. The odometry system is supposed to have an uncertainty $\sigma_{V_l}^2 = 10$ and $\sigma_{\Omega}^2 = 1$ in linear and angular speed respectively. The initialization trajectory is shown in Figure 8. The image measurements are considered polluted with a Gaussian process of $\sigma_v^2 = 10$, independently of each measurement and image coordinate.

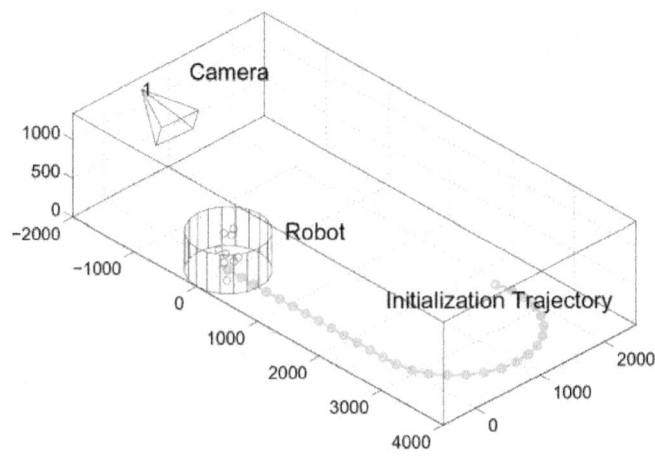

Figure 8. Initialization trajectory of the robot in the experiment based on synthetic data.

To compare the performance of the initialization method proposed in Section 3., the following error magnitudes are described:

$$\epsilon_M = \frac{\sqrt{\sum_{i=1}^{N} \|M^i - M^i_{gth}\|^2}}{\sqrt{\sum_{i=1}^{N} \|M^i_{gth}\|^2}} \quad \epsilon_T = \|T_0 - T_{0,gth}\| \quad \epsilon_\alpha = \|\alpha_0 - \alpha_{0,gth}\| \tag{43}$$

where $T_{0,gth}$, $\alpha_{0,gth}$ and M^i_{gth} correspond to the ground truth values of the initialization vector X^a_0. The following two experiments are proposed:

- *Initialization errors in function of the odometry error.* In this experiment the value of σ^2_{vl} and σ_Ω^2 are multiplied by the following multiplicative factor:

$$\rho \in \begin{pmatrix} 0.01 & 0.1 & 0.5 & 1 & 5 & 10 \end{pmatrix} \tag{44}$$

The different errors of Equation 43 can be viewed in Figure 9a-c in terms of ρ. As it can be observed in the results, the M.C. (Complete correlated matrix Σ_L) method outperforms the rest of approximations of Σ_L, specially the full diagonal method M.I., which means that the statistical modelling proposed in this paper is effective.

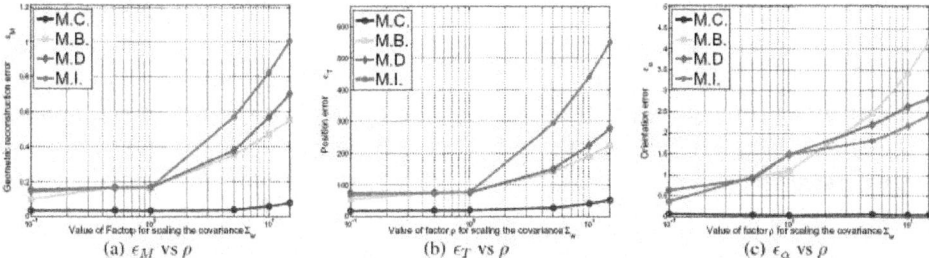

(a) ϵ_M vs ρ (b) ϵ_T vs ρ (c) ϵ_α vs ρ

Figure 9. Experiment showing the different initialization errors in function of the amount of error in odometry readings.

- *Initialization errors in function of the trajectory length.* The trajectory used in the experiment and displayed in Figure 8 is uniformly scaled by parameter

$$\rho_t \in \begin{pmatrix} 0.2 & 0.4 & 0.6 & 0.8 & 1 & 1.2 & 1.4 \end{pmatrix} \tag{45}$$

so that it can be guessed the relationship between trajectory length and initialization errors. In Figure 10 the initialization errors are displayed versus ρ_t.

In light of the results shown in Figure 10a-c, the M.C. method is capable of reducing the error no matter how large is the trajectory chosen. However, in the rest of the approximations of Σ_L there is an optimal point where the initialization errors are minimum. This results make sense, as without statistical modelling large trajectories contain accumulative errors which usually affects the final solution of X^a_0.

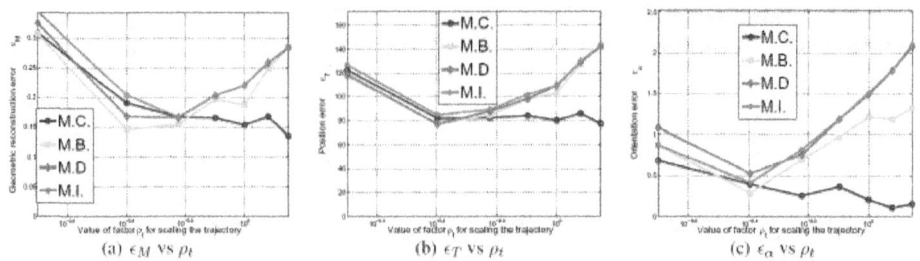

Figure 10. Experiment showing the different initialization errors in function of the trajectory length performed by the robot.

Both experiments clearly manifest that the complete matrix Σ_L, with all its cross-correlated terms (M.C.), outperforms the rest of proposals, especially when it is compared to the case where Σ_L is the identity matrix (M.I.), which means no statistical modeling.

5.2. Real Data

The initialization experiment proposed in this paper consists of a robot performing a short trajectory from which its 3D model and initial position is obtained using the results of Section 3.. We present a comparison of the initialization results using three different trajectories. Each one of the trajectories is displayed in Figure 11 as a colored interval of the whole trajectory performed by the robot in the experiment.

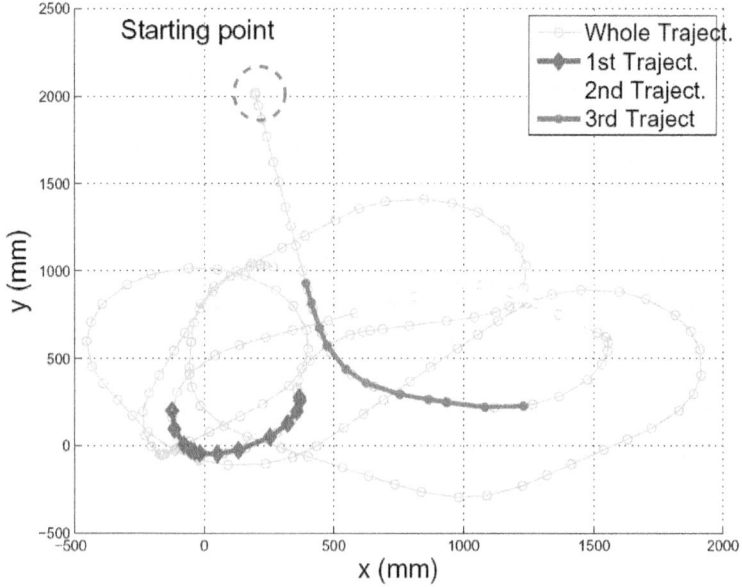

Figure 11. Intervals of robot's trajectory used for initialization.

The results of the initialization method on each of the 3 trajectories selected is shown in Figure 12. Depending on the trajectory used, the features viewed by the camera vary and thus the corresponding initialized 3D model. As it can be seen in Figure 12, the 3D model is accurate and its projection matches with points in the robot's structure in all cases. It must be remarked that on each case, the position obtained as a result of the initialization (*i.e.*, X_0) corresponds to the first position of each interval.

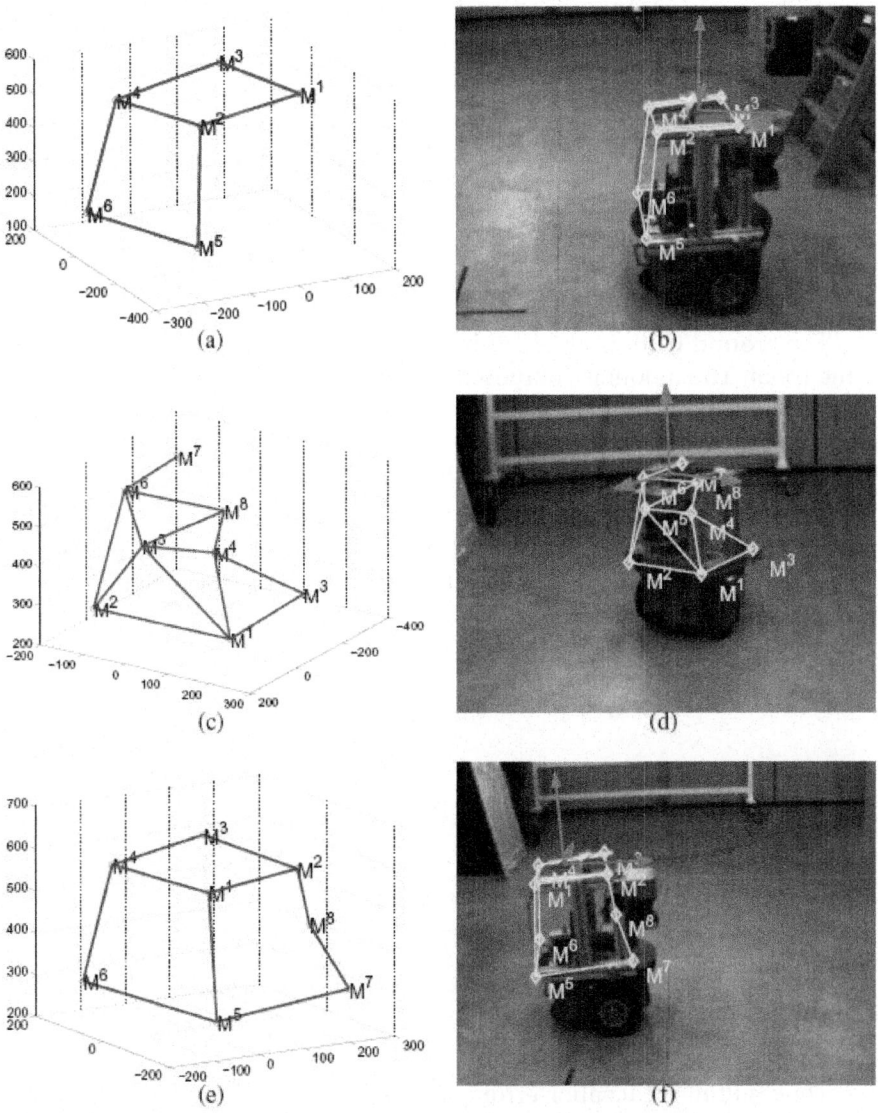

Figure 12. Single camera initialization results. On each row it is shown the resulting reconstruction and its projection in the image plane of the camera.

The sequential algorithm is tested using the whole trajectory shown in Figure 11. The initial pose and 3D model are the result of the initialization results shown in Figure 12e,f. The estimated position of the robot, compared to a "ground truth" measurement, is presented in Figure 13a,b.

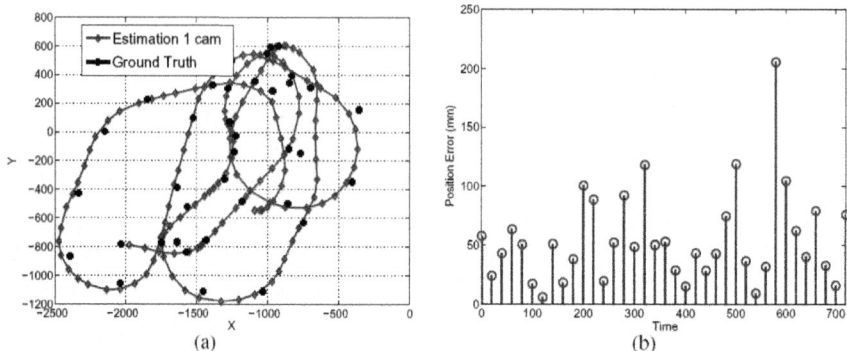

(a) (b)

Figure 13. Online robot's pose compared with the "ground-truth".

The ground truth is obtained by means of a manually measured 3D model of the robot. The model is composed of 3D points, that are easily recognized by a human observer. By manually clicking points of the 3D model in the image plane, and by using the method proposed in [18], the reference pose of the robot is obtained in some of the frames of the experiment.

Another experiment is proposed to test the online algorithm with occlusions and a larger path followed by the robot. In Figure 14a it is shown the geometric placement of the camera and in Figure 14b it is shown the geometric model obtained during initialization.

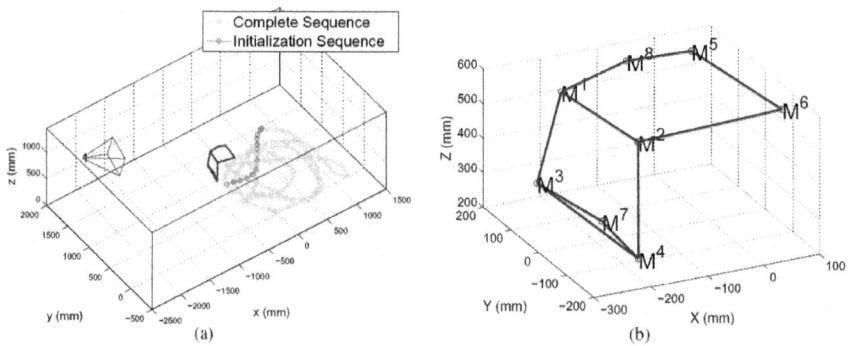

(a) (b)

Figure 14. Scene geometry and 3D model obtained during initialization.

The resulting localization error is shown in Figure 15b. In Figure 16 it is shown several frames, indexed by the time sample number k, presenting hard occlusions between the camera and the robot without losing the tracking. The RANSAC method used in the Kalman loop avoid erroneous matches in the occluded parts of the object. Besides, in those frames with completely occluded features, only the estimation step of the Kalman filter is done, which is accurate enough for short periods of time.

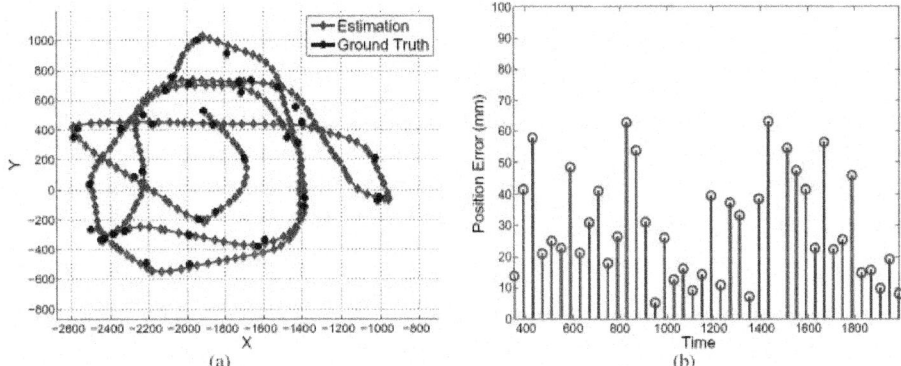

Figure 15. Localization results in an experiment with occlusions.

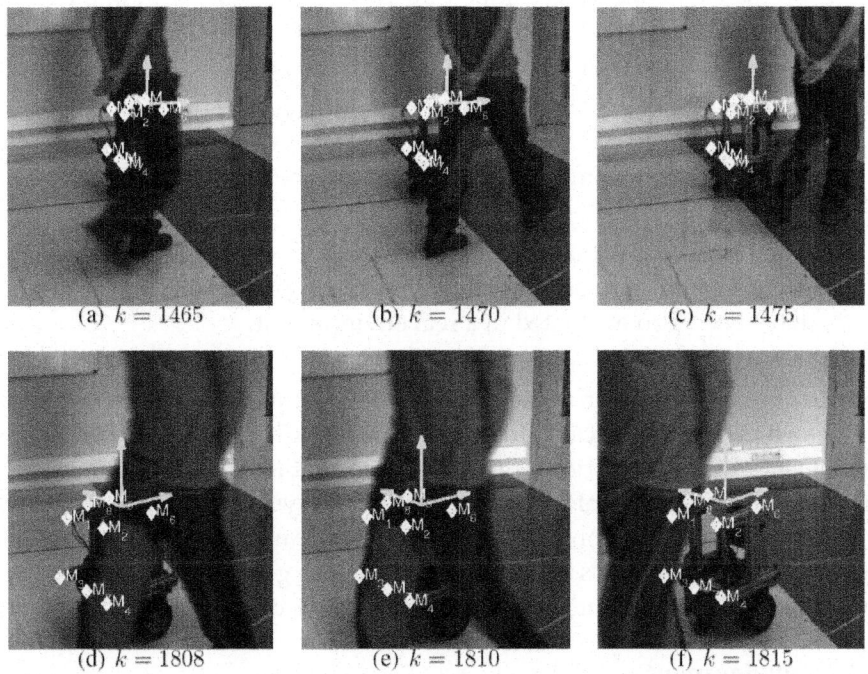

Figure 16. Occlusions and its influence in the localization accuracy.

In both sequential experiments the 3D model is composed of 8 to 10 points and no extra geometrical information is introduced in the state vector. As a future line of study a simultaneous reconstruction and localization approach can be adopted to introduce online extra 3D points in the state vector as the robot is tracked. A very similar approach is done in the Simultaneous Localization and Mapping (SLAM) problem, and some of their solutions [5] are perfectly applicable to the problem assessed in this paper.

6. CONCLUSIONS

This paper has presented a complete localization system for mobile robots inside intelligent environments with a single external camera. The objectives of obtaining the pose of the robot based on its natural appearance has been tackled in the paper using a reconstruction approach followed by a sequential localization approach.

The main contributions of this paper are summarized in the following list:

- The initialization step provides a non-supervised method for obtaining the initial pose and structure of the robot previously to its sequential localization. A Maximum Likelihood cost function is proposed, which obtains the pose and geometry given a trajectory performed by the robot. The proposal of this paper allows to compensate for the odometry drift in the solution. Also an exact initialization method has been proposed and the degenerate configurations have been identified theoretically.

- The online approach of the algorithm obtains the robot's pose by using a sequential Bayesian inference approach. A robust step, based on the RANSAC algorithm, is proposed to clean the measurement vector out of outliers.

- The results show that the proposed method is suitable to be used in real environments. The accuracy and non-drifting nature of the pose estimation have been also evaluated in a real environment.

The future research must be oriented to scale the problem for large configurations of non-overlapped cameras and multiple robots, where extra problems arise, such as to automatically detect the transitions between cameras and online refinement of the geometric models. It is important from our point of view to, in the future, combine the information given by the system proposed in this paper with information sensed onboard the robots using cameras. This approach allows to jointly build large maps attached to information given by the cameras, so that robots can be localized and controlled to perform a complex task.

Acknowledgements

This work was supported by the Ministry of Science and Technology under RESELAI project (references TIN2006-14896-C02-01) and also by the Spanish Ministry of Science and Innovation (MICINN) under the VISNU project (REF-TIN2009-08984).

REFERENCES

1. Lee, J.; Hashimoto, H. Intelligent space concept and contents. *Advanced Robotics* 2002, *16*, 265–280.

2. Aicon 3D. Available online: www.aicon.de (accessed January 2010).

3. ViconPeak. Available Online: www.vicon.com (accessed January 2010).

4. Se, S.; Lowe, D.; Little, J. Vision-based global localization and mapping for mobile robots. *IEEE Trans. Robotics Automat.* 2005, *21*, 364–375.

5. Davison, A.J.; Murray, D.W. Simultaneous Localization and Map-Building Using Active Vision. *IEEE Trans. Patt. Anal.* 2002, *24*, 865–880.

6. Hada, Y.; Hemeldan, E.; Takase, K.; Gakuhari, H. Trajectory tracking control of a nonholonomic mobile robot using iGPS and odometry. In *Proceedings of IEEE International Conference on Multisensor Fusion and Integration for Intelligent Systems*, Tokyo, Japan, 2003; pp. 51–57.

7. Fernandez, I.; Mazo, M.; Lazaro, J.; Pizarro, D.; Santiso, E.; Martin, P.; Losada, C. Guidance of a mobile robot using an array of static cameras located in the environment. *Auton. Robots* 2007, *23*, 305–324.

8. Morioka, K.; Mao, X.; Hashimoto, H. Global color model based object matching in the multi-camera environment. In *Proceedings of IEEE/RSJ International Conference on Intelligent Robots and Systems*, Beijing, China, 2006; pp. 2644–2649.

9. Chung, J.; Kim, N.; Kim, J.; Park, C.M. POSTRACK: a low cost real-time motion tracking system for VR application. In *Proceedings of Seventh International Conference on Virtual Systems and Multimedia*, Berkeley, CA, USA, 2001; pp. 383–392.

10. Sogo, T.; Ishiguro, H.; Ishida, T. Acquisition of qualitative spatial representation by visual observation. In *Proceedings of International Joint Conference on Artificial Intelligence*, Stockholm, Sweden, 1999.

11. Kruse, E.; Wahl, F.M. Camera-based observation of obstacle motions to derive statistical data for mobile robot motion planning. In *Proceedings of IEEE International Conference on Robotics and Automation*, Leuven, Belgium, 1998; pp. 662–667.

12. Hoover, A.; Olsen, B.D. Sensor network perception for mobile robotics. In *Proceedings of IEEE International Conference on Robotics and Automation*, Krakow, Poland, 2000; pp. 342–347.

13. Steinhaus, P.; Ehrenmann, M.; Dillmann, R. MEPHISTO. A Modular and extensible path planning system using observation. *Lect. Notes Comput. Sci.* 1999, *1542*, 361–375.

14. Shi, J.; Tomasi, C. Good features to track. In *Proceedings of IEEE Conference on Computer Vision and Pattern Recognition*, Seattle, WA, USA, 1994; pp. 593–600.

15. Lowe, D.G. Distinctive image features from scale-invariant keypoints. *Int. J. Comput. Vision* 2004, *60*, 91–110.

16. Triggs, B.; McLauchlan, P.; Hartley, R.; Fitzgibbon, A. Bundle adjustment–a modern synthesis. *Lect. Notes Comput. Sci.* 1999, *1883*, 298–372.

17. Hartley, R.; Zisserman, A. *Multiple View Geometry in Computer Vision*, 2nd ed.; Cambridge University Press: Cambridge, UK, 2003.

18. Moreno-Noguer, F.; Lepetit, V.; Fua, P. Accurate non-iterative o (n) solution to the PnP problem. In *Proceedings of IEEE International Conference on Computer Vision*, Rio de Janeiro, Brasil, 2007; pp. 1–8.

19. Schweighofer, G.; Pinz, A. Globally optimal o (n) solution to the pnp problem for general camera models. In *Proceedings of BMVC*, Leeds, UK, 2008.

20. Van der Merwe, R.; de Freitas, N.; Doucet, A.; Wan, E. *The Unscented Particle Filter*; Technical Report CUED/F-INFENG/TR380; Engineering Department of Cambridge University: Cambridge, UK, 2000.

21. Pizarro, D.; Mazo, M.; Santiso, E.; Hashimoto, H. Mobile robot geometry initialization from single camera. Field and Service Robotics; Springer Tracts in Advanced Robotics: Heidelberg, Germany, 2008; pp. 93–102.

22. Fischler, M.A.; Bolles, R.C. Random sample consensus: a paradigm for model fitting with applications to image analysis and automated cartography. *Comm. Assoc. Comp. Match.* 1981, 24, 381–395.

23. GEINTRA. Group of Electronic Engineering Applied to Intelligent Spaces and Transport. Available online: http://www.geintra-uah.org/en (accessed December 2009).

24. Bouguet, J.Y. Matlab Calibration Toolbox. Available online: http://www.vision.caltech.edu/bouguetj/calib-doc (accessed December 2008).

INDEX